T0134616

Intelligent Systems Reference Library

Volume 125

Series editors

Janusz Kacprzyk, Polish Academy of Sciences, Warsaw, Poland
e-mail: kacprzyk@ibspan.waw.pl

Lakhmi C. Jain, University of Canberra, Canberra, Australia;
Bournemouth University, UK;
KES International, UK
e-mails: jainlc2002@yahoo.co.uk; Lakhmi.Jain@canberra.edu.au
URL: http://www.kesinternational.org/organisation.php

About this Series

The aim of this series is to publish a Reference Library, including novel advances and developments in all aspects of Intelligent Systems in an easily accessible and well structured form. The series includes reference works, handbooks, compendia, textbooks, well-structured monographs, dictionaries, and encyclopedias. It contains well integrated knowledge and current information in the field of Intelligent Systems. The series covers the theory, applications, and design methods of Intelligent Systems. Virtually all disciplines such as engineering, computer science, avionics, business, e-commerce, environment, healthcare, physics and life science are included.

More information about this series at http://www.springer.com/series/8578

Gerasimos G. Rigatos

State-Space Approaches for Modelling and Control in Financial Engineering

Systems Theory and Machine Learning Methods

Gerasimos G. Rigatos
Unit of Industrial Automation
Industrial Systems Institute
Rion, Patras
Greece

ISSN 1868-4394 ISSN 1868-4408 (electronic)
Intelligent Systems Reference Library
ISBN 978-3-319-85004-7 ISBN 978-3-319-52866-3 (eBook)
DOI 10.1007/978-3-319-52866-3

Printed on acid-free paper

This Springer imprint is published by Springer Nature
The registered company is Springer International Publishing AG
The registered company address is: Gewerbestrasse 11, 6330 Cham, Switzerland

To Elektra

Foreword

In the recent years, there has been significant research interest in modelling and control of financial systems through state-space representation of their dynamics. This is because such approaches eliminate the use of heuristics in financial decision-making while assuring stability and in several cases optimality in the functioning of economic systems. As one delves into the complexity of the financial dynamics, he perceives that deterministic modelling is unlikely to work and that variability, parametric uncertainty and stochasticity are factors that should be seriously taken into account for the efficient management of economic systems. Through a synergism of systems theory and machine learning methods, this monograph develops modelling and control approaches which finally assure that the monitored financial systems will evolve according to the specifications and optimality objectives, while the risk of wrong decision-making in the management of these systems will be also minimized.

The use of state-space models in financial engineering allows to eliminate heuristics and empirical methods currently in use in decision-making procedures for finance. On the other side, it permits to establish methods of fault-free performance and optimality in the management of assets and capitals and methods assuring stability in the functioning of financial systems (e.g. of several financial institutions and of the banking sector). The systems theory-based and machine learning methods developed by the monograph stand for a genuine and significant contribution to the field of financial engineering. First, the monograph solves in a conclusive manner problems associated with the control and stabilization of nonlinear and chaotic dynamics in financial systems, when these are described in the form of nonlinear ordinary differential equations. Next, it solves in a conclusive manner problems associated with the control and stabilization of financial systems governed by spatiotemporal dynamics, that is systems described by partial differential equations (e.g. the Black–Scholes PDE and its variants). Moreover, the monograph solves the problem of filtering for the aforementioned types of financial models, that is of estimation of the entire dynamics of the financial systems when using limited information (partial observations) obtained from them. Finally, the monograph solves in a conclusive and optimal manner the problem of statistical validation of computational models and tools used to support financial engineers in decision taking. Through the methods it develops, the monograph enables to identify inconsistent and inappropriately parameterized financial models and to take necessary actions for their update.

The topics studied in the monograph are of primary importance for financial engineering and for the profitable management of financial systems. It is a common sense that decision-making in finance should stop being based on intuition, heuristics and empirical rules and should move progressively to systematic methods of assured performance. To this end, the monograph demonstrates first that it is possible to identify the complete dynamics of financial systems using limited information out of them, and next, it shows that the estimated dynamics can be used for the control and stabilization of such systems. The monograph's estimation, forecasting and control methods are not only addressed to financial systems described by nonlinear ordinary differential equations, but also extended to financial systems exhibiting spatiotemporal dynamics, as in the case of the Black–Scholes PDE. Offering solution to estimation and control problems in PDE models met often in finance is one of the main contributions of the monograph, and this can be useful both for the academic community and for financial engineers working in practice. Another major contribution of the monograph is in statistical validation of decision-making tools used in financial engineering. Taking into account the need for reliable functioning of software developed for decision support in finance, one can easily understand the significance of the monograph's results about validation of the computational models of financial systems and of the associated forecasting tools. The monograph offers to financial engineers optimal statistical methods for determining whether the models used in estimation of the state of financial systems are accurate or whether they contain inconsistent parameters which result in forecasting of low precision.

The contents of the monograph cover the following key areas for financial engineering: (i) control and stabilization of financial systems dynamics, (ii) state estimation and forecasting and (iii) statistical validation of decision-making tools. The monograph is primarily addressed to the academic community. The content of the monograph can be used for teaching undergraduate or postgraduate courses in financial engineering. Therefore, it can be used by both academic tutors and students as a reference book for such a course. A significant part of the monographs readership is also expected to come from the engineering and computer science community, as well as from the finance and economics community. The nonlinear PDE control and estimation methods analysed in the proposed monograph can be a powerful tool and useful companion for people working on applied financial engineering.

Athens, Greece Gerasimos G. Rigatos
March 2017

Preface

The present monograph contains new results and findings on control and estimation problems for financial systems and for statistical validation of computational tools used for financial decision-making. The use of state-space models in financial engineering will allow to eliminate heuristics and empirical methods currently in use in decision-making procedures for finance. On the other side, it will permit to establish methods of fault-free performance and optimality in the management of assets and capitals and methods assuring stability in the functioning of financial systems (e.g. of several financial institutions and of the banking sector). As it can be confirmed from an overview of the relevant bibliography, the systems theory-based and machine learning methods developed by the monograph stand for a genuine and significant contribution to the field of financial engineering. First, the monograph solves in a conclusive manner problems associated with the control and stabilization of nonlinear and chaotic dynamics in financial systems, when these are described in the form of nonlinear ordinary differential equations. Next, it solves in a conclusive manner problems associated with the control and stabilization of financial systems governed by spatiotemporal dynamics, that is systems described by partial differential equations (e.g. the Black–Scholes PDE and its variants). Moreover, the monograph solves the problem of filtering for the aforementioned types of financial models, that is of estimation of the entire dynamics of the financial systems when using limited information (partial observations) obtained from them. Finally, the monograph solves in a conclusive and optimal manner the problem of statistical validation of computational models and tools used to support financial engineers in decision taking. Through the methods it develops, the monograph enables to identify inconsistent and inappropriately parameterized financial models and to take necessary actions for their update.

The monograph comes to address the need about decision-making in finance that will be no longer based on heuristics and intuition but will make use of computational methods and tools characterized by fault-free performance and optimality. Through the synergism of systems theory and machine learning methods, the monograph offers solutions, in a conclusive manner, to the following key problems met in financial engineering: (i) control and stabilization of financial systems exhibiting nonlinear and chaotic dynamics; (ii) control and stabilization of financial systems exhibiting spatiotemporal dynamics described by partial differential equations; (iii) solution to the associated filtering problems, that is estimation of the complete dynamics of the aforementioned complex types of financial models with the use of limited information extracted out of them; (iv) elaborated computational tools for the assessment of risk in financial systems and for the optimized management of capitals and assets; and (v) statistical validation of decision support tools used in finance, such as forecasting models and models of financial systems dynamics. The monograph is primarily addressed to the academic and research community of financial engineering as well as to tutors of relevant university courses. It can also be a useful reference for students of financial engineering, at both undergraduate and postgraduate level, helping them to get acquainted with established approaches for control, estimation and forecasting in finance as well as with methods for validating the precision of computational tools used in decision support. Finally, it is addressed to financial engineers working on practical problems of risk-free decision-making and aiming at profitable management of funds, commodities and financial resources.

The management of financial systems has to address the following issues: (i) stability, (ii) modelling and forecasting and (iii) validation and update of decision-making tools. About (i), it is noted that although the dynamics of financial systems has been described efficiently by the Black–Scholes PDE and its variants, little has been done about its stabilization. The problem of control and stabilization of diffusion PDEs of this type is a non-trivial one and has to be implemented using as control inputs only the PDEs boundary conditions. The monograph offers solution of assured convergence and performance for this difficult control problem. Additionally, there are several types of financial systems described by nonlinear ODEs which exhibit chaotic dynamics. The monograph provides stabilizing control methods for such systems too. About (ii), it is easy to understand that forecasting in financial systems is significant for risk assessment and successful decision-making. By being in position to predict future states of the financial system, early warning indications are handled and profitable actions are taken for asset and capital management. The monograph's method contributes to this direction. About (iii), it is apparent that the effectiveness of all decision-making processes in finance is dependent on the sufficiency of the information collected from the financial system and on the accuracy and credibility of decision support tools. The statistical validation of decision-making software and of the models used by it is important for the maximization of profits in financial systems management and for the minimization of risks. Clearly, the monograph solves the statistical validation problems in a conclusive manner.

The monograph comprises the following chapters:

In Chap. 1, systems theory and stability concepts are overviewed. This chapter analyses the basics of systems theory which can be used in the modelling of nonlinear dynamics. To understand the oscillatory behaviour of nonlinear systems that can exhibit such dynamics, benchmark examples of oscillators are given. Moreover, using examples from state-space models, the following properties are analysed: phase diagram, isoclines, attractors, local stability, bifurcations of fixed points and chaos properties.

In Chap. 2, main approaches to nonlinear control with potential application to financial systems are analysed. In control and stabilization of the dynamics of financial systems, one can distinguish three main research axes: (i) methods based on global linearization, (ii) methods based on asymptotic linearization and (iii) Lyapunov methods. As far as approach (i) is concerned, these are methods for the transformation of the nonlinear dynamics of the system to equivalent linear state-space descriptions for which one can design controllers using state feedback and can also solve the associated state estimation (filtering) problem. One can classify here methods based on the theory of differentially flat systems and methods based on Lie algebra. As far as approach (ii) is concerned, solutions are pursued to the problem of nonlinear control with the use of local linear models (obtained at local equilibria). For such local linear models, feedback controllers of proven stability can be developed. One can select the parameters of such local controllers in a manner that assures the robustness of the control loop to both external perturbations and model parametric uncertainty. As far as approach (iii) is concerned, that is methods of nonlinear control of the Lyapunov type, one comes against the problems of minimization of Lyapunov functions so as to assure the asymptotic stability of the control loop. For the development of Lyapunov-type controllers, one can either exploit a model about the financial systems dynamics or proceed in a model-free manner, as in the case of indirect adaptive control.

In Chap. 3, main approaches to nonlinear estimation with potential application to financial systems are analysed. To treat the filtering problem for nonlinear dynamics in financial systems, the Extended Kalman Filter is an established approach. However, since this is based on approximate linearization of the system's state-space description and in the truncation of higher order terms in the associated Taylor series expansion, the Unscented Kalman Filter is frequently used in its place. The latter filter performs state estimation by averaging on state vectors that are selected at each iteration of the filtering algorithm and being defined by the columns of the estimation error vector covariance matrix. Additionally, to handle the case of non-Gaussian noises in the filtering procedure, the particle filter has been proposed. A number of potential state vector values (particles) are updated in time through elitism criteria, and out of this set, the estimate of the state vector is computed. The topic of nonlinear estimation is completed by a new nonlinear filtering approach under the name Derivative-free nonlinear Kalman Filter. This filter based on linearizing transformation of the monitored financial system is proven to conditionally maintain the optimality features of the standard Kalman Filter and to be computationally faster than other nonlinear estimation methods. Moreover, to treat the distributed filtering and state estimation in financial systems, one can apply established methods for decentralized state estimation, such as the Extended Information Filter (EIF) and the Unscented Information Filter (UIF). EIF stands for the distributed implementation of the Extended Kalman Filter, while UIF stands for the distributed implementation of the Unscented Kalman Filter. Additionally, to obtain a distributed filtering scheme in this monograph, the Derivative-free

Extended Information Filter (DEIF) is implemented. This stands for the distributed implementation of a differential flatness theory-based filtering method under the name Derivative-free distributed nonlinear Kalman Filter. The improved performance of DEIF compared to the EIF and UIF is confirmed both in terms of improved estimation accuracy and in terms of improved speed of computation. Finally, one can note distributed filtering with the use of the distributed particle filter. This consists of multiple particle filters running at distributed computation units, while a consensus criterion is used to fuse the local state estimates.

In Chap. 4, linearizing control and filtering for nonlinear dynamics in financial systems is explained. A flatness-based adaptive fuzzy control is applied to the problem of stabilization of the dynamics of a chaotic finance system, describing interaction between the interest rate, the investment demand and the price exponent. First, it is proven that the system is differentially flat. This implies that all its state variables and its control inputs can be expressed as differential functions of a specific state variable, which is a so-called flat output. It also implies that the flat output and its derivatives are differentially independent which means that they are not connected to each other through an ordinary differential equation. By proving that the finance system is differentially flat and by applying differential flatness diffeomorphisms, its transformation to the linear canonical (Brunovsky) is performed. For the latter description of the system, the design of a stabilizing state feedback controller becomes possible. A first problem in the design of such a controller is that the dynamic model of the finance system is unknown, and thus, it has to be identified with the use neurofuzzy approximators. The estimated dynamics provided by the approximators is used in the computation of the control input, thus establishing an indirect adaptive control scheme. The learning rate of the approximators is chosen from the requirement the system's Lyapunov function to have always a negative first-order derivative. Another problem that has to be dealt with is that the control loop is implemented only with the use of output feedback. To estimate the non-measurable state vector elements of the finance system, a state observer is implemented in the control loop. The computation of the feedback control signal requires the solution of two algebraic Riccati equations at each iteration of the control algorithm. Lyapunov stability analysis demonstrates first that an H-infinity tracking performance criterion is satisfied. This signifies elevated robustness against modelling errors and external perturbations. Moreover, global asymptotic stability is proven for the control loop.

In Chap. 5, nonlinear optimal control and filtering for financial systems is explained. A new nonlinear optimal control approach is proposed for the stabilization of the dynamics of a chaotic finance model. The dynamic model of the financial system, which expresses interaction between the interest rate, the investment demand, the price exponent and the profit margin, undergoes approximate linearization round local operating points. These local equilibria are defined at each iteration of the control algorithm and consist of the present value of the system's state vector and the last value of the control inputs vector that was exerted on it. The approximate linearization makes use of Taylor series expansion and of the computation of the associated Jacobian matrices. The truncation of higher order terms in the Taylor series expansion is considered to be a modelling error that is compensated by the robustness of the control loop. As the control algorithm runs, the temporary equilibrium is shifted towards the reference trajectory and finally converges to it. The control method needs to compute an H-infinity feedback control law at each iteration and requires the repetitive solution of an algebraic Riccati equation. Through Lyapunov stability analysis, it is shown that an H-infinity tracking performance criterion holds for the control loop. This implies elevated robustness against model approximations and external perturbations. Moreover, under moderate conditions, the global asymptotic stability of the finance system's feedback control is proven.

In Chap. 6, a Kalman Filtering approach for the detection of option mispricing in the Black–Scholes PDE is introduced. Financial derivatives and option pricing models are usually described with the use of stochastic differential equations and diffusion-type partial differential equations (e.g. Black–Scholes models). Considering the latter case in this chapter, a new filtering method for distributed parameter systems is developed for estimating option price variations without the knowledge of initial conditions. The proposed filtering method is the so-called Derivative-free nonlinear Kalman Filter and is based on a decomposition of the nonlinear partial differential equation model into a set of ordinary differential equations with respect to time. Next, each one of the local models associated with the ordinary differential equations is transformed into a model of the linear canonical (Brunovsky) form through a change of coordinates (diffeomorphism) which is based on differential flatness theory. This transformation provides an extended model of the nonlinear dynamics of the option pricing model for which state estimation is possible by applying the standard Kalman Filter recursion. Based on the provided state estimate, validation of the Black–Scholes PDE model can be performed and the existence of inconsistent parameters in the Black–Scholes PDE model can be concluded.

In Chap. 7, a Kalman Filtering approach to the detection of option mispricing in electric power markets is analysed. As mentioned in the previous chapter, option pricing models are usually described with the use of stochastic differential equations and diffusion-type partial differential equations (e.g. Black–Scholes models). In case of electric power markets these models are complemented with integral terms which describe the effects of jumps and changes in the diffusion process and which are associated with variations in the production rates, in the condition of the transmission and distribution system, in the pay-off capability, etc. Considering the latter case, that is a partial integrodifferential equation for the option's price, a new filtering method, is developed for estimating option price variations without knowledge of initial conditions. The proposed filtering method is the so-called Derivative-free nonlinear Kalman Filter and is based on a transformation of the initial option price dynamics into a state-space model of the linear canonical form. The transformation is shown to be based on differential flatness theory and finally provides a model of the option price dynamics for which state estimation is possible by applying the standard Kalman Filter recursion. Based on the provided state estimate, validation of the Black–Scholes partial integrodifferential equation can be performed and the existence of inconsistent parameters in the electricity market pricing model can be concluded.

In Chap. 8, corporations' default probability forecasting using the Derivative-free nonlinear Kalman Filter is explained. This chapter proposes a systematic method for forecasting default probabilities for financial firms with particular interest in electric power corporations. According to the credit risk theory, a company's closeness to default is determined by the distance of its assets' value from its debts. The assets' value depends primarily on the company's market (option) value through a complex nonlinear relation. By forecasting with accuracy the enterprise's option value, it becomes also possible to estimate the future value of the enterprise's assets and the associated probability of default. This chapter proposes a systematic method for forecasting the proximity to default for companies (option/asset value forecasting methods) using the new nonlinear Kalman Filtering method under the name Derivative-free nonlinear Kalman Filter. The firm's option value is considered to be described by the Black–Scholes nonlinear partial differential equation. Using differential flatness theory, the partial differential equation is transformed into an equivalent state-space model in the so-called canonical form. Using the latter model and by redesigning the Derivative-free nonlinear Kalman Filter as a m-step ahead predictor, estimates are obtained of the company's future option values. By forecasting the company's market (option) values, it becomes finally possible to forecast the associated asset value and volatility and also to estimate the company's future default risk.

In Chap. 9, validation of financial options models using neural networks with invariance to Fourier transform is explained. It is known that numerical solution of the Black–Scholes PDE enables to compute with precision the values of financial options, within a finite-time horizon. It is also known that solutions to the option pricing problem can be obtained in closed form using Fourier methods, such as the Fast Fourier Transform, the expansion in Fourier-cosine series or the expansion in Fourier–Hermite series. In this chapter, modelling of financial options' dynamics is performed, using a neural network with 2D Gauss-Hermite basis functions that remain invariant to Fourier transform. Knowing that the Gauss-Hermite basis functions satisfy the orthogonality property and remain unchanged under the Fourier transform, subjected only to a change of scale, one has that the considered neural network provides the spectral analysis of the options' dynamics model. Actually, the squares of the weights of the output layer of the neural network denote the spectral components for the monitored options' dynamics. By observing changes in the amplitude of the aforementioned spectral components, one can have also an indication about deviations from nominal values, for parameters that affect the options' dynamics, such as interest rate, dividend payment and volatility. Moreover, since specific parametric changes are associated with amplitude changes of specific spectral components of the options' model, isolation of the distorted parameters can be also performed.

In Chap. 10, statistical validation of financial forecasting tools with generalized likelihood ratio approaches is analysed. The local statistical approach for fault detection and isolation is applied to the problem of validation of a fuzzy model which can be used in forecasting. The method detects the inconsistencies between a fuzzy rule base and the modelled system. It can also identify which are the faulty parameters of the fuzzy model. The Fisher information matrix explains the detectability of changes in the parameters of the fuzzy model. Simulation tests illustrate the method's credibility. As a case study, statistical validation of a neurofuzzy model of chaotic time series is considered.

In Chap. 11, distributed Kalman Filtering for risk assessment in interconnected financial markets is analysed. In financial decision-making, such as in the trading of options, it is important to regularly validate the accuracy and reliability of decision support tools. In this context, this chapter introduces a distributed scheme for the validation of option price forecasting models enabling early diagnosis of options mispricing. It is considered that N independent agents monitor and forecast the variation of option prices through locally parameterized Kalman Filters. It is also assumed that final decision about the options' price is taken through a fuzzy consensus scheme, that is the individual forecasts of the distributed agents, provided by local Kalman Filters are fused with a fuzzy weighting process. Thus, forecasting is finally performed by a fuzzy Kalman Filter. It is likely, though, that some of the distributed models are improperly parametrized and fail to describe accurately the real dynamics of the option's market. To this end, a statistical method is developed capable of (i) detecting if the estimation about the options's price that is provided by the multi-agent system is sufficiently precise or not and (ii) isolating the ith agent that makes use of an improperly parameterized model. This chapter provides one of the few approaches for testing the accuracy of distributed Kalman Filters for financial decision-making and the only one that permits to detect parametric changes that are of magnitude of less than 1% of the nominal value of the monitored financial system.

In Chap. 12, stabilization of financial systems dynamics through feedback control of the Black–Scholes PDE is analysed. The objective of this chapter was to develop a boundary control method for the Black–Scholes PDE which describes option dynamics. It is shown that the procedure for numerical solution of Black–Scholes PDE results in a set of nonlinear ordinary differential equations (ODEs) and an associated state equations model. For the local subsystems, into which a Black–Scholes PDE is decomposed, it becomes possible to apply boundary-based feedback control. The controller design proceeds by showing that the state-space model of the Black–Scholes PDE stands for a differentially flat system. Next, for each subsystem which is related to a nonlinear ODE, a virtual control input is computed, which can invert the subsystem's dynamics and can eliminate the subsystem's tracking error. From the last row of the state-space description, the control input (boundary condition) that is actually applied to the Black–Scholes PDE is found. This control input contains recursively all virtual control inputs which were computed for the individual ODE subsystems associated with the previous rows of the state-space equation. Thus, by tracing the rows of the state-space model backwards, at each iteration of the control algorithm, one can finally obtain the control input that should be applied to the Black–Scholes PDE so as to assure that all its state variables will converge to the desirable setpoints.

In Chap. 13, stabilization of the multi-asset Black–Scholes PDE using differential flatness theory is analysed. A method for feedback control of the multi-asset Black–Scholes PDE is developed. By applying once more semi-discretization and a finite differences scheme, the multi-asset Black–Scholes PDE is transformed into a state-space model consisting of ordinary nonlinear differential equations. For this set of differential equations, it is shown that differential flatness properties hold. This enables to solve the associated control problem and to succeed stabilization of the options' dynamics. It is shown that the previous procedure results in a set of nonlinear ordinary differential equations (ODEs) and to an associated state equations model. For the local subsystems, into which a multi-asset Black–Scholes PDE is decomposed, it becomes possible to apply boundary-based feedback control. The controller design proceeds by showing that the state-space model of the multi-asset Black–Scholes PDE stands for a differentially flat system. Next, for each subsystem which is related to a nonlinear ODE, a virtual control input is computed, which can invert the subsystem's dynamics and can eliminate the subsystem's tracking error. From the last row of the state-space description, the control input (boundary condition) that is actually applied to the multi-asset Black–Scholes PDE system is found. This control input contains recursively all virtual control inputs which were computed for the individual ODE subsystems associated with the previous rows of the state-space equation. Thus, by tracing the rows of the state-space model backwards, at each iteration of the control algorithm, one can finally obtain the control input that should be applied to the multi-asset Black–Scholes PDE so as to assure that all its state variables will converge to the desirable setpoints.

In Chap. 14, stabilization of commodities pricing PDE using differential flatness theory is explained. Pricing of commodities (e.g. oil, carbon, mining products, electric power and agricultural crops) is vital for the majority of transactions taking place in financial markets. A method for feedback control of commodities pricing dynamics is developed. The PDE model of the commodities price dynamics is shown to be equivalent to a multi-asset Black–Scholes PDE. Actually, it is a diffusion process evolving in a 2D assets space, where the first asset is the commodity's spot price and the second asset is the convenience yield. As in the previous chapters, by applying semi-discretization and a finite differences scheme, this multi-asset PDE is transformed into a state-space model consisting of ordinary nonlinear differential equations. For the local subsystems, into which the commodities PDE is decomposed, it becomes possible to apply boundary-based feedback control. The controller design proceeds by showing that the state-space model of the commodities PDE stands for a differentially flat system. Next, for each subsystem which is related to a nonlinear ODE, a virtual control input is computed, which can invert the subsystem's dynamics and can eliminate the subsystem's tracking error. From the last row of the state-space description, the control input (boundary condition) that is actually applied to the multi-factor commodities' PDE system is found. This control input contains recursively all virtual control inputs which were computed for the individual ODE subsystems associated with the previous rows of the state-space equation. Thus, by tracing the rows of the state-space model backwards, at each iteration of the control algorithm, one can finally obtain the control input that should be applied to the commodities PDE system so as to assure that all its state variables will converge to the desirable setpoints. By demonstrating the feasibility of such a control method it is also proven that through selected purchase and sales during the trading procedure, the price of the negotiated commodities can be made to converge and stabilize at specific reference values.

In Chap. 15, stabilization of mortgage price dynamics using differential flatness theory is analysed. Pricing of mortgages (loans for the purchase of residences, land or farms) is vital for the majority of transactions taking place in financial markets. In this chapter, a method for stabilization of mortgage price dynamics is developed. It is considered that mortgage prices follow a PDE model which is equivalent to a multi-asset Black–Scholes PDE. Actually, it is a diffusion process evolving in a 2D assets space, where the first asset is the residence price and the second asset is the interest rate. By applying semi-discretization and a finite differences scheme, this multi-asset PDE is transformed into a state-space model consisting of ordinary nonlinear differential equations. For the local subsystems, into which the mortgage PDE is decomposed, it becomes possible to apply boundary-based feedback control. The controller design proceeds by showing that the state-space model of the mortgage price PDE stands for a differentially flat system. Next, for each subsystem which is related to a nonlinear ODE, a virtual control input is computed, which can invert the subsystem's dynamics and can eliminate the subsystem's tracking error. From the last row of the state-space description, the control input (boundary condition) that is actually applied to the multi-factor mortgage price PDE system is found. This control input contains recursively all virtual control inputs which were computed for the individual ODE subsystems associated with the previous rows of the state-space equation. Thus, by tracing the rows of the state-space model backwards, at each iteration of the control algorithm, one can finally obtain the control input that should be applied to the mortgage price PDE system so as to assure that all its state variables will converge to the desirable setpoints. By showing the feasibility of such a control method, it is also proven that through selected modification of the PDE boundary conditions, the price of the mortgage can be made to converge and stabilize at specific reference values.

The main purpose of this book was to disseminate new findings useful for academic teaching and research in the area of financial engineering and to develop systematic methods for management and risk minimization in financial systems. Methods for solving control and estimation problems in financial systems become progressively part of the curriculum of several academic departments at undergraduate level. This is because there is a need to acquaint future engineers with technologies that enable the functioning of financial systems according to the desirable specifications, even under uncertainty and partial information about their dynamic model. The present book contains teaching material which can be used for independent courses on financial engineering. This book can also serve perfectly the needs of postgraduate courses on financial engineering where more emphasis can be given to advanced computational and the mathematical techniques for the profitable and risk-free management of financial systems. The title of the course can be the same as the title of the book, i.e. state-space approaches to modelling and control in financial engineering: systems theory and machine learning methods. Starting from the analysis of dynamical systems theory and of established approaches for control and estimation in nonlinear dynamical systems, the monograph moves progressively to the solution of key problems met in financial engineering, such as (i) nonlinear control and filtering for financial systems exhibiting complex and chaotic dynamics, (ii) control and estimation for the PDE dynamics of financial systems, and (iii) statistical validation of decision support tools used in financial engineering. Through the balanced interaction between the theoretical and the application part, students can assimilate the new knowledge and can become efficient in control and estimation of financial systems and in methods for the optimized management of capitals and assets.

However, this book and is not only addressed to the academic community but also targets people working in practical problems and applications of financial engineering. There is continuous demand for developing elaborated software tools that will enable optimal decision-making about financial systems. To this end, there is a need to eliminate heuristics and intuition-based approaches in financial engineering and to establish methods that assure stabilization and convergence of financial systems to desirable performance indexes. The monograph's contribution to this direction is clear.

Athens, Greece
March 2017

Gerasimos G. Rigatos Ph.D.
Electrical and Computer Engineer

Acknowledgements

The author of this monograph would like to thank researchers in the area of financial systems, as well as in the area of dynamical systems modelling and control for contributing to the development and completion of this research work, through reviews, comments and meaningful remarks.

Contents

Chapter 1
Systems Theory and Stability Concepts

1.1 Outline

The chapter analyzes the basics of systems theory which can be used in the modeling of financial systems. Financial systems may exhibit complex dynamics characterized by oscillations or chaos. Parametric variations in the models of financial systems may also affect and modify their stability properties. The following properties are analyzed: phase diagram, isoclines, attractors, local stability, bifurcations of fixed points and chaos properties.

1.2 Characteristics of the Dynamics of Nonlinear Systems

Main features characterizing the stability of nonlinear dynamical systems are defined as follows [121, 274]:

1. *Finite escape time*: It is the finite time within which the state-vector of the nonlinear system converges to infinity.
2. *Multiple isolated equilibria*: A linear system can have only one equilibrium to which converges the state vector of the system in steady-state. A nonlinear system can have more than one isolated equilibria (fixed points). Depending on the initial state of the system, in steady-state the state vector of the system can converge to one of these equilibria.
3. *Limit cycles*: For a linear system to exhibit oscillations it must have eigenvalues on the imaginary axis. The amplitude of the oscillations depends on initial conditions. In nonlinear systems one may have oscillations of constant amplitude and frequency, which do not depend on initial conditions. This type of oscillations is known as *limit cycles*.
4. *Sub-harmonic, harmonic and almost periodic oscillations*: A stable linear system under periodic input produces an output of the same frequency. A nonlinear system,

G.G. Rigatos, *State-Space Approaches for Modelling and Control in Financial Engineering*, Intelligent Systems Reference Library 125,
DOI 10.1007/978-3-319-52866-3_1

under periodic excitation can generate oscillations with frequencies which are several times smaller (subharmonic) or multiples of the frequency of the input (harmonic). It may also generate almost periodic oscillations with frequencies which are not necessarily multiples of a basis frequency (almost periodic oscillations).
5. *Chaos*: A nonlinear system in steady-state can exhibit a behavior which is not characterized as equilibrium, periodic oscillation or almost periodic oscillation. This behavior is characterized as chaos. As time advances the behavior of the system changes in a random-like manner, and this depends on the initial conditions. Although the dynamic system is deterministic it exhibits randomness in the way it evolves in time.
6. *Multiple modes of behavior*: It is possible the same dynamical system to exhibit simultaneously more than one of the aforementioned characteristics (1)–(5). Thus, a system without external excitation may exhibit simultaneously more than one limit cycles. A system receiving a periodic external input may exhibit harmonic or subharmonic oscillations, or an even more complex behavior in steady state which depends on the amplitude and frequency of the excitation.

1.3 Computation of Isoclines

An autonomous second order system is described by two differential equations of the form

$$\begin{aligned} \dot{x}_1 &= f_1(x_1, x_2) \\ \dot{x}_2 &= f_2(x_1, x_2) \end{aligned} \tag{1.1}$$

The method of the isoclines consists of computing the slope (ratio) between f_2 and f_1 for every point of the trajectory of the state vector (x_1, x_2).

$$s(x) = \frac{f_2(x_1,x_2)}{f_1(x_1,x_2)} \tag{1.2}$$

The case $s(x) = c$ describes a curve in the $x_1 - x_2$ plane along which the trajectories $\dot{x}_1 = f_1(x_1, x_2)$ and $\dot{x}_2 = f_2(x_1, x_2)$ have a constant slope.

The curve $s(x) = c$ is drawn in the $x_1 - x_2$ plane and along this curve one also draws small linear segments of length c. The curve $s(x) = c$ is known as isocline. The direction of these small linear segments is according to the sign of the ratio $f_2(x_1, x_2)/f_1(x_1, x_2)$.

Example 1:

The following simplified nonlinear dynamical system is considered

$$\begin{aligned} \dot{x}_1 &= x_2 \\ \dot{x}_2 &= -sin(x_1) \end{aligned} \tag{1.3}$$

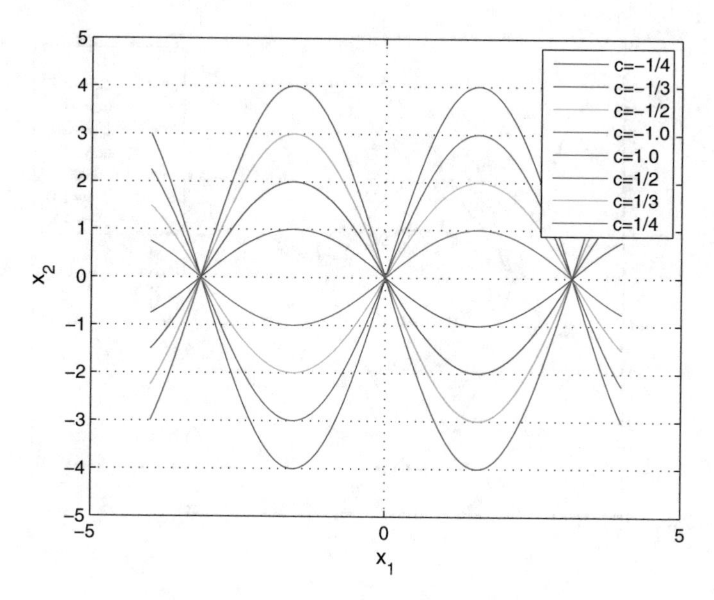

Fig. 1.1 Isoclines diagram for $s(x) = \frac{-sin(x_1)}{x_2}$

The slope $s(x)$ is given by the relation

$$s(x) = \frac{f_2(x_1,x_2)}{f_1(x_1,x_2)} \Rightarrow s(x) = -\frac{sin(x_2)}{x_2} \tag{1.4}$$

Setting $s(x) = c$ it holds that the isoclines are given by the relation

$$x_2 = -\frac{1}{c} sin(x_1) \tag{1.5}$$

For different values of c one has the following isoclines diagram depicted in Fig. 1.1.
Example 2:

The following oscillator model is considered, being free of friction and with state-space equations

$$\begin{aligned} \dot{x}_1 &= x_2 \\ \dot{x}_2 &= -0.5x_2 - sin(x_1) \end{aligned} \tag{1.6}$$

To compute isoclines one has

$$\begin{aligned} s(x) = \frac{-0.5x_2 - sin(x_1)}{x_2} = c \Rightarrow -0.5x_2 - sin(x_1) = cx_2 \Rightarrow \\ (0.5 + c)x_2 = sin(x_1) \Rightarrow x_2 = -\frac{1}{0.5+c} sin(x_1) \end{aligned} \tag{1.7}$$

For different values of parameter c the isoclines are depicted in Fig. 1.2.

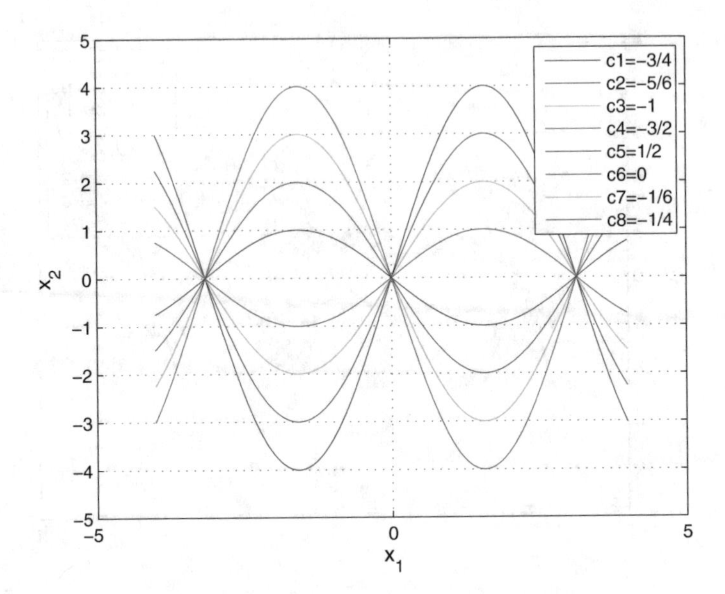

Fig. 1.2 Isoclines diagram for $s(x) = \frac{-0.5x_2 - sin(x_1)}{x_2}$

1.4 Stability Features of Dynamical Systems

Basic features that are important in the study of nonlinear dynamical systems are (i) equilibria (fixed points), (ii) limit cycles, (iii) phase diagrams, (iv) periodic orbits and (v) bifurcations of fixed points [121, 222, 274]. The definition of these features will be given through examples

1.4.1 The Phase Diagram

One can consider the following nonlinear model with the two state variables, V and η. The dynamics of this model can be written as

$$\begin{aligned} \frac{dV}{dt} &= f(V, t) \\ \frac{d\eta}{dt} &= g(V, \eta) \end{aligned} \tag{1.8}$$

The phase diagram consists of the points on the trajectories of the solution of the associated differential equation, i.e. $(V(t_k), \eta(t_k))$.

At a fixed point or equilibrium it holds $f(V_R, \eta_R) = 0$ and $g(V_R, \eta_R) = 0$. The closed trajectories are associated with periodic solutions. If there are closed trajectories then $\exists T > 0$ such that $(V(t_k), \eta(t_k)) = (V(t_k + T), \eta(t_k + T))$.

Another useful parameter is the nullclines. The V-nullcline is characterized by the relation $\dot{V} = f(V, \eta) = 0$. The η-nullcline is characterized by the relation $\dot{\eta} = g(V, \eta) = 0$. The fixed points (or equilibria) are found on the intersection of nullclines.

1.4.2 *Stability Analysis of Nonlinear Systems*

1.4.2.1 Local Linearization

A manner to examine stability in nonlinear dynamical systems is to perform local linearization around an equilibrium. Assume the nonlinear system $\dot{x} = f(x, u)$ and $f(x_0, u) = 0$ that is x_0 is the local equilibrium. The linearization of $f(x, u)$ with respect to u round x_0 (Taylor series expansion) gives the equivalent description

$$\dot{x} = Ax + Bu \tag{1.9}$$

where

$$A = \nabla_x f = \begin{pmatrix} \frac{\partial f_1}{\partial x_1} & \frac{\partial f_1}{\partial x_2} & \cdots & \frac{\partial f_1}{\partial x_n} \\ \frac{\partial f_2}{\partial x_1} & \frac{\partial f_2}{\partial x_2} & \cdots & \frac{\partial f_2}{\partial x_n} \\ \cdots & \cdots & \cdots & \cdots \\ \frac{\partial f_n}{\partial x_1} & \frac{\partial f_n}{\partial x_2} & \cdots & \frac{\partial f_n}{\partial x_n} \end{pmatrix} |_{x=x_0} \tag{1.10}$$

and

$$B = \nabla_u f = \begin{pmatrix} \frac{\partial f_1}{\partial u_1} & \frac{\partial f_1}{\partial u_2} & \cdots & \frac{\partial f_1}{\partial u_n} \\ \frac{\partial f_2}{\partial u_1} & \frac{\partial f_2}{\partial u_2} & \cdots & \frac{\partial f_2}{\partial u_n} \\ \cdots & \cdots & \cdots & \cdots \\ \frac{\partial f_n}{\partial u_1} & \frac{\partial f_n}{\partial u_2} & \cdots & \frac{\partial f_n}{\partial u_n} \end{pmatrix} |_{x=x_0} \tag{1.11}$$

The eigenvalues of matrix A define the local stability features of the system:

Example 1:

Assume the nonlinear system

$$\frac{d^2x}{dt^2} + 2x\frac{dx}{dt} + 2x^2 - 4x = 0 \tag{1.12}$$

By defining the state variables $x_1 = x$, $x_2 = \dot{x}$ the system can be written in the following form

$$\begin{aligned} \dot{x}_1 &= x_2 \\ \dot{x}_2 &= -2x_1x_2 - 2x_1^2 + 4x_1 \end{aligned} \tag{1.13}$$

It holds that $\dot{x} = 0$ if $f(x) = 0$ that is $(x_1^{*,1}, x_2^{*,1}) = (0, 0)$ and $(x_1^{*,2}, x_2^{*,2}) = (2, 0)$. Round the first equilibrium $(x_1^{*,1}, x_2^{*,1}) = (0, 0)$ the system's dynamics is written as

$$A = \nabla_x f = \begin{pmatrix} 0 & 1 \\ 4 & 0 \end{pmatrix} \text{ or } \begin{pmatrix} \dot{x}_1 \\ \dot{x}_2 \end{pmatrix} = \begin{pmatrix} 0 & 1 \\ 4 & 0 \end{pmatrix} \begin{pmatrix} x_1 \\ x_2 \end{pmatrix} \tag{1.14}$$

The eigenvalues of the system are $\lambda_1 = 2$ and $\lambda_2 = -2$. This means that the fixed point $(x_1^{*,1}, x_2^{*,1}) = (0, 0)$ is an unstable one.

Next, the fixed point $(x_1^{*,2}, x_2^{*,2}) = (2, 0)$ is analyzed. The associated Jacobian matrix is computed again. It holds that

$$A = \nabla_x f = \begin{pmatrix} 0 & 1 \\ -4 & -4 \end{pmatrix} \tag{1.15}$$

The eigenvalues of the system are $\lambda_1 = -2$ and $\lambda_2 = -2$. Consequently, the fixed point $(x_1^{*,2}, x_2^{*,2}) = (2, 0)$ is a stable one.

1.4.2.2 Lyapunov Stability Approach

The Lyapunov method analyzes the stability of a dynamical system without the need to compute explicitly the trajectories of the state vector $x = [x_1, x_2, \cdots, x_n]^T$.

Theorem: The system described by the relation $\dot{x} = f(x)$ is asymptotically stable in the vicinity of the equilibrium $x_0 = 0$ if there is a function $V(x)$ such that

(i) $V(x)$ to be continuous and to have a continuous first order derivative at x_0
(ii) $V(x) > 0$ if $x \neq 0$ and $V(0) = 0$
(iii) $\dot{V}(x) < 0, \forall x \neq 0$.

The Lyapunov function is usually chosen to be a quadratic (and thus positive) energy function of the system however there in no systematic method to define it.

Assume now, that $\dot{x} = f(x)$ and $x_0 = 0$ is the equilibrium. Then the system is globally asymptotically stable if for every $\varepsilon > 0$, $\exists \delta(\varepsilon) > 0$, such that if $||x(0)|| < \delta$ then $||x(t)|| < \varepsilon, \forall t \geq 0$.

This means that if the state vector of the system starts in a disc of radius δ then as time advances it will remain in the disc of radius ε, as shown in Fig. 1.3. Moreover, if $lim_{t \to \infty} ||x(t)|| = x_0 = 0$ then the system is globally asymptotically stable.

Example 1: Consider the system

$$\begin{aligned} \dot{x}_1 &= x_2 \\ \dot{x}_2 &= -x_1 - x_3^2 \end{aligned} \tag{1.16}$$

The following Lyapunov function is defined

$$V(x) = x_1^2 + x_2^2 \tag{1.17}$$

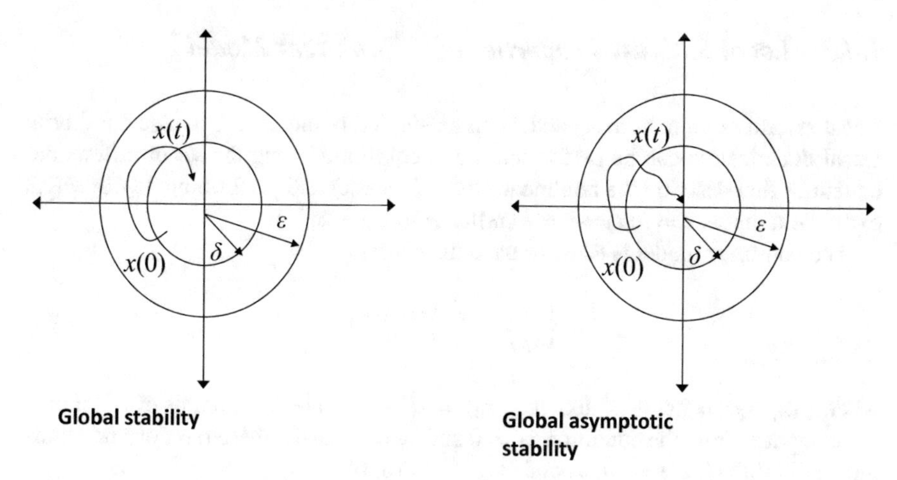

Fig. 1.3 Global stability and global asymptotic stability

The equilibrium point is $(x_1 = 0, x_2 = 0)$. It holds that $V(x) > 0\ \forall\ (x_1, x_2) \neq (0, 0)$ and $V(x) = 0$ for $(x_1, x_2) = (0, 0)$. Moreover, it holds

$$\begin{aligned} \dot{V}(x) = 2x_1\dot{x}_1 + 2x_2\dot{x}_2 = 2x_1x_2 + 2x_2(-x_1 - x_2^3) \Rightarrow \\ \dot{V}(x) = -2x_2^4 < 0\ \forall\ (x_1, x_2) \neq (0, 0) \end{aligned} \tag{1.18}$$

Therefore, the system is asymptotically stable and $lim_{t\to\infty}(x_1, x_2) = (0, 0)$.

Example 2: Consider the system

$$\begin{aligned} \dot{x}_1 = -x_1(1 + 2x_1x_2^2) \\ \dot{x}_2 = x_1^3x_2 \end{aligned} \tag{1.19}$$

The following Lyapunov function is considered

$$V(x) = \tfrac{1}{2}x_1^2 + x_2^2 \tag{1.20}$$

The equilibrium point is $x_1 = 0, x_2 = 0$. It holds that

$$\begin{aligned} \dot{V}(x) = x_1\dot{x}_1 + 2x_2\dot{x}_2 = -x_1^2(1 + 2x_1x_2^2 + 2x_2(x_1^3x_2)) \Rightarrow \\ \dot{V}(x) = -x_1^2 < 0\ \forall\ (x_1, x_2) \neq (0, 0) \end{aligned} \tag{1.21}$$

Therefore, the system is asymptotically stable and $lim_{t\to\infty}(x_1, x_2) = (0, 0)$.

1.4.3 Local Stability Properties of a Nonlinear Model

Local stability of a nonlinear model can be studied round the associated equilibria. Local linearization can be performed round equilibria, using the set of differential equations that describe the nonlinear model $\dot{x} = h(x)$ and performing Taylor series expansion, that is $\dot{x} = h(x) \Rightarrow \dot{x} = h(x_0)|_{x_0} + \nabla_x h(x - x_0) + \cdots$.

The nonlinear model is taken to have the generic form

$$\begin{pmatrix} \dot{x}_1 \\ \dot{x}_2 \end{pmatrix} = \begin{pmatrix} f(x_1, x_2) \\ g(x_1, x_2) \end{pmatrix} \tag{1.22}$$

where $f(x_1, x_2) = 2x_1 + x_2^2$ and $g(x_1, x_2) = x_1^2 + 2x_2$. The fixed points of this model are computed from the condition $\dot{x}_1 = 0$ and $\dot{x}_2 = 0$. Using these relations one finds the equilibria $(x_1^*, x_2^*) = (0, 0)$ and $(x_1^*, x_2^*) = (0, 0)$

The Jacobian matrix $\nabla_x h = M$ is given by

$$M = \begin{pmatrix} \frac{\partial f}{\partial x_1} & \frac{\partial f}{\partial x_2} \\ \frac{\partial g}{\partial x_1} & \frac{\partial g}{\partial x_2} \end{pmatrix} \tag{1.23}$$

which results into the matrix

$$J = \begin{pmatrix} 2 & 2x_2 \\ 2x_1 & 2 \end{pmatrix} \tag{1.24}$$

The eigenvalues of matrix M define stability round fixed points (stable or unstable fixed point). To this end, one has to find the roots of the associated characteristic polynomial that is given by $det(\lambda I - J) = 0$ where I is the identity matrix. Thus for the fixed point $(x_1^*, x_2^*) = (0, 0)$ one obtains

$$det(\lambda I - M) = \begin{pmatrix} \lambda - 2 & -2x_2 \\ -2x_1 & \lambda - 2 \end{pmatrix}_{|((x_1^*, x_2^*) = (0,0))} \tag{1.25}$$

which results into the characteristic polynomial $p(\lambda) = (\lambda - 2)^2$ and has the positive eigenvalues $\lambda_1 = 2$, $\lambda_2 = 2$. Therefore, the equilibrium $(x_1^*, x_2^*) = (0, 0)$ is an unstable one. For the fixed point $(x_1^*, x_2^*) = (-2, -2)$ one has

$$det(\lambda I - M) = \begin{pmatrix} \lambda - 2 & -2x_2 \\ -2x_1 & \lambda - 2 \end{pmatrix}_{|((x_1^*, x_2^*) = (-2,2))} \tag{1.26}$$

which results into the characteristic polynomial $p(\lambda) = (\lambda - 2)^2 - 16$ and has the eigenvalues $\lambda_1 = 6$, $\lambda_2 = -2$. Therefore, the equilibrium $(x_1^*, x_2^*) = (-2, -2)$ is again an unstable one.

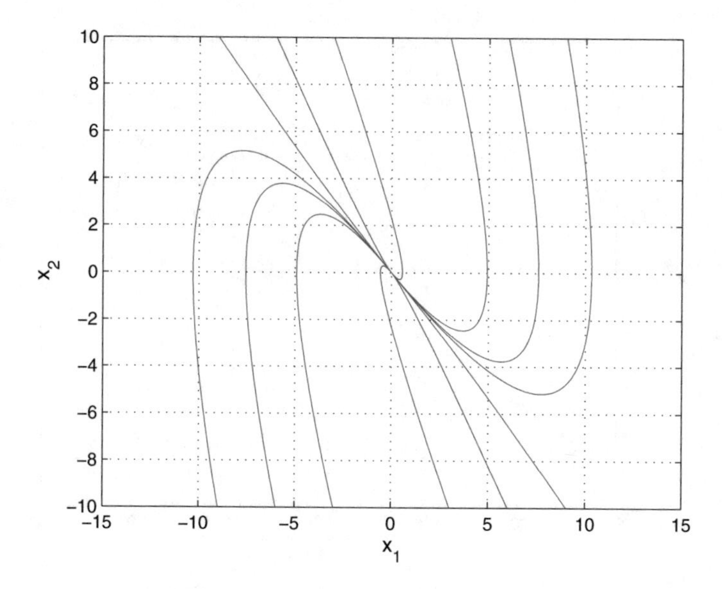

Fig. 1.4 Phase diagram of initial state variables x_1, x_2 of a 2nd order linear autonomous system with negative eigenvalues, where $\lambda_1 < \lambda_2 < 0$

1.5 Phase Diagrams and Equilibria

1.5.1 Phase Diagrams for Linear Dynamical Systems

The following autonomous linear system is considered

$$\dot{x} = Ax \tag{1.27}$$

The eigenvalues of matrix A define the system dynamics. Some terminology associated with fixed points is as follows:

A fixed point for the system of Eq. (1.27) is called hyperbolic if none of the eigenvalues of matrix A has zero real part. A hyperbolic fixed point is called a saddle if some of the eigenvalues of matrix A have real parts greater than zero and the rest of the eigenvalues have real parts less than zero. If all of the eigenvalues have negative real parts then the hyperbolic fixed point is called a stable node or sink. If all of the eigenvalues have positive real parts then the hyperbolic fixed point is called an unstable node or source. If the eigenvalues are purely imaginary then one has an elliptic fixed point which is said to be a center.

Case 1: Both eigenvalues of matrix A are real and unequal, that is $\lambda_1 \neq \lambda_1 \neq 0$.
For $\lambda_1 < 0$ and $\lambda_2 < 0$ the phase diagram for z_1 and z_2 is shown in Fig. 1.4.
In case that λ_2 is smaller than λ_1 the term $e^{\lambda_2 t}$ decays faster than $e^{\lambda_1 t}$.
For $\lambda_1 > 0 > \lambda_2$ the phase diagram of Fig. 1.5 is obtained.

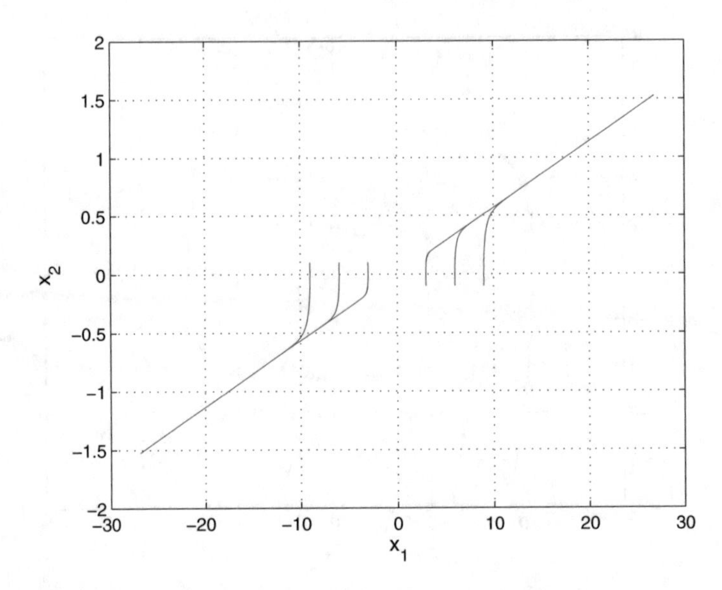

Fig. 1.5 Phase diagram of initial state variables x_1, x_2 of a 2nd order linear autonomous system with real eigenvalues, where $\lambda_1 > 0 > \lambda_2$

In the latter case there are stable trajectories along eigenvector v_1 and unstable trajectories along eigenvector v_2 of matrix A. The stability point (0, 0) is said to be a saddle point.

When $\lambda_1 > \lambda_2 > 0$ then one has the phase diagrams of Fig. 1.6.

Case 2: Complex eigenvalues:
Typical phase diagrams in the case of stable complex eigenvalues are given in Fig. 1.7.

Typical phase diagrams in the case of unstable complex eigenvalues are given in Fig. 1.8.

Typical phase diagrams in the case of imaginary eigenvalues are given in Fig. 1.9.

Case 3: Matrix A has nonzero eigenvalues which are equal to each other. The associated phase diagram is given in Fig. 1.10.

1.5.2 Multiple Equilibria for Nonlinear Dynamical Systems

A nonlinear system can have multiple equilibria as shown in the following example. Consider, for instance the model of a pendulum under friction

$$\begin{aligned} \dot{x}_1 &= x_2 \\ \dot{x}_2 &= -\tfrac{g}{l} sin(x_1) - \tfrac{K}{m} x_2 \end{aligned} \tag{1.28}$$

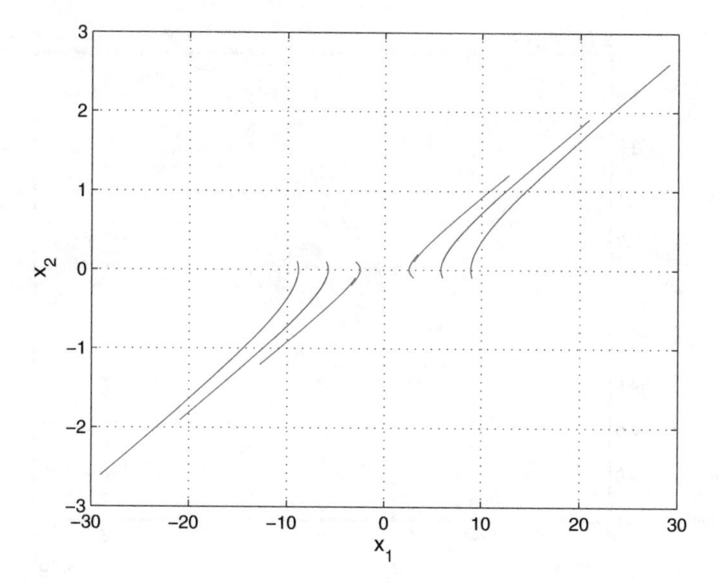

Fig. 1.6 Phase diagram of state variables x_1, x_2 of a 2nd order linear autonomous system with real eigenvalues $\lambda_1 > \lambda_2 > 0$

The associated phase diagram is designed for different initial conditions and is given in Fig. 1.11.

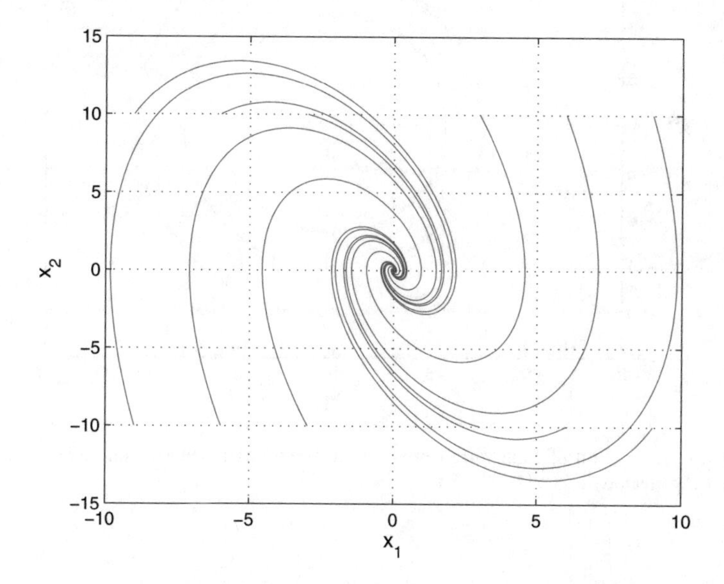

Fig. 1.7 Phase diagram of state variables x_1, x_2 of a 2nd order linear autonomous system with complex stable eigenvalues

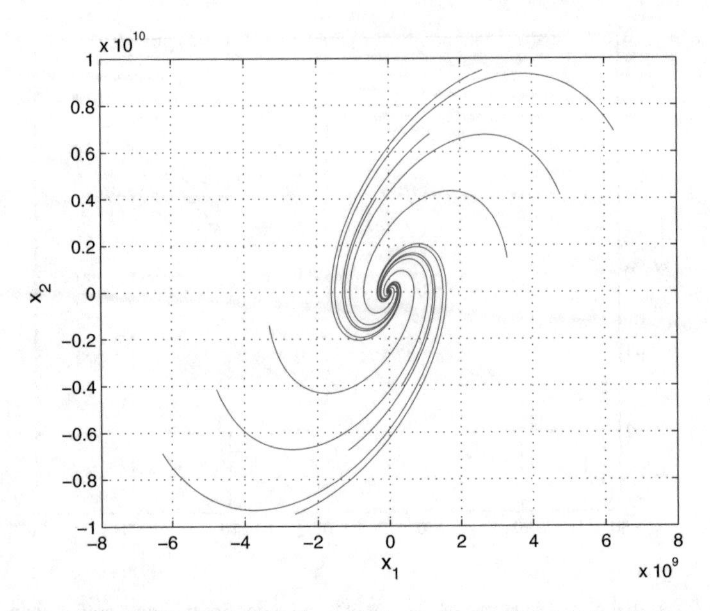

Fig. 1.8 Phase diagram of state variables x_1, x_2 of a 2nd order linear autonomous system with complex unstable eigenvalues

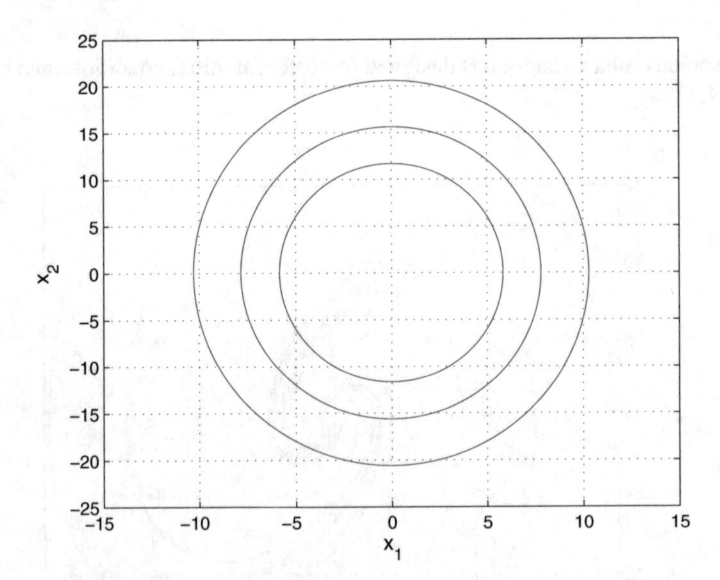

Fig. 1.9 Phase diagram of state variables x_1, x_2 of a 2nd order linear autonomous system with imaginary eigenvalues

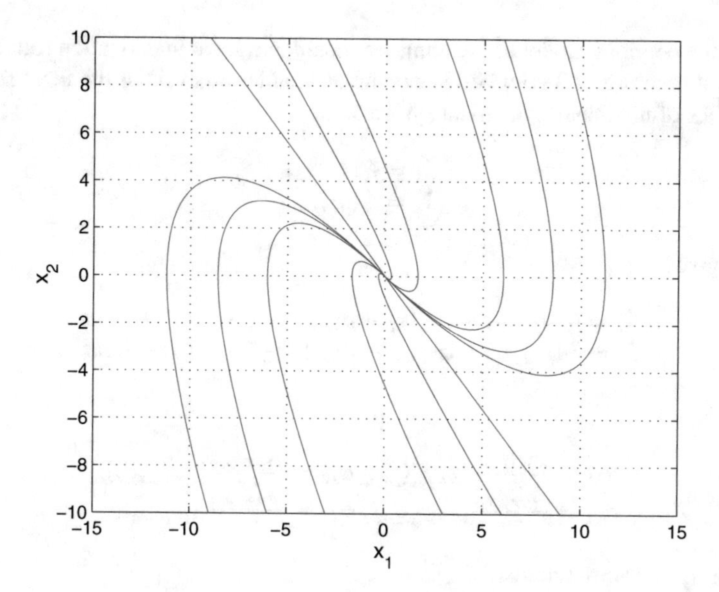

Fig. 1.10 Phase diagram of state variables x_1, x_2 of a 2nd order linear autonomous system with identical stable eigenvalues

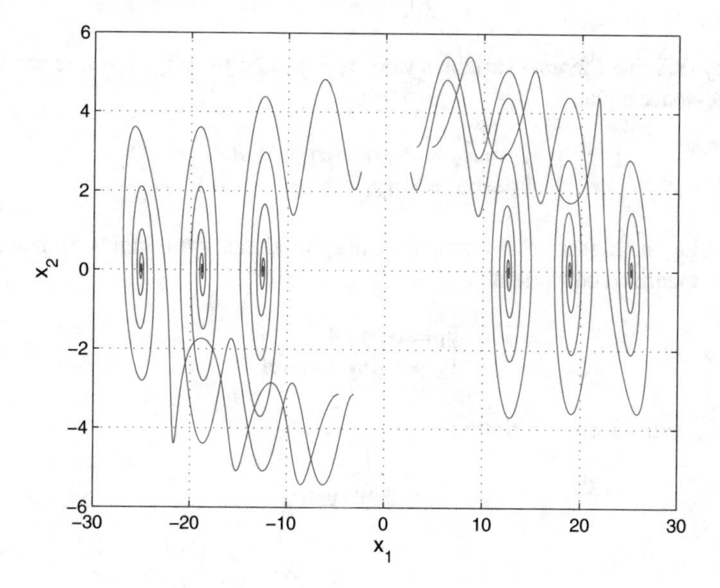

Fig. 1.11 Phase diagram of state variables x_1, x_2 of a 2nd order nonlinear oscillator that exhibits multiple equilibria

For the previous model of the nonlinear oscillator, local linearization round equilibria with the use of Taylor series expansion enables analysis of the local stability properties of nonlinear dynamical systems

$$\begin{aligned} \dot{x}_1 &= f_1(x_1, x_2) \\ \dot{x}_2 &= f_2(x_1, x_2) \end{aligned} \tag{1.29}$$

which gives

$$\begin{aligned} \dot{x}_1 &= f_1(p_1, p_2) + \alpha_{11}(x_1 - p_1) + \alpha_{12}(x_2 - p_2) + h.o.t \\ \dot{x}_2 &= f_2(p_1, p_2) + \alpha_{21}(x_1 - p_1) + \alpha_{22}(x_2 - p_2) + h.o.t \end{aligned} \tag{1.30}$$

where

$$\begin{aligned} \alpha_{11} &= \frac{\partial f_1(x_1,x_2)}{\partial x_1}|_{x_1=p_1,x_2=p_2} \ \alpha_{12} = \frac{\partial f_1(x_1,x_2)}{\partial x_2}|_{x_1=p_1,x_2=p_2} \\ \alpha_{21} &= \frac{\partial f_2(x_1,x_2)}{\partial x_1}|_{x_1=p_1,x_2=p_2} \ \alpha_{22} = \frac{\partial f_2(x_1,x_2)}{\partial x_2}|_{x_1=p_1,x_2=p_2} \end{aligned} \tag{1.31}$$

For the equilibrium it holds:

$$\begin{aligned} f_1(p_1, p_2) &= 0 \\ f_2(p_1, p_2) &= 0 \end{aligned} \tag{1.32}$$

Next, by defining the new variables $y_1 = x_1 - p_1$ and $y_2 = x_2 - p_2$ one can rewrite the state-space equation

$$\begin{aligned} \dot{y}_1 &= \dot{x}_1 = \alpha_{11}y_1 + \alpha_{12}y_2 + h.o.t. \\ \dot{y}_2 &= \dot{x}_2 = \alpha_{21}y_1 + \alpha_{22}y_2 + h.o.t. \end{aligned} \tag{1.33}$$

By omitting the higher order terms one can approximate the initial nonlinear system with its linearized equivalent

$$\begin{aligned} \dot{y}_1 &= \alpha_{11}y_1 + \alpha_{12}y_2 \\ \dot{y}_2 &= \alpha_{21}y_1 + \alpha_{22}y_2 \end{aligned} \tag{1.34}$$

which in matrix form is written as

$$\dot{y} = Ay \tag{1.35}$$

where

$$A = \begin{pmatrix} \alpha_{11} & \alpha_{12} \\ \alpha_{21} & \alpha_{22} \end{pmatrix} = \begin{pmatrix} \frac{\partial f_1}{\partial x_1} & \frac{\partial f_1}{\partial x_2} \\ \frac{\partial f_2}{\partial x_1} & \frac{\partial f_2}{\partial x_2} \end{pmatrix}|_{x=p} = \frac{\partial f}{\partial x}|_{x=p} \tag{1.36}$$

Matrix $A = \frac{\partial f}{\partial x}$ is the Jacobian matrix of the system that is computed at point $(x_1, x_2) = (p_1, p_2)$. It is anticipated that the trajectories of the phase diagram of the linearized system in the vicinity of the equilibrium point will be also close to the trajectories of the phase diagram of the nonlinear system.

Therefore, if the origin (equilibrium) in the phase diagram of the linearized system is (i) a stable node (matrix A has stable linear eigenvalues), (ii) a stable focus (matrix A has stable complex eignevalues), (iii) a saddle point (matrix A has some eigenvalues with negative real part while the rest of the eigenvalues have positive real part) then one concludes that the same properties hold for phase diagram of the nonlinear system.

Example: A nonlinear oscillator of the following form is considered:

$$\dot{x} = f(x) \Rightarrow$$

$$\begin{pmatrix} \dot{x}_1 \\ \dot{x}_2 \end{pmatrix} = \begin{pmatrix} x_2 \\ -sin(x_1) - 0.5x_2 \end{pmatrix} \tag{1.37}$$

The associated Jacobian matrix is

$$\frac{\partial f}{\partial x} = \begin{pmatrix} 0 & 1 \\ -cos(x_1) & -0.5 \end{pmatrix} \tag{1.38}$$

There are two equilibrium points $(0, 0)$ and $(\pi, 0)$. The linearization round the equilibria gives

$$A_1 = \begin{pmatrix} 0 & 1 \\ -1 & -0.5 \end{pmatrix} \quad A_2 = \begin{pmatrix} 0 & 1 \\ 1 & -0.5 \end{pmatrix}$$

$$\text{with eigenvalues} \qquad \text{with eigenvalues} \tag{1.39}$$

$$\lambda_{1,2} = -0.25 \pm j0.97 \ \lambda_1 = -1.28 \ \lambda_2 = 0.78$$

Consequently, the equilibrium $(0, 0)$ is a stable focus (matrix A_1 has stable complex eignevalues) and the equilibrium $(\pi, 0)$ is a saddle point (matrix A_2 has an unstable eigenvalue).

1.5.3 Limit Cycles

A dynamical system is considered to exhibit limit cycles when it admits the nontrivial periodic solution

$$x(t + T) = x(t) \ \forall \ t \geq 0 \tag{1.40}$$

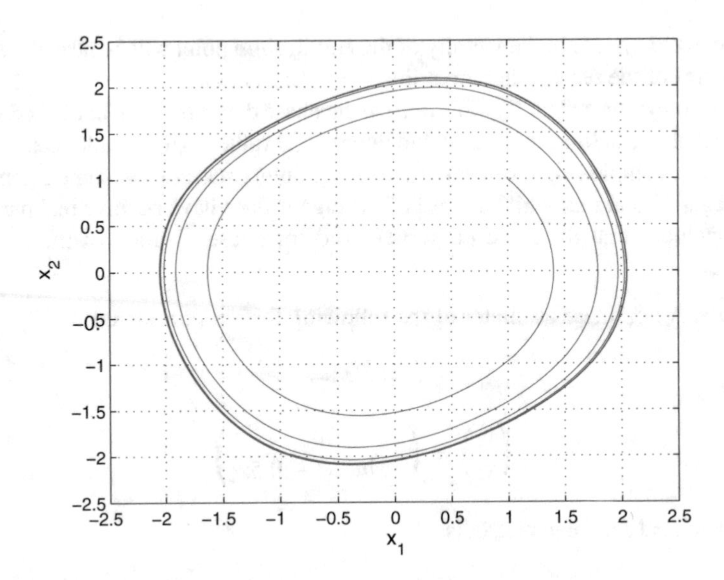

Fig. 1.12 Phase diagram of the Van der Pol oscillator for $\varepsilon = 0.2$

for some $T > 0$ (the trivial periodic solution is the one associated with $x(t)$=constant). An example about the existence of limit cycles is examined in the case of the Van der Pol oscillator. The state equations of the oscillator are

$$\begin{aligned} \dot{x}_1 &= x_2 \\ \dot{x}_2 &= -x_1 + \varepsilon(1 - x_1^2)x_2 \end{aligned} \tag{1.41}$$

Next the phase diagram of the Van der Pol oscillator is designed for two different values of parameter ε, namely $\varepsilon = 0.2$, $\varepsilon = 1$. In Figs. 1.12 and 1.13, it can be observed that in all cases there is a closed trajectory to which converge all curves of the phase diagram that start from points far from it.

To study conditions under which dynamical systems exhibit limit cycles, a second order autonomous nonlinear system is considered next, given by

$$\begin{pmatrix} \dot{x}_1 \\ \dot{x}_2 \end{pmatrix} = \begin{pmatrix} f_1(x_1, x_2) \\ f_2(x_1, x_2) \end{pmatrix} \tag{1.42}$$

Next, the following theorem defines the appearance of limit cycles in the phase diagram of dynamical systems [121, 274].

Theorem 1 (Poincaré-Bendixson): If a trajectory of the nonlinear system of Eq. (1.42) remains in a finite region Ω, then one of the following is true: (i) the trajectory goes to an equilibrium point, (ii) the trajectory tends to a limit cycle, (iii) the trajectory is itself a limit cycle.

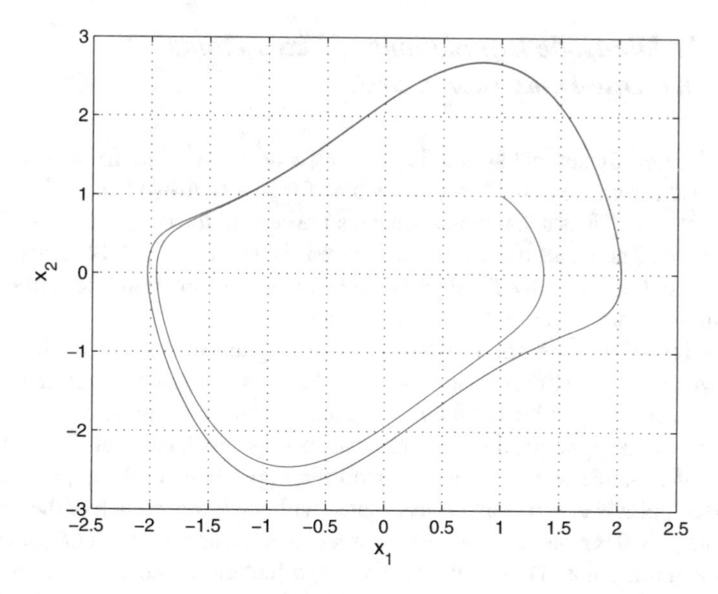

Fig. 1.13 Phase diagram of the Van der Pol oscillator for $\varepsilon = 1$

Moreover, the following theorem provides a sufficient condition for the nonexistence of limit cycles:

Theorem 2 (Bendixson): For the nonlinear system of Eq. (1.42), no limit cycle can exist in a region Ω of the phase plane in which $\frac{\partial f_1}{\partial x_1} + \frac{\partial f_2}{\partial x_2}$ does not vanish and does not change sign.

1.6 Bifurcations

1.6.1 Bifurcations of Fixed Points

As the parameters of the nonlinear model of the system are changed, the stability of the equilibrium point can also change and also the number of equilibria may vary. Values of these parameters at which the locus of the equilibrium (as a function of the parameter) changes and different branches appear, are known as critical or bifurcation values. The phenomenon of bifurcation, has to do with quantitative changes of parameters leading to qualitative changes of the system's properties [48, 70, 103, 126, 135].

Another issue in bifurcation analysis has to do with the study of the segments of the bifurcation branches in which the fixed points are no longer stable but either become unstable or are associated with limit cycles. The latter case is called Hopf bifurcation and the system's Jacobian matrix has a pair of complex imaginary eigenvalues.

1.6.2 Saddle-Node Bifurcations of Fixed Points in a One-Dimensional System

The considered dynamical system is given by $\dot{x} = \mu - x^2$. The fixed points of the system result from the condition $\dot{x} = 0$ which for $\mu > 0$ gives $x^* = \pm\sqrt{\mu}$. The first fixed point $x = \sqrt{\mu}$ is a stable one whereas the second fixed point $x = -\sqrt{\mu}$ is an unstable one. The phase diagram of the system is given in Fig. 1.14. Since there is one stable and one unstable fixed point the associated bifurcation (locus of the fixed points in the phase plane) will be a saddle-node one.

The bifurcations diagram is given next. The diagram shows how the fixed points of the dynamical system vary with respect to the values of parameter μ. In the above case it represents a parabola in the $\mu - x$ plane as shown in Fig. 1.15.

For $\mu > 0$ the dynamical system has two fixed points located at $\pm\sqrt{\mu}$. The one fixed point is stable and is associated with the upper branch of the parabola. The other fixed point is unstable and is associated with the lower branch of the parabola. The value $\mu = 0$ is considered to be a bifurcation value and the point $(x, \mu) = (0, 0)$ is a bifurcation point. This particular type of bifurcation where the one branch is associated with fixed points and the other branch is not associated to any fixed points is known as saddle-node bifurcation.

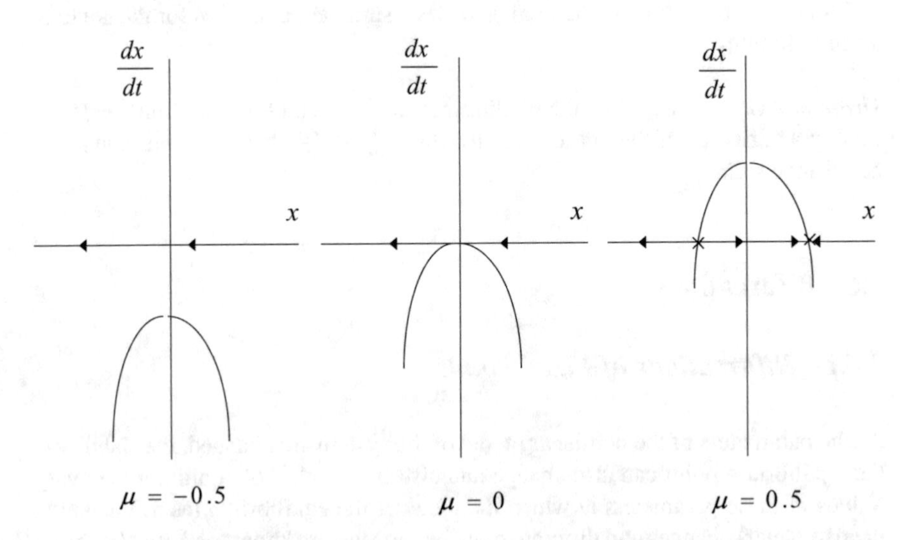

Fig. 1.14 Phase diagram and fixed points of the system $\dot{x} = \mu - x^2$. Converging arrows denote a stable fixed point

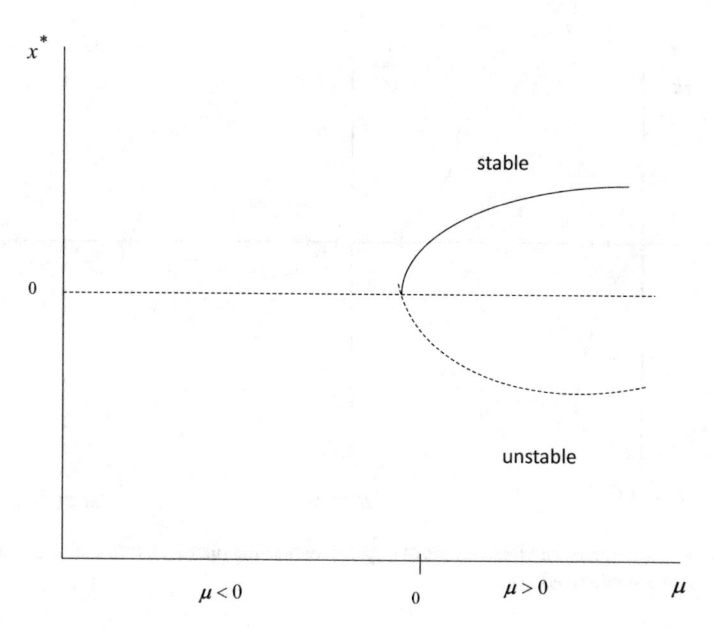

Fig. 1.15 Saddle-node bifurcation diagram

1.6.3 Pitchfork Bifurcation of Fixed Points

In pitchfork bifurcations the number of fixed points varies with respect to the values of the bifurcation parameter. The dynamical system $\dot{x} = x(\mu - x^2)$ is considered. The associated fixed points are found by the condition $\dot{x} = 0$. For $\mu < 0$ there is one fixed point at zero which is stable. For $\mu = 0$ there is still one fixed point at zero which is still stable. For $\mu > 0$ there are three fixed points, one at $x = 0$, one at $x = +\sqrt{\mu}$ which is stable and one at $x = -\sqrt{\mu}$ which is also stable. The associated phase diagrams and fixed points are presented in Fig. 1.16.

The bifurcations diagram is given next. The diagram shows how the fixed points of the dynamical system vary with respect to the values of parameter μ. In the above case it represents a parabola in the $\mu - x$ plane as shown in Fig. 1.17.

1.6.4 The Hopf Bifurcation

Bifurcation of the equilibrium point means that the locus of the equilibrium on the phase plane changes due to variation of the system's parameters. Equilibrium x^* is a hyperbolic equilibrium point if the real parts of all its eigenvalues are non-zero. A Hopf bifurcation appears when the hyperbolicity of the equilibrium point is lost due to variation of the system parameters and the eigenvalues become purely imaginary.

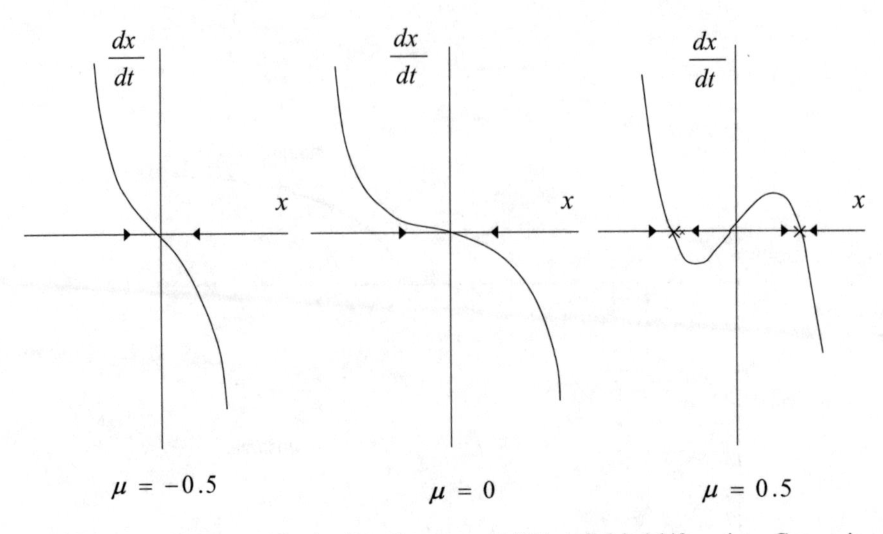

Fig. 1.16 Phase diagram and fixed points of a system exhibiting pitchfork bifurcations. Converging arrows denote a stable fixed point

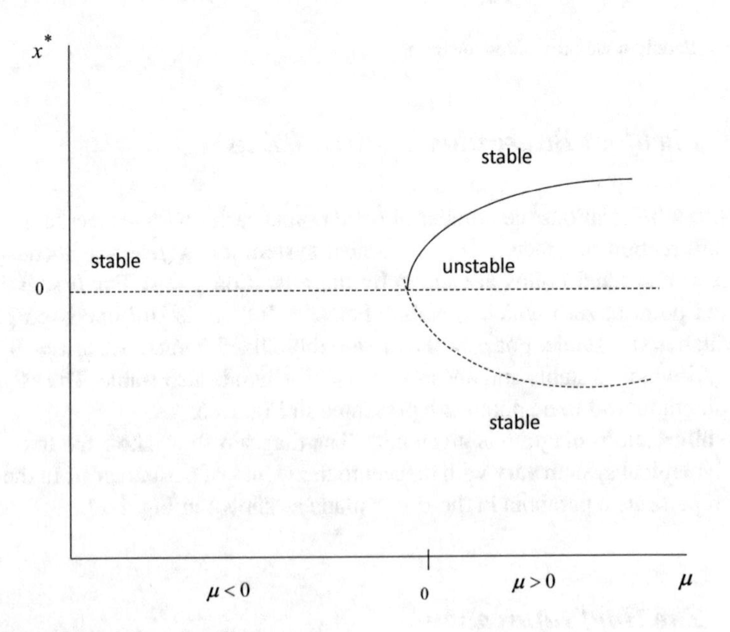

Fig. 1.17 Pitchfork bifurcation diagram

By changing the values of the parameters at a Hopf bifurcation, an oscillatory solution appears [175].

The stages for finding Hopf bifurcations in nonlinear dynamical systems are described next. The following autonomous differential equation is considered:

$$\frac{dx}{dt} = f_1(x, \lambda) \quad (1.43)$$

where x is the state vector $x \in R^n$ and $\lambda \in R^m$ is the vector of the system parameters. In Eq. (1.43) a point x^* satisfying $f_1(x^*) = 0$ is an equilibrium point. Therefore from the condition $f_1(x^*) = 0$ one obtains a set of equations which provide the equilibrium point as function of the bifurcating parameter. The stability of the equilibrium point can be evaluated by linearizing the system's dynamic model around the equilibrium point and by computing eigenvalues of the Jacobian matrix. The Jacobian matrix at the equilibrium point can be written as

$$J_{f_1(x^*)} = \frac{\partial f_1(x)}{\partial x}|_{x=x^*} \quad (1.44)$$

and the determinant of the Jacobian matrix provides the characteristic equation which is given by

$$det(\mu_i I_n - J_{f_1(x^*)}) = \lambda_1^n + \alpha_1 \lambda_i^{n-1} + \cdots + \alpha_{n-1}\lambda_i + \alpha_n = 0 \quad (1.45)$$

where I_n is the $n \times n$ identity matrix, μ_i with $i = 1, 2, \cdots, n$ denotes the eigenvalues of the Jacobian matrix $Df_1(x^*)$. From the requirement the eigenvalues of the system to be purely imaginary one obtains a condition, i.e. values that the bifurcating parameter should take, for the appearance of Hopf bifurcations.

As example, the following nonlinear system is considered:

$$\begin{aligned} \dot{x}_1 &= x_2 \\ \dot{x}_2 &= -x_1 + (m - x_1^2)x_1 \end{aligned} \quad (1.46)$$

Setting $\dot{x}_1 = 0$ and $\dot{x}_2 = 0$ one obtains the system's fixed points. For $m \leq 1$ the system has the fixed point $(x_1^*, x_2^*) = (0, 0)$. For $m > 1$ the system has the fixed points $(x_1^*, x_2^*) = (0, 0)$, $(x_1^*, x_2^*) = (\sqrt{m-1}, 0)$, $(x_1^*, x_2^*) = (-\sqrt{m-1}, 0)$. The system's Jacobian is

$$J = \begin{pmatrix} 0 & 1 \\ -1 + m - 3x_1^2 & 0 \end{pmatrix} \quad (1.47)$$

If $m \leq 1$, the eigenvalues of the Jacobian at the fixed point $(x_1^*, x_2^*) = (0, 0)$ will be imaginary. Thus the system exhibits Hopf bifurcations, as shown in Fig. 1.18. If $m > 1$ then the fixed point $(x_1^*, x_2^*) = (0, 0)$ is an unstable one. On the other hand, if $m > 1$ the eigenvalues of the Jacobian at the fixed points $(x_1^*, x_2^*) = (\sqrt{m-1}, 0)$

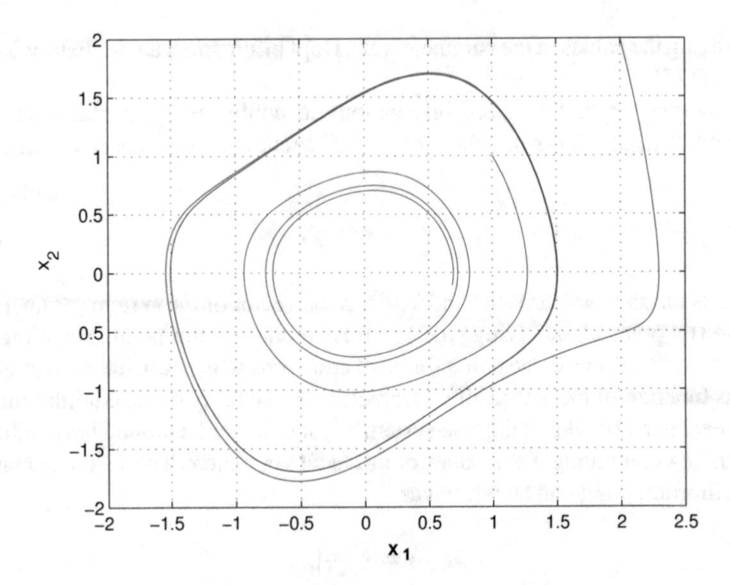

Fig. 1.18 Phase diagram and fixed points of a system exhibiting Hopf bifurcations

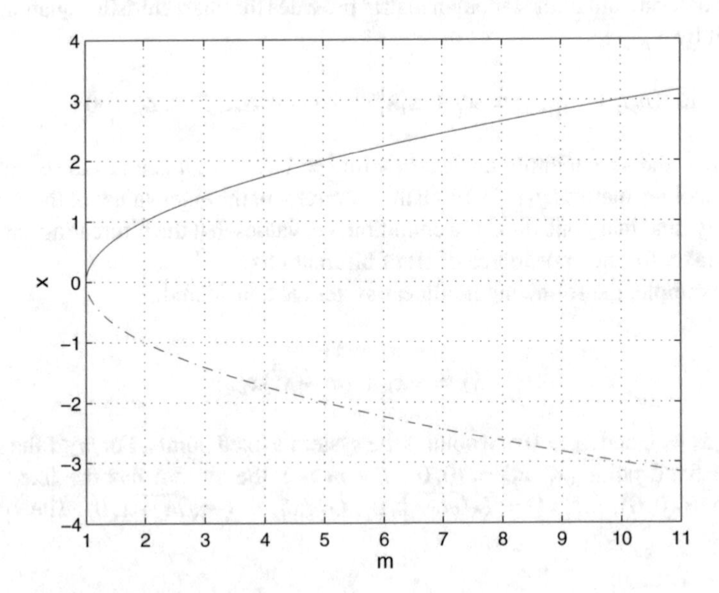

Fig. 1.19 Hopf bifurcation diagram

and $(x_1^*, x_2^*) = (-\sqrt{m-1}, 0)$ will be imaginary and the system exhibits again Hopf bifurcations.

The bifurcations diagram is given next. The diagram shows how the fixed points of the dynamical system vary with respect to the values of parameter μ. In the above case it represents a parabola in the $\mu - x$ plane as shown in Fig. 1.19.

1.7 Chaos in Dynamical Systems

1.7.1 Chaotic Dynamics

Nonlinear dynamical systems, can exhibit a phenomenon called chaos, by which it is meant that the system's output becomes extremely sensitive to initial conditions. The essential feature of chaos is the unpredictability of the system's output. Even if the model of the chaotic system is known, the system's response in the long run cannot be predicted. Chaos is distinguished from random motion. In the latter case, the system's model or input contain uncertainty and as a result, the time variation of the output cannot be predicted exactly (only statistical measures can be computed). In chaotic systems, the involved dynamical model is deterministic and there is little uncertainty about the system's model, input and initial conditions. In such systems, by slightly varying initial conditions or parameters values, a completely different phase diagram can be produced [284, 285].

A known chaotic dynamical systems are the Van der Pol oscillator which exhibits limit cycles and which has been analyzed in subsection 1.5.3 [273, 302]. Other chaotic models are the Duffing oscillator, the Lorenz oscillator, the Genesio-Tesi oscillator or Chen's system.

1.7.2 Examples of Chaotic Dynamical Systems

1. The Lorenz chaotic system

The Lorenz system is obtained from studying a fluid layer heated from below and cooled from above, such that a temperature difference is established across it. The convection motion is described by the Navier-Stokes equations. Taking Fourier expansions of these equations along two spatial directions and truncating the remaining expressions after the third mode, the following simplified model is obtained:

$$\begin{aligned} \dot{x} &= \sigma(y - z) \\ \dot{y} &= \rho x - y - xz \\ \dot{z} &= -\beta z + xy \end{aligned} \tag{1.48}$$

where σ, ρ and β are real parameters denoting the Prandtl number, the Rayleigh number and a geometric factor, respectively. The state variables x, y and z represent measures of fluid velocities and the spatial temperature distribution in the fluid layer under gravity. The Rayleigh number can be manipulated by changing the heat transfer to the fluid from below. This parameter is considered to be a control input $u = \rho$.

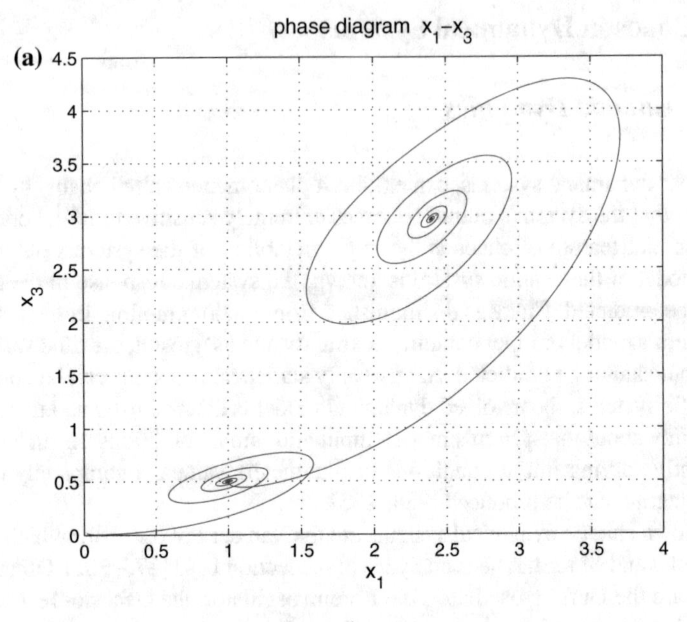

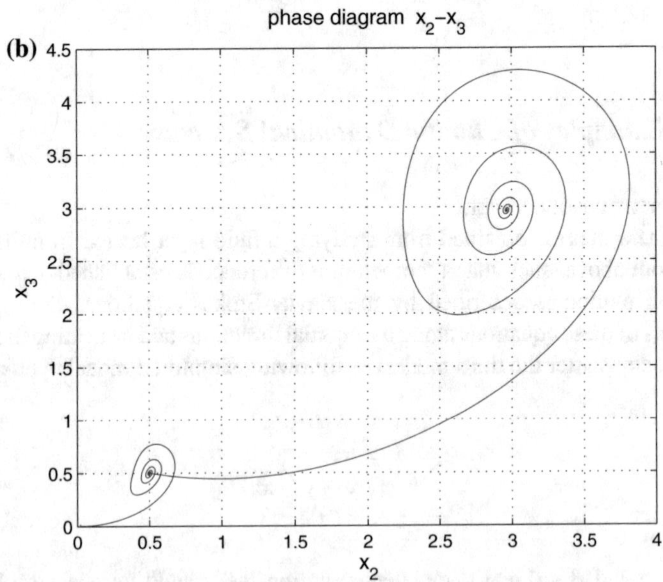

Fig. 1.20 **a** indicative phase diagram for the Lorenz system for states x_1 and x_3 **b** indicative phase diagram for the Lorenz system for states x_2 and x_3

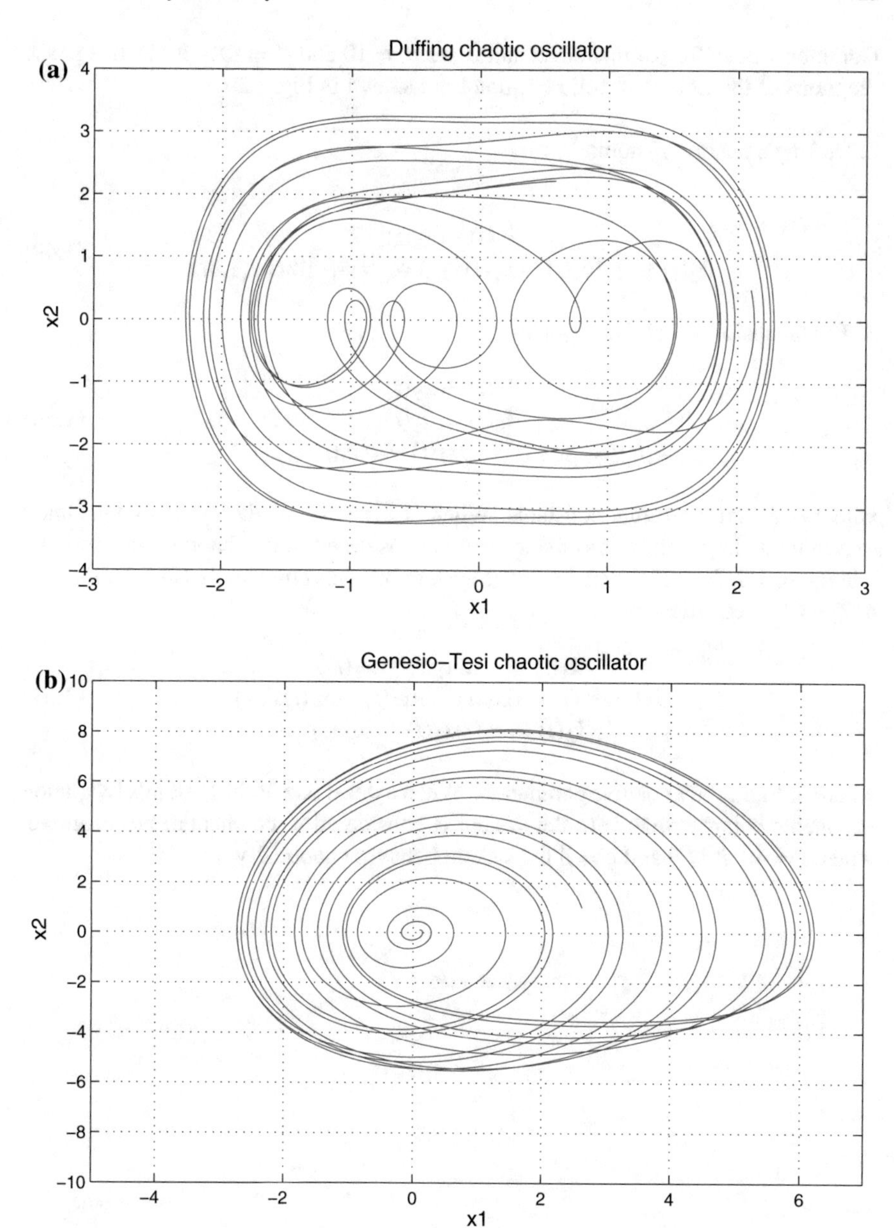

Fig. 1.21 Phase diagram of typical chaotic oscillators **a** Duffing's oscillator **b** Genesio–Tesi's oscillator

Common values for parameters σ and β are $\sigma = 10$ and $\beta = 8/3$. Indicative phase diagrams of the Lorenz chaotic oscillator are shown in Fig. 1.20.

2. Duffing's chaotic system:

$$\begin{aligned} \dot{x}_1(t) &= x_2(t) \\ \dot{x}_2(t) &= 1.1x_1(t) - x_1^3(t) - 0.4x_2(t) + 1.8cos(1.8t) \end{aligned} \tag{1.49}$$

3. The Genesio-Tesi chaotic system:

$$\begin{aligned} \dot{x}_1(t) &= x_2(t) \\ \dot{x}_2(t) &= x_3(t) \\ \dot{x}_3(t) &= -cx_1(t) - bx_2(t) - \alpha x_3(t) + x_1^2(t) \end{aligned} \tag{1.50}$$

where a, b and c are real constants. When at least one of the system's Lyapunov exponents is larger than zero the system is considered to be chaotic. For example, when $a = 1.2$, $b = 2.92$ and $c = 6$ the system behaves chaotically (Fig. 1.21).
4. The Chen chaotic system

$$\begin{aligned} \dot{x}_1(t) &= -a(x_1(t) - x_2(t)) \\ \dot{x}_2(t) &= (c - a)x_1(t) + cx_2(t) - x_1(t)x_3(t) \\ \dot{x}_3(t) &= x_1(t)x_2(t) - bx_3(t) \end{aligned} \tag{1.51}$$

where a, b and c are positive parameters. When at least one of the system's Lyapunov exponents is larger than zero the system is considered to be chaotic. For example, when $a = 40$, $b = 3$ and $c = 3$ the system behaves chaotically.

Chapter 2
Main Approaches to Nonlinear Control

2.1 Outline

In control and stabilization of the dynamics of financial systems, one can distinguish three main research axes: (i) Methods of global linearization, (ii) Methods of asymptotic linearization, and (iii) Lyapunov methods, As far as approach (i) is concerned, that is methods of global linearization, these aim at the transformation of the nonlinear dynamics of the system to equivalent linear state-space descriptions for which one can design controllers using state feedback and can also solve the associated state estimation (filtering) problem. One can classify here methods based on the theory of differentially flat systems and methods based on Lie algebra. As far as approach (ii) is concerned, solutions are pursued to the problem of nonlinear control with the use of local linear models (obtained at local equilibria). For such local linear models, feedback controllers of proven stability can be developed. One can select the parameters of such local controllers in a manner that assures the robustness of the control loop to both external perturbations and to model parametric uncertainty. As far as approach (iii) is concerned, that is methods of nonlinear control of the Lyapunov type one comes against problems of minimization of Lyapunov functions so as to assure the asymptotic stability of the control loop. For the development of Lyapunov type controllers one can either exploit a model about the systems dynamics or can proceed in a model-free manner, as in the case of indirect adaptive control.

2.2 Overview of Main Approaches to Nonlinear Control

One can note three axes in the development of nonlinear control systems based on state-space representation of the system's dynamics: (1) Methods of global linearization-based control (2) Methods of asymptotic linearization-based control, and (3) Lyapunov theory-based methods of control [121, 216, 222, 225, 228, 248].

G.G. Rigatos, *State-Space Approaches for Modelling and Control in Financial Engineering*, Intelligent Systems Reference Library 125,
DOI 10.1007/978-3-319-52866-3_2

As far as approach (i) is concerned, that is methods of global linearization these aim at the transformation of the nonlinear dynamics of the system to equivalent linear state-space descriptions for which one can design controllers using state feedback and can also solve the associated state estimation (filtering) problem. One can classify here methods based on the theory of differentially flat systems and methods based on Lie algebra. These approaches avoid approximate modelling errors and arrive at controllers of elevated precision and robustness. Using global linearization methods one can also solve nonlinear state estimation (filtering) problems.

As far as approach (ii) is concerned that is methods of asymptotic linearization, one can note results on robust and adaptive control with the use of a decomposition of the systems dynamics into local linear models. Solution to the problem of nonlinear control with the use of local linear models (obtained at local equilibria) is often pursued. For such local linear models, feedback controllers of proven stability can be developed. One can select the parameters of such local controllers in a manner that assures the robustness of the control loop to both external perturbations and to model parametric uncertainty. These controllers succeed asymptotically (that is as time advances) the compensation of the systems nonlinear dynamics and the stabilization of the closed control loops. In this area, one can apply robust control which is based on local approximate linearization of the systems dynamics and which requires the computation of Jacobian matrices, Such a method can solve the nonlinear H-infinity control problem.

As far as approach (iii) is concerned, that is methods of nonlinear control of the Lyapunov type the minimization of Lyapunov functions is pursued so as to assure the asymptotic stability of the control loop. For the development of Lyapunov type controllers one can either exploit a model about the systems dynamics or can proceed in a model-free manner, as in the case of indirect adaptive control. In the latter approach, the systems dynamics is taken to be completely unknown and can be approximated by adaptive algorithms which are suitably designed so as to assure the stabilization and robustness of the control loop.

2.3 Control Based on Global Linearization Methods

2.3.1 Overview of Differential Flatness Theory

As far as methods of global linearization are concerned, these are methods for the transformation of the nonlinear dynamics of the system to equivalent linear state-space descriptions for which one can design controllers using state feedback and can also solve the associated state estimation (filtering) problem. One can classify here methods based on the theory of differentially flat systems and methods based on Lie algebra. These approaches avoid approximate modelling errors and arrive at controllers of elevated precision and robustness.

Differential flatness theory and flatness-based control were introduced in the late 80's by Michel Fliess and co-researchers and since then they keep on being developed and on providing efficient solutions to advanced control and state estimation problems [83].

The definition of a differentially flat system is as follows: A system $\dot{x} = f(x, u)$ with state vector $x \in R^n$, input vector $u \in R^m$ where f is a smooth vector field, is differentially flat if there exists a vector $y \in R^m$ in the form

$$y = h(x, u, \dot{u}, \ldots, u^{(r)}) \tag{2.1}$$

such that

$$\begin{aligned} x &= \phi(y, \dot{y}, \ldots, y^{(q)}) \\ u &= \alpha(y, \dot{y}, \ldots, y^{(q)}) \end{aligned} \tag{2.2}$$

where h, ϕ and α are smooth functions. This means that the new system's description is given by the m algebraic variables y_i, $i = 1, 2, \ldots, m$. The definition of the flat output given above was $y = h(x, u, \dot{u}, \ldots, u^{(r)})$. If the flat output is exclusively a function of the state vector x then the system is a 0-flat one. However, there may be need to express the flat output as a function of not only the state vector x but also as a function of the control u and of its derivatives. For instance, the latter holds in the case of dynamic feedback linearization and in the application of the so-called dynamic extension. This means that the state vector of the system is extended by considering as additional state variables the control inputs and its derivatives.

Equation (2.2) shows that the state vector of the differentially flat system and its control inputs can be expressed as function of the flat output and of the flat output's derivatives. The basic question that arises in the study of differential flatness is whether, given the differential equations that describe the nonlinear system dynamics $\dot{x} = f(x, u)$, there exists a function $y = h()$ given by $y = h(x, u, \dot{u}, \ldots, u^{(r)}$, such that the state vector of the system x and the control input u can be expressed as functions of y and of its derivatives, as in Eq. (2.2).

Next a formal definition will be given about the analogy in terms of the control theory, of what was described above as under-determined differential-equation systems which can be solved without integration.

2.3.2 Differential Flatness for Finite Dimensional Systems

As noted in Eqs. (2.1) and (2.2) differential flatness is a structural property of a class of nonlinear dynamical systems, denoting that all system variables (such as state vector elements and control inputs) can be written in terms of a set of specific variables (the so-called flat outputs) and their derivatives. The following nonlinear system is considered:

$$\dot{x}(t) = f(x(t), u(t)) \tag{2.3}$$

The time variable is $t \in R$, the state vector is $x(t) \in R^n$ with initial conditions $x(0) = x_0$, and the input variable is $u(t) \in R^m$. Next, the main principles of differentially flat systems are given [231, 265]:

The finite dimensional system of Eq. (2.3) can be written in the general form of an ordinary differential equation (ODE), i.e. $S_i(w, \dot{w}, \ddot{w}, \ldots, w^{(i)})$, $i = 1, 2, \ldots, q$. The entity w is a generic notation for the system variables (these variables are for instance the elements of the system's state vector $x(t)$ and the elements of the control input $u(t)$) while $w^{(i)}$, $i = 1, 2, \ldots, q$ are the associated derivatives. Such a system is said to be differentially flat if there is a collection of m functions $y = (y_1, \ldots, y_m)$ of the system variables and of their time-derivatives, i.e. $y_i = \phi(w, \dot{w}, \ddot{w}, \ldots, w^{(\alpha_i)})$, $i = 1, \ldots, m$ satisfying the following two conditions [77, 158, 167, 169, 222]:

1. There does not exist any differential relation of the form $R(y, \dot{y}, \ldots, y^{(\beta)}) = 0$ which implies that the derivatives of the flat output are not coupled in the sense of an ODE, or equivalently it can be said that the flat output is differentially independent.
2. All system variables (i.e. the elements of the system's state vector w and the control input) can be expressed using only the flat output y and its time derivatives $w_i = \psi_i(y, \dot{y}, \ldots, y^{(\gamma_i)})$, $i = 1, \ldots, s$. An equivalent definition of differentially flat systems is as follows:

Definition: The system $\dot{x} = f(x, u)$, $x \in R^n$, $u \in R^m$ is differentially flat if there exist relations

$$\begin{aligned} h &: R^n \times (R^m)^{r+1} \rightarrow R^m, \\ \phi &: (R^m)^r \rightarrow R^n \text{ and} \\ \psi &: (R^m)^{r+1} \rightarrow R^m \end{aligned} \tag{2.4}$$

such that

$$\begin{aligned} y &= h(x, u, \dot{u}, \ldots, u^{(r)}), \\ x &= \phi(y, \dot{y}, \ldots, y^{(r-1)}), \text{ and} \\ u &= \psi(y, \dot{y}, \ldots, y^{(r-1)}, y^{(r)}). \end{aligned} \tag{2.5}$$

This means that all system dynamics can be expressed as a function of the flat output and its derivatives, therefore the state vector and the control input can be written as

$$\begin{aligned} x(t) &= \phi(y(t), \dot{y}(t), \ldots, y^{(r-1)}(t)), \text{ and} \\ u(t) &= \psi(y(t), \dot{y}(t), \ldots, y^{(r)}(t)) \end{aligned} \tag{2.6}$$

Next, an example is given to explain the design of a differentially flat controller for finite dimensional systems of known parameters.

Example 1: Flatness-based control for a nonlinear system of known parameters [128]. Consider the following model:

$$\begin{aligned} \dot{x}_1 &= x_3 - x_2 u \\ \dot{x}_2 &= -x_2 + u \\ \dot{x}_3 &= x_2 - x_1 + 2x_2(u - x_2) \end{aligned} \tag{2.7}$$

The flat output is chosen to be $y_1 = x_1 + \frac{x_2^2}{2}$. Thus one gets:

$$\begin{array}{l} y_1 = x_1 + \frac{x_2^2}{2} \\ y_2 = \dot{y}_1 = (x_3 - x_2 u) + x_2(u - x_2) = x_3 - x_2^2 \\ y_3 = \dot{y}_2 = \ddot{y}_1 = x_2 - x_1 + 2x_2(u - x_2) - 2x_2(u - x_2) = -x_1 + x_2 \\ v = \dot{y}_3 = y_1^{(3)} = -x_3 + x_2 u - x_2 + u = -x_2 - x_3 + u(1 + x_2) \end{array} \tag{2.8}$$

It can be verified that property (1) holds, i.e. there does not exist any differential relation of the form $R(y, \dot{y}, \ldots, y^{(\beta)}) = 0$, and this implies that the derivatives of the flat output are not coupled. Moreover, it can be shown that property (2) also holds i.e. the components w of the system (elements of the system's state vector and control input) can be expressed using only the flat output y and its time derivatives $w_i = \psi_i(y, \dot{y}, \ldots, y^{(\gamma_i)}), i = i, \ldots, s$.

For instance to calculate x_1 with respect to $y_1, \dot{y}_1, \ddot{y}_1$ and $y_1^{(3)}$ the relation of $\ddot{y}_1$ is used, i.e.:

$$x_1^2 + 2x_1(1 + \ddot{y}_1) + \ddot{y}_1^2 - 2y_1 = 0 \tag{2.9}$$

from which two possible solutions are derived, i.e.: $x_1 = -(1 + \ddot{y}_1 - \sqrt{1 + 2(y_1 + \ddot{y}_1)})$ and $x_1 = -(1 + \ddot{y}_1 + \sqrt{1 + 2(y_1 + \ddot{y}_1)})$. Keeping the biggest out of these two solutions one obtains:

$$\begin{array}{l} x_1 = -(1 + \ddot{y}_1) + \sqrt{1 + 2(y_1 + \ddot{y}_1)} \\ x_2 = \ddot{y}_1 + x_1 \\ x_3 = \dot{y}_1 + \ddot{y}_1^2 + 2x_1\ddot{y}_1 + x_1^2 \\ \\ u = \dfrac{y_1^{(3)} + \ddot{y}_1^2 + \ddot{y}_1 + \dot{y}_1 + x_1 + 2x_1\ddot{y}_1 + x_1^2}{1 + x_1 + \ddot{y}_1} \end{array} \tag{2.10}$$

The computation of the equivalent model of the system in the linear canonical form is summarized as follows: by finding the derivatives of the flat output one gets a set of equations which can be solved with respect to the state variables and the control input of the initial state-space description of the system. First, the binomial of variable x_1 given in Eq. (2.9) is solved providing x_1 as a function of the flat output and its derivatives. Next, using the expression for x_1 and Eq. (2.10), state variable x_2 is also written as a function of the flat output and its derivatives. Finally, using the expressions for both x_1 and x_2 and Eq. (2.10), state variable x_3 is written as a function of the flat output and its derivatives. Thus one can finally express the state vector elements and the control input as function of the flat output and its derivatives, which completes the proof about differential flatness of the system.

From Eq. (2.10) it can be concluded that the initial system of Eq. (2.7) is indeed differentially flat. Using the flat output and its derivatives, the system of Eq. (2.7) can be written in Brunovsky (canonical) form:

$$\frac{d}{dt}\begin{pmatrix} y_1 \\ y_2 \\ y_3 \end{pmatrix} = \begin{pmatrix} 0 & 1 & 0 \\ 0 & 0 & 1 \\ 0 & 0 & 0 \end{pmatrix} \begin{pmatrix} y_1 \\ y_2 \\ y_3 \end{pmatrix} + \begin{pmatrix} 0 \\ 0 \\ 1 \end{pmatrix} v \tag{2.11}$$

where the new control input is $v = f(x) + g(x)u$. Therefore, a transformation of the system into a linear equivalent description is obtained and then it is straightforward to design a a controller based on linear control theory. Thus, given the reference trajectory $[x_1^*, x_2^*, x_3^*]^T$ one can find the transformed reference trajectory $[y_1^*, \dot{y}_1^*, \ddot{y}_1^*]^T$ and select the appropriate control input v that succeeds tracking of the reference setpoints. Knowing v, the control u of the initial system can be found. Knowing v the control input that is actually applied to the system is $u = g^{-1}(x)[v - f(x)]$.

It is noted that for linear systems the property of differential flatness is equivalent to that of controllability.

2.4 Control Based on Approximate Linearization Methods

In this research area solutions are pursued to the problem of nonlinear control with the use of local linear models (obtained at local equilibria). For such local linear models, feedback controllers of proven stability can be developed. One can select the parameters of such local controllers in a manner that assures the robustness of the control loop to both external perturbations and to model parametric uncertainty. These controllers succeed asymptotically (that is as time advances) the compensation of the systems nonlinear dynamics and the stabilization of the closed control loops. In this area, research focuses on a new method of robust control which is based on local approximate linearization of the systems dynamics and which requires the computation of Jacobian matrices, The method solves the nonlinear H-infinity control problem.

2.4.1 Approximate Linearization Round Temporary Equilibria

A nonlinear system in the form

$$\dot{x} = f(x, u)\ x \in R^n,\ f \in R^n,\ u \in R^m \tag{2.12}$$

is considered. Linearization of the model round a temporary equilibrium is performed, where the equilibrium is defined by the present value of the system's state vector and the last value of the control inputs vector exerted on it. After such a linearization round its present operating point, the system's model is written as

$$\dot{x} = Ax + Bu + d_1 \tag{2.13}$$

where

$$\begin{aligned} A &= \nabla_x f(x, u)|_{(x^*, u^*)} \\ B &= \nabla_u f(x, u)|_{(x^*, u^*)} \end{aligned} \tag{2.14}$$

Parameter d_1 stands for the linearization error in the system's model appearing in Eq. (2.13). The desirable trajectory of the system is denoted by $\mathbf{x_d} = [\mathbf{x_d}, \mathbf{y_d}, \boldsymbol{\theta}_\mathbf{d}]$. Tracking of this trajectory is succeeded after applying the control input u^*. At every time instant the control input u^* is assumed to differ from the control input u appearing in Eq. (2.13) by an amount equal to Δu, that is $u^* = u + \Delta u$

$$\dot{x}_d = Ax_d + Bu^* + d_2 \tag{2.15}$$

The dynamics of the controlled system described in Eq. (2.13) can be also written as

$$\dot{x} = Ax + Bu + Bu^* - Bu^* + d_1 \tag{2.16}$$

and by denoting $d_3 = -Bu^* + d_1$ as an aggregate disturbance term one obtains

$$\dot{x} = Ax + Bu + Bu^* + d_3 \tag{2.17}$$

By subtracting Eq. (2.15) from Eq. (2.17) one has

$$\dot{x} - \dot{x}_d = A(x - x_d) + Bu + d_3 - d_2 \tag{2.18}$$

By denoting the tracking error as $e = x - x_d$ and the aggregate disturbance term as $\tilde{d} = d_3 - d_2$, the tracking error dynamics becomes

$$\dot{e} = Ae + Bu + \tilde{d} \tag{2.19}$$

The above linearized form of the nonlinear system can be efficiently controlled after applying an H-infinity feedback control scheme.

2.4.2 *The Nonlinear H-Infinity Control*

2.4.2.1 Mini-max Control and Disturbance Rejection

The initial nonlinear system is assumed to be in the form

$$\dot{x} = f(x, u) \quad x \in R^n, \ u \in R^m \tag{2.20}$$

Linearization of the system is performed at each iteration of the control algorithm round its present operating point $(x^*, u^*) = (x(t), u(t - T_s))$. The linearized equivalent of the system is described by

$$\dot{x} = Ax + Bu + L\tilde{d} \quad x \in R^n, \ u \in R^m, \ \tilde{d} \in R^q \tag{2.21}$$

where matrices A and B are obtained from the computation of the Jacobians

$$A = \begin{pmatrix} \frac{\partial f_1}{\partial x_1} & \frac{\partial f_1}{\partial x_2} & \cdots & \frac{\partial f_1}{\partial x_n} \\ \frac{\partial f_2}{\partial x_1} & \frac{\partial f_2}{\partial x_2} & \cdots & \frac{\partial f_2}{\partial x_n} \\ \cdots & \cdots & \cdots & \cdots \\ \frac{\partial f_n}{\partial x_1} & \frac{\partial f_n}{\partial x_2} & \cdots & \frac{\partial f_n}{\partial x_n} \end{pmatrix} |_{(x^*,u^*)} \tag{2.22}$$

$$B = \begin{pmatrix} \frac{\partial f_1}{\partial u_1} & \frac{\partial f_1}{\partial u_2} & \cdots & \frac{\partial f_1}{\partial u_m} \\ \frac{\partial f_2}{\partial u_1} & \frac{\partial f_2}{\partial u_2} & \cdots & \frac{\partial f_2}{\partial u_m} \\ \cdots & \cdots & \cdots & \cdots \\ \frac{\partial f_n}{\partial u_1} & \frac{\partial f_n}{\partial u_2} & \cdots & \frac{\partial f_n}{\partial u_m} \end{pmatrix} |_{(x^*,u^*)} \tag{2.23}$$

and vector $\tilde{d}$ denotes disturbance terms due to linearization errors. The problem of disturbance rejection for the linearized model that is described by

$$\begin{aligned} \dot{x} &= Ax + Bu + Ld \\ y &= Cx \end{aligned} \tag{2.24}$$

where $x \in R^n$, $u \in R^m$, $\tilde{d} \in R^q$ and $y \in R^p$, cannot be handled efficiently if the classical LQR control scheme is applied. This is because of the existence of the perturbation term d. The disturbance term d apart from modeling (parametric) uncertainty and external perturbation terms can also represent noise terms of any distribution.

In the H_∞ control approach, a feedback control scheme is designed for trajectory tracking by the system's state vector and simultaneous disturbance rejection, considering that the disturbance affects the system in the worst possible manner. The disturbances' effects are incorporated in the following quadratic cost function:

$$\begin{aligned} J(t) = \tfrac{1}{2}\int_0^T [y^T(t)y(t) + \\ + ru^T(t)u(t) - \rho^2\tilde{d}^T(t)\tilde{d}(t)]dt, \quad r, \rho > 0 \end{aligned} \tag{2.25}$$

The significance of the negative sign in the cost function's term that is associated with the perturbation variable $\tilde{d}(t)$ is that the disturbance tries to maximize the cost function $J(t)$ while the control signal $u(t)$ tries to minimize it. The physical meaning of the relation given above is that the control signal and the disturbances compete to each other within a mini-max differential game. This problem of mini-max optimization can be written as

$$min_u max_{\tilde{d}} J(u, \tilde{d}) \tag{2.26}$$

The objective of the optimization procedure is to compute a control signal $u(t)$ which can compensate for the worst possible disturbance, that is externally imposed to the system. However, the solution to the mini-max optimization problem is directly related to the value of the parameter ρ. This means that there is an upper bound in the disturbances magnitude that can be annihilated by the control signal.

2.4.2.2 H-Infinity Feedback Control

For the linearized system given by Eq. (2.24) the cost function of Eq. (2.25) is defined, where the coefficient r determines the penalization of the control input and the weight coefficient ρ determines the reward of the disturbances' effects.

It is assumed that (i) The energy that is transferred from the disturbances signal $d(t)$ is bounded, that is

$$\int_0^{\infty}\tilde{d}^T(t)\tilde{d}(t)dt < \infty \tag{2.27}$$

(ii) the matrices $[A\ B]$ and $[A\ L]$ are stabilizable, (iii) the matrix $[AC]$ is detectable. Then, the optimal feedback control law is given by

$$u(t) = -Kx(t) \tag{2.28}$$

with

$$K = \frac{1}{r}B^T P \tag{2.29}$$

where P is a positive semi-definite symmetric matrix which is obtained from the solution of the Riccati equation

$$A^T P + PA + Q - P\left(\frac{1}{r}BB^T - \frac{1}{2\rho^2}LL^T\right)P = 0 \tag{2.30}$$

where Q is also a positive definite symmetric matrix. The worst case disturbance is given by

$$\tilde{d}(t) = \frac{1}{\rho^2}L^T Px(t) \tag{2.31}$$

2.4.2.3 The Role of Riccati Equation Coefficients in H_∞ Control Robustness

The parameter ρ in Eq. (2.25), is an indication of the closed-loop system robustness. If the values of $\rho > 0$ are excessively decreased with respect to r, then the solution of the Riccati equation is no longer a positive definite matrix. Consequently there is a lower bound ρ_{min} of ρ for which the H_∞ control problem has a solution. The acceptable values of ρ lie in the interval $[\rho_{min}, \infty)$. If ρ_{min} is found and used in the design of the H_∞ controller, then the closed-loop system will have increased robustness. Unlike this, if a value $\rho > \rho_{min}$ is used, then an admissible stabilizing H_∞ controller will be derived but it will be a suboptimal one. The Hamiltonian matrix

$$H = \begin{pmatrix} A & -(\frac{1}{r}BB^T - \frac{1}{\rho^2}LL^T) \\ -Q & -A^T \end{pmatrix} \tag{2.32}$$

provides a criterion for the existence of a solution of the Riccati equation Eq. (2.30). A necessary condition for the solution of the algebraic Riccati equation to be a positive semi-definite symmetric matrix is that H has no imaginary eigenvalues [225].

2.4.2.4 Lyapunov Stability Analysis

Through Lyapunov stability analysis it will be shown that the proposed nonlinear control scheme assures H_∞ tracking performance, and that in case of bounded disturbance terms asymptotic convergence to the reference setpoints is succeeded.

The tracking error dynamics for the nonlinear system is written in the form

$$\dot{e} = Ae + Bu + L\tilde{d} \tag{2.33}$$

where in this example $L = I \in R^n$ with I being the identity matrix. The following Lyapunov equation is considered

$$V = \frac{1}{2}e^T Pe \tag{2.34}$$

where $e = x - x_d$ is the tracking error. By differentiating with respect to time one obtains

$$\begin{gathered} \dot{V} = \frac{1}{2}\dot{e}^T Pe + \frac{1}{2}eP\dot{e} \Rightarrow \\ \dot{V} = \frac{1}{2}[Ae + Bu + L\tilde{d}]^T P + \frac{1}{2}e^T P[Ae + Bu + L\tilde{d}] \Rightarrow \end{gathered} \tag{2.35}$$

$$\begin{gathered} \dot{V} = \frac{1}{2}[e^T A^T + u^T B^T + \tilde{d}^T L^T]Pe + \\ + \frac{1}{2}e^T P[Ae + Bu + L\tilde{d}] \Rightarrow \end{gathered} \tag{2.36}$$

$$\begin{gathered} \dot{V} = \frac{1}{2}e^T A^T Pe + \frac{1}{2}u^T B^T Pe + \frac{1}{2}\tilde{d}^T L^T Pe + \\ \frac{1}{2}e^T PAe + \frac{1}{2}e^T PBu + \frac{1}{2}e^T PL\tilde{d} \end{gathered} \tag{2.37}$$

The previous equation is rewritten as

$$\begin{gathered} \dot{V} = \frac{1}{2}e^T(A^T P + PA)e + \left(\frac{1}{2}u^T B^T Pe + \frac{1}{2}e^T PBu\right) + \\ + \left(\frac{1}{2}\tilde{d}^T L^T Pe + \frac{1}{2}e^T PL\tilde{d}\right) \end{gathered} \tag{2.38}$$

Assumption: For given positive definite matrix Q and coefficients r and ρ there exists a positive definite matrix P, which is the solution of the following matrix equation

$$A^T P + PA = -Q + P\left(\frac{1}{r}BB^T - \frac{1}{2\rho^2}LL^T\right)P \tag{2.39}$$

Moreover, the following feedback control law is applied to the system

$$u = -\frac{1}{r}B^T Pe \tag{2.40}$$

By substituting Eq. (2.39) and Eq. (2.40) one obtains

$$\dot{V} = \frac{1}{2}e^T[-Q + P(\frac{1}{r}BB^T - \frac{1}{2\rho^2}LL^T)P]e + \\ + e^T PB(-\frac{1}{r}B^T Pe + e^T PL\tilde{d} \Rightarrow \tag{2.41}$$

$$\dot{V} = -\frac{1}{2}e^T Qe + (\frac{1}{r}PBB^T Pe - \frac{1}{2\rho^2}e^T PLL^T)Pe \\ -\frac{1}{r}e^T PBB^T Pe + e^T PL\tilde{d} \tag{2.42}$$

which after intermediate operations gives

$$\dot{V} = -\frac{1}{2}e^T Qe - \frac{1}{2\rho^2}e^T PLL^T Pe + e^T PL\tilde{d} \tag{2.43}$$

or, equivalently

$$\dot{V} = -\frac{1}{2}e^T Qe - \frac{1}{2\rho^2}e^T PLL^T Pe + \\ + \frac{1}{2}e^T PL\tilde{d} + \frac{1}{2}\tilde{d}^T L^T Pe \tag{2.44}$$

Lemma: The following inequality holds

$$\frac{1}{2}e^T L\tilde{d} + \frac{1}{2}\tilde{d}L^T Pe - \frac{1}{2\rho^2}e^T PLL^T Pe \leq \frac{1}{2}\rho^2\tilde{d}^T\tilde{d} \tag{2.45}$$

Proof: The binomial $(\rho\alpha - \frac{1}{\rho}b)^2$ is considered. Expanding the left part of the above inequality one gets

$$\rho^2 a^2 + \frac{1}{\rho^2}b^2 - 2ab \geq 0 \Rightarrow \frac{1}{2}\rho^2 a^2 + \frac{1}{2\rho^2}b^2 - ab \geq 0 \Rightarrow \\ ab - \frac{1}{2\rho^2}b^2 \leq \frac{1}{2}\rho^2 a^2 \Rightarrow \frac{1}{2}ab + \frac{1}{2}ab - \frac{1}{2\rho^2}b^2 \leq \frac{1}{2}\rho^2 a^2 \tag{2.46}$$

The following substitutions are carried out: $a = \tilde{d}$ and $b = e^T PL$ and the previous relation becomes

$$\frac{1}{2}\tilde{d}^T L^T Pe + \frac{1}{2}e^T PL\tilde{d} - \frac{1}{2\rho^2}e^T PLL^T Pe \leq \frac{1}{2}\rho^2\tilde{d}^T\tilde{d} \tag{2.47}$$

Equation (2.47) is substituted in Eq. (2.44) and the inequality is enforced, thus giving

$$\dot{V} \leq -\frac{1}{2}e^T Qe + \frac{1}{2}\rho^2\tilde{d}^T\tilde{d} \tag{2.48}$$

Equation (2.48) shows that the H_∞ tracking performance criterion is satisfied. The integration of $\dot{V}$ from 0 to T gives

$$\int_0^T \dot{V}(t)dt \leq -\frac{1}{2}\int_0^T ||e||_Q^2 dt + \frac{1}{2}\rho^2\int_0^T ||\tilde{d}||^2 dt \Rightarrow \\ 2V(T) + \int_0^T ||e||_Q^2 dt \leq 2V(0) + \rho^2\int_0^T ||\tilde{d}||^2 dt \tag{2.49}$$

Moreover, if there exists a positive constant $M_d > 0$ such that

$$\int_0^{\infty} ||\tilde{d}||^2 dt \leq M_d \tag{2.50}$$

then one gets

$$\int_0^{\infty} ||e||_Q^2 dt \leq 2V(0) + \rho^2 M_d \tag{2.51}$$

Thus, the integral $\int_0^{\infty} ||e||_Q^2 dt$ is bounded. Moreover, $V(T)$ is bounded and from the definition of the Lyapunov function V in Eq. (2.34) it becomes clear that $e(t)$ will be also bounded since $e(t) \in \Omega_e = \{e|e^T Pe \leq 2V(0) + \rho^2 M_d\}$.

According to the above and with the use of Barbalat's Lemma one obtains $lim_{t\to\infty} e(t) = 0$.

2.4.3 Approximate Linearization with Local Fuzzy Models

This approach commonly makes use of Takagi-Sugeno fuzzy modelling and the solution of Linear Matrix Inequalities (LMIs) [282].

After linearization round local operating points the initial nonlinear dynamical system of the form

$$\dot{x} = f(x, u) \tag{2.52}$$

is written in the form of Takagi–Sugeno fuzzy model, that is [72]

$$\begin{array}{l} \text{Rule i}: IF\ x_1(t)\ is\ M_1^i\ AND\ x_2(t)\ is\ M_2^i\ AND\ \cdots\ AND\ x_n(t)\ is\ M_n^i \\ \qquad THEN\ \delta x(t) = A_i x(t) + B_i u(t)\ i = 1, 2, \ldots, r \end{array} \tag{2.53}$$

where x_j is the j-th variable of the state vector, M_j^i is the i-th fuzzy set into which the value range of the j-th input variable is partitioned, $x(t) = [x_1(t), \ldots, x_n(t)]^T\ in\ R^n$ is the state vector, $u(t) = [u_1(t), \ldots, u_m(t)]^T \in R^m$ is the input vector, while it holds $A_i \in R^{n\times n}$ and $B_i \in R^{n\times m}$. It is noted that the model described in Eq. (2.53) includes also the autonomous system dynamics, that is when $u(t) = 0$.

The output of the Takagi–Sugeno fuzzy model is

$$\delta x(t) = \frac{\sum_{i=1}^r w_i [A_i x(t) + B_i u(t)]}{\sum_{i=1}^r w_i} \tag{2.54}$$

where

$$w_i = \prod_{k=1}^r \mu_K^i(x_k(t)) \tag{2.55}$$

A condition that assures that the aforementioned system remains stable is stated as follows [72]:

Condition 1: The equilibrium of an autonomous Takagi–Sugeno fuzzy model is globally asymptotically stable if there exists a common symmetric positive definite matrix P, such that

$$A_i^T P + PA_i < 0 \; \forall \, i = 1, 2, \ldots, r \tag{2.56}$$

Next, stability conditions are formulated for the case of a non-autonomous system. A feedback fuzzy controller is designed by using the antecedent part of the Takagi–Sugeno fuzzy model of the system. The fuzzy controller consists of the following rules

$$\begin{array}{l} \text{Rule j}: \textit{IF } x_1(t) \textit{ is } M_1^j \textit{ AND } x_2(t) \textit{ is } M_2^j \textit{ AND } \cdots \textit{ AND } x_n(t) \textit{ is } M_n^j \\ \qquad \textit{THEN } u(t) = K_j x(t) \; j = 1, 2, \ldots, r \end{array} \tag{2.57}$$

The aggregate control signal that is applied to the nonlinear system becomes

$$u(t) = \frac{\sum_{j=1}^r w_j K_j x(t)}{\sum_{j=1}^r w_j} \tag{2.58}$$

By applying the above feedback control the output of the closed-loop system becomes

$$\delta x(t) = \frac{\sum_{i=1}^r \sum_{j=1}^r w_i w_j (A_i + B_i K_j) x(t)}{\sum_{i=1}^r \sum_{j=1}^r w_i w_j} \tag{2.59}$$

Stability conditions for the continuous-time Takagi–Sugeno fuzzy system under state feedback can be also formulated [72].

Condition 2: The equilibrium of the fuzzy system is globally asymptotically stable if there exists a common symmetric positive definite matrix P such that

$$(A_i + B_i K_j)^T P + P(A_i + B_i K_j) < 0 \; \forall \, i, j = 1, 2, \ldots, r \tag{2.60}$$

Finally, a less conservative condition about the stability of the continuous-time Takagi–Sugeno system is formulated as follows [72].

Condition 3: The equilibrium state for the continuous-time fuzzy system is globally asymptotically stable if there exists a common positive definite matrix P such that

$$\begin{array}{c} (A_i + B_i K_i)^T P + P(A_i + B_i K_i) < 0, \; i = 1, 2, \ldots, r \\ (G_{ij}^T P + PG_{ij}) < 0, \; i < j \leq r \end{array} \tag{2.61}$$

where

$$G_{ij} = \frac{(A_i + B_i K_j) + (A_j + B_j K_i)}{2}, \; i, j \leq r \tag{2.62}$$

2.5 Control Based on Lyapunov Stability Analysis

2.5.1 *Transformation of Nonlinear Systems into a Canonical Form*

As far as methods of nonlinear control of the Lyapunov type are concerned, one comes against problems of minimization of Lyapunov functions so as to assure the asymptotic stability of the control loop. For the development of Lyapunov type controllers one can either exploit a model about the system's dynamics or can proceed in a model-free manner, as in the case of indirect adaptive control. In the latter case, the system's dynamics is taken to be completely unknown and can be approximated by adaptive algorithms which are suitably designed so as to assure the stabilization and robustness of the control loop.

A single-input differentially flat dynamical system is considered next:

$$\dot{x} = f_s(x,t) + g_s(x,t)(u+\tilde{d}),\ x \in R^n,\ u \in R,\ \tilde{d} \in R \tag{2.63}$$

where $f_s(x,t)$, $g_s(x,t)$ are nonlinear vector fields defining the system's dynamics, u denotes the control input and $\tilde{d}$ denotes additive input disturbances. Knowing that the system of Eq. (2.63) is differentially flat, the next step is to try to write it into a Brunovsky form. It has been shown that, in general, transformation into the Brunovsky (canonical) form can be succeeded for systems that admit static feedback linearization [158]. Single input differentially flat systems, admit static feedback linearization, and can be transformed into the Brunovksy form.

The selected flat output is denoted by y. Then, for the state variables x_i of the system of Eq. (2.63) it holds

$$x_i = \phi_i(y, \dot{y}, \dots, y^{(r-1)}), i = 1, \dots, n \tag{2.64}$$

while for the control input it holds

$$u = \psi(y, \dot{y}, \dots, y^{(r-1)}, y^{(r)}) \tag{2.65}$$

Introducing the new state variables $y_1 = y$ and $y_i = y^{(i-1)}$, $i = 2, \dots, n$, the initial system of Eq. (2.63) can be written in the Brunovsky form:

$$\begin{pmatrix} \dot{y}_1 \\ \dot{y}_2 \\ \cdots \\ \cdots \\ \dot{y}_{n-1} \\ \dot{y}_n \end{pmatrix} = \begin{pmatrix} 0 & 1 & 0 & \cdots & 0 \\ 0 & 0 & 1 & \cdots & 0 \\ \cdots & \cdots & \cdots & \cdots & \cdots \\ \cdots & \cdots & \cdots & \cdots & \cdots \\ 0 & 0 & 0 & \cdots & 1 \\ 0 & 0 & 0 & \cdots & 0 \end{pmatrix} \begin{pmatrix} y_1 \\ y_2 \\ \cdots \\ \cdots \\ y_{n-1} \\ y_n \end{pmatrix} + \begin{pmatrix} 0 \\ 0 \\ \cdots \\ \cdots \\ 0 \\ 1 \end{pmatrix} v \tag{2.66}$$

where $v = f(x, t) + g(x, t)(u + \tilde{d})$ is the control input for the linearized model, and $\tilde{d}$ denotes additive input disturbances. Thus one can use that

$$y^{(n)} = f(x, t) + g(x, t)(u + \tilde{d}) \tag{2.67}$$

where $f(x, t)$, $g(x, t)$ are unknown nonlinear functions, while as mentioned above $\tilde{d}$ is an unknown additive disturbance. It is possible to make the system's state vector x follow a given bounded reference trajectory x_d. In the presence of model uncertainties and external disturbances, denoted by w_d, successful tracking of the reference trajectory is provided by the H_∞ criterion [140, 206, 258]:

$$\int_0^T e^T Q e dt \leq \rho^2 \int_0^T {w_d}^T w_d dt \tag{2.68}$$

where ρ is the attenuation level and corresponds to the maximum singular value of the transfer function $G(s)$ of the linearized model associated to Eqs. (2.66) and (2.67).

The concept of H_∞ control comes from the need to compensate for the worst type of disturbances that may affect a nonlinear dynamical system. This was initially formulated for linear dynamics described with the use of transfer functions. The H_∞ norm of a linear system with transfer function $G(s)$, is denoted by $||G||_\infty$ and is defined as $||G||_\infty = sup_\omega \sigma_{max}[G(j\omega)]$ where σ_{max} is the system's maximum singular value [196, 199, 206]. The term *sup* denotes the supremum or least upper bound of the function $\sigma_{max}[G(j(\omega)]$, and thus the H_∞ norm of $G(s)$ is the maximum value of $\sigma_{max}[G(j(\omega)]$ over all frequencies ω. H_∞ norm has a physically meaningful interpretation when considering the system $y(s) = G(s)u(s)$. When this system is driven with a unit sinusoidal input at a specific frequency, $\sigma_{max}|G(j\omega)|$ is the largest possible output for the corresponding sinusoidal input. Thus, the H_∞ norm is the largest possible amplification over all frequencies of a sinusoidal input.

2.5.2 *Adaptive Control Law for Nonlinear Systems*

For the measurable state vector x of the system of Eqs. (2.66) and (2.67), and for uncertain functions $f(x, t)$ and $g(x, t)$ an appropriate control law is

$$u = \frac{1}{\hat{g}(x, t)}[y_d^{(n)} - \hat{f}(x, t) - K^T e + u_c] \tag{2.69}$$

with $e = [e, \dot{e}, \ddot{e}, \ldots, e^{(n-1)}]^T$ and $e = y - y_d$, $K^T = [k_n, k_{n-1}, \ldots, k_1]$, such that the polynomial $e^{(n)} + k_1 e^{(n-1)} + k_2 e^{(n-2)} + \cdots + k_n e$ is Hurwitz. The term u_c denotes the supervisory (supplementary) control input that is used for unmodeled dynamics and external perturbations. The control law of Eq. (2.69) results into

$$e^{(n)} = -K^T e + u_c + [f(x,t) - \hat{f}(x,t)] + [g(x,t) - \hat{g}(x,t)]u + g(x,t)\tilde{d} \quad (2.70)$$

where the supervisory control term u_c aims at the compensation of the approximation error

$$w = [f(x,t) - \hat{f}(x,t)] + [g(x,t) - \hat{g}(x,t)]u \quad (2.71)$$

as well as of the additive disturbance term $d_1 = g(x,t)\tilde{d}$. The above relation can be written in a state-equation form. The state vector is taken to be $e^T = [e, \dot{e}, \ldots, e^{(n-1)}]$, which after some operations yields

$$\dot{e} = (A - BK^T)e + Bu_c + B\{[f(x,t) - \hat{f}(x,t)] + [g(x,t) - \hat{g}(x,t)]u + d_1\} \quad (2.72)$$

where

$$A = \begin{pmatrix} 0 & 1 & 0 & \cdots\cdots & 0 \\ 0 & 0 & 1 & \cdots\cdots & 0 \\ \cdots & \cdots & \cdots & \cdots & \cdots \\ \cdots & \cdots & \cdots & \cdots & \cdots \\ 0 & 0 & 0 & \cdots\cdots & 1 \\ 0 & 0 & 0 & \cdots\cdots & 0 \end{pmatrix}, \quad B = \begin{pmatrix} 0 \\ 0 \\ \cdots \\ \cdots \\ 0 \\ 1 \end{pmatrix} \quad (2.73)$$

and $K = [k_n, k_{n-1}, \ldots, k_2, k_1]^T$. This is again a canonical form for the dynamics of the tracking error. As explained above, the control signal u_c is an auxiliary control term, used for the compensation of $\tilde{d}$ and w, which can be selected according to H_∞ control theory:

$$u_c = -\frac{1}{r}B^T P e \quad (2.74)$$

2.5.3 Approximators of System Unknown Dynamics

The approximation of functions $f(x,t)$ and $g(x,t)$ of Eq. (2.67) can be carried out with neuro-fuzzy networks (Fig. 2.1). The estimation of $f(x,t)$ and $g(x,t)$ can be written as [267, 268]:

$$\hat{f}(x|\theta_f) = \theta_f^T \phi(x), \quad \hat{g}(x|\theta_g) = \theta_g^T \phi(x), \quad (2.75)$$

where $\phi(x)$ are kernel functions with elements

$$\phi^l(x) = \frac{\prod_{i=1}^n \mu_{A_i}^l(x_i)}{\sum_{l=1}^N \prod_{i=1}^n \mu_{A_i}^l(x_i)} \quad l = 1, 2, \ldots, N \quad (2.76)$$

It is assumed that the weights θ_f and θ_g vary in the bounded areas M_{θ_f} and M_{θ_g} which are defined as

$$\begin{aligned} M_{\theta_f} &= \{\theta_f \in R^h : ||\theta_f|| \leq m_{\theta_f}\}, \\ M_{\theta_g} &= \{\theta_g \in R^h : ||\theta_g|| \leq m_{\theta_g}\} \end{aligned} \quad (2.77)$$

with m_{θ_f} and m_{θ_g} positive constants. The values of θ_f and θ_g that give optimal approximation are:

$$\begin{aligned}\theta_f^* &= \arg\min_{\theta_f \in M_{\theta_f}} [sup_{x \in U_x} |f(x) - \hat{f}(x|\theta_f)|] \\ \theta_g^* &= \arg\min_{\theta_g \in M_{\theta_g}} [sup_{x \in U_x} |g(x) - \hat{g}(x|\theta_g)|]\end{aligned} \tag{2.78}$$

The approximation error of $f(x,t)$ and $g(x,t)$ is given by

$$\begin{aligned}w &= [f(x,t) - \hat{f}(x,|\theta_f^*)] + [g(x,t) - \hat{g}(x|\theta_f^*)]u \Rightarrow \\ w &= \{[\hat{f}(x|\theta_f) - \hat{f}(x|\theta_f^*)] + [f(x,t) - \hat{f}(x|\theta_f)]\} + \\ &+ \{[\hat{g}(x|\theta_g) - \hat{g}(x|\theta_g^*)] + [g(x,t) - \hat{g}(x|\theta_g)]\}u\end{aligned} \tag{2.79}$$

where: (i) $\hat{f}(x|\theta_f^*)$ is the approximation of f for the best estimation θ_f^* of the weights' vector θ_f, (ii) $\hat{g}(x|\theta_g^*)$ is the approximation of g for the best estimation θ_g^* of the weights' vector θ_g.

The approximation error w can be decomposed into w_a and w_b, where

$$\begin{aligned}w_a &= [\hat{f}(x|\theta_f) - \hat{f}(x|\theta_f^*)] + [\hat{g}(x|\theta_g) - \hat{g}(x|\theta_g^*)]u \\ w_b &= [\hat{f}(x|\theta_f^*) - f(x,t)] + [\hat{g}(x|\theta_g^*) - g(x,t)]u\end{aligned} \tag{2.80}$$

Finally, the following two parameters are defined:

$$\begin{aligned}\tilde{\theta}_f &= \theta_f - \theta_f^* \\ \tilde{\theta}_g &= \theta_g - \theta_g^*.\end{aligned} \tag{2.81}$$

A difference between the neurofuzzy approximator given in Fig. 2.1 and a radial basis functions (RBF) neural network is the normalization layer that appears between the Gaussian basis functions layer and the weights output layer. After normalization, the sum (over the complete set of rules) of the fuzzy membership values of each input pattern becomes equal to 1.

2.5.4 *Lyapunov Stability Analysis for Dynamical Systems*

The adaptation law of the weights θ_f and θ_g as well as of the supervisory control term u_c is derived by the requirement for negative derivative of the quadratic Lyapunov function

$$V = \frac{1}{2}e^T Pe + \frac{1}{2\gamma_1}\tilde{\theta}_f^T \tilde{\theta}_f + \frac{1}{2\gamma_2}\tilde{\theta}_g^T \tilde{\theta}_g \tag{2.82}$$

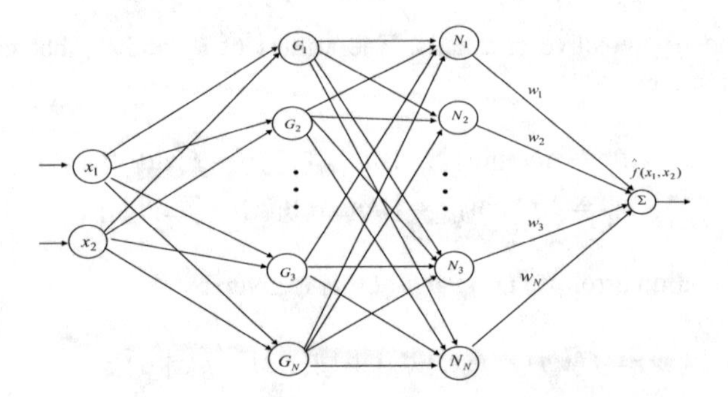

Fig. 2.1 Neuro-fuzzy approximator: G_i Gaussian basis function, N_i: normalization unit

Differentiating Eq. (2.82) gives

$$\dot{V} = \frac{1}{2}\dot{e}^T Pe + \frac{1}{2}e^T P\dot{e} + \frac{1}{\gamma_1}\tilde{\theta}_f^T \dot{\tilde{\theta}}_f + \frac{1}{\gamma_2}\tilde{\theta}_g^T \dot{\tilde{\theta}}_g \Rightarrow \tag{2.83}$$

Next, substituting Eq. (2.79) into Eq. (2.83) gives

$$\begin{aligned} \dot{V} &= \frac{1}{2}e^T\{(A - BK^T)^T P + P(A - BK^T)\}e + \\ &+ B^T Pe(u_c + w + d_1) + \frac{1}{\gamma_1}\tilde{\theta}_f^T \dot{\tilde{\theta}}_f + \frac{1}{\gamma_2}\tilde{\theta}_g^T \dot{\tilde{\theta}}_g \end{aligned} \tag{2.84}$$

Assumption 1: For given positive definite matrix Q and coefficients r and ρ there exists a positive definite matrix P, which is the solution of the following matrix equation

$$(A - BK^T)^T P + P(A - BK^T) - PB\left(\frac{2}{r} - \frac{1}{\rho^2}\right)B^T P + Q = 0 \tag{2.85}$$

Substituting Eq. (2.85) into $\dot{V}$ yields after some operations

$$\begin{aligned} \dot{V} &= -\frac{1}{2}e^T Qe + \frac{1}{2}e^T PB\left(\frac{2}{r} - \frac{1}{\rho^2}\right)B^T Pe + B^T Pe\left(-\frac{1}{r}e^T PB\right) + \\ &+ B^T Pe(w + d_1) + \frac{1}{\gamma_1}\tilde{\theta}_f^T \dot{\tilde{\theta}}_f + \frac{1}{\gamma_2}\tilde{\theta}_g^T \dot{\tilde{\theta}}_g \end{aligned} \tag{2.86}$$

It holds that

$$\dot{\tilde{\theta}}_f = \dot{\theta}_f - \dot{\theta}_f^* = \dot{\theta}_f \\ \dot{\tilde{\theta}}_g = \dot{\theta}_g - \dot{\theta}_g^* = \dot{\theta}_g \tag{2.87}$$

The following weight adaptation laws are considered [267]

$$\dot{\theta}_f = \begin{cases} -\gamma_1 e^T PB\phi(x) \ \ if \ ||\theta_f|| < m_{\theta_f} \\ 0 \ ||\theta_f|| \geq m_{\theta_f} \end{cases} \tag{2.88}$$

$$\dot{\theta}_g = \begin{cases} -\gamma_2 e^T PB\phi(x)u_c \ \ if \ ||\theta_g|| < m_{\theta_g} \\ 0 \ ||\theta_g|| \geq m_{\theta_g} \end{cases} \tag{2.89}$$

$\dot{\theta}_f$ and $\dot{\theta}_g$ are set to 0, when

$$||\theta_f|| \geq m_{\theta_f}, \ \ ||\theta_g|| \geq m_{\theta_g}. \tag{2.90}$$

The update of θ_f stands for a gradient algorithm on the cost function $\frac{1}{2}(f - \hat{f})^2$. The update of θ_g is also of the gradient type, while u_c implicitly tunes the adaptation gain γ_2. Substituting Eqs. (2.88) and (2.89) in $\dot{V}$ finally gives

$$\dot{V} = -\frac{1}{2}e^T Qe - \frac{1}{2\rho^2}e^T PBB^T Pe + e^T PB(w + d_1) - \\ - e^T PB(\theta_f - \theta_f^*)^T \phi(x) - e^T PB(\theta_g - \theta_g^*)^T \phi(x)u_c \Rightarrow \\ \dot{V} = -\frac{1}{2}e^T Qe - \frac{1}{2\rho^2}e^T PBB^T Pe + e^T PB(w + d_1) + e^T PBw_\alpha. \tag{2.91}$$

Denoting $w_1 = w + d_1 + w_\alpha$ one gets

$$\dot{V} = -\frac{1}{2}e^T Qe - \frac{1}{2\rho^2}e^T PBB^T Pe + e^T PBw_1 \text{ or equivalently,} \\ \dot{V} = -\frac{1}{2}e^T Qe - \frac{1}{2\rho^2}e^T PBB^T Pe + \frac{1}{2}e^T PBw_1 + \frac{1}{2}w_1^T B^T Pe \tag{2.92}$$

Lemma: The following inequality holds:

$$\frac{1}{2}e^T PBw_1 + \frac{1}{2}w_1^T B^T Pe - \frac{1}{2\rho^2}e^T PBB^T Pe \leq \frac{1}{2}\rho^2 w_1^T w_1 \tag{2.93}$$

Proof: The binomial $(\rho a - \frac{1}{\rho}b)^2 \geq 0$ is considered. Expanding the left part of the above inequality one gets

$$\rho^2 a^2 + \frac{1}{\rho^2}b^2 - 2ab \geq 0 \Rightarrow \frac{1}{2}\rho^2 a^2 + \frac{1}{2\rho^2}b^2 - ab \geq 0 \Rightarrow \\ ab - \frac{1}{2\rho^2}b^2 \leq \frac{1}{2}\rho^2 a^2 \Rightarrow \frac{1}{2}ab + \frac{1}{2}ab - \frac{1}{2\rho^2}b^2 \leq \frac{1}{2}\rho^2 a^2 \tag{2.94}$$

The following substitutions are carried out: $a = w_1$ and $b = e^T PB$ and the previous relation becomes

$$\tfrac{1}{2}w_1^T B^T Pe + \tfrac{1}{2}e^T PBw_1 - \tfrac{1}{2\rho^2}e^T PBB^T Pe \leq \tfrac{1}{2}\rho^2 w_1^T w_1 \tag{2.95}$$

The previous inequality is used in $\dot{V}$, and thus one arrives at

$$\dot{V} \leq -\frac{1}{2}e^T Qe + \frac{1}{2}\rho^2 w_1^T w_1 \tag{2.96}$$

The attenuation coefficient ρ can be chosen such that the right part of Eq. (2.96) is always upper bounded by 0. For instance, it suffices at any iteration of the control algorithm to have

$$\begin{array}{c} -\tfrac{1}{2}e^T Qe + \tfrac{1}{2}\rho^2 ||w||_1^2 \leq 0 \Rightarrow \\ -\tfrac{1}{2}||e||_Q^2 + \tfrac{1}{2}\rho^2 ||w||_1^2 \leq 0 \Rightarrow \\ \rho^2 \leq \frac{||e||_Q^2}{||w||_1^2} \end{array} \tag{2.97}$$

Without knowledge of the uncertainty and disturbances term $||w||_1$ a sufficiently small value of ρ can assure that the above inequality holds and thus that the control loop stability can be assured. Actually ρ should be given the least value which permits to obtain a solution for the Riccati equation of Eq. (2.85).

Equation (2.96) can be used to show that the H_∞ performance criterion is satisfied. The integration of $\dot{V}$ from 0 to T gives

$$\begin{array}{c} \int_0^T \dot{V}(t)dt \leq -\tfrac{1}{2}\int_0^T ||e||^2 dt + \tfrac{1}{2}\rho^2 \int_0^T ||w_1||^2 dt \Rightarrow \\ 2V(T) + \int_0^T ||e||_Q^2 dt \leq 2V(0) + \rho^2 \int_0^T ||w_1||^2 dt \end{array} \tag{2.98}$$

Moreover, if there exists a positive constant $M_w > 0$ such that

$$\int_0^\infty ||w_1||^2 dt \leq M_w \tag{2.99}$$

then one gets

$$\int_0^\infty ||e||_Q^2 dt \leq 2V(0) + \rho^2 M_w \tag{2.100}$$

Thus, the integral $\int_0^\infty ||e||_Q^2 dt$ is bounded. Moreover, $V(T)$ is bounded and from the definition of the Lyapunov function V in Eq. (2.82) it becomes clear that $e(t)$ will be also bounded since $e(t) \in \Omega_e = \{e|e^T Pe \leq 2V(0) + \rho^2 M_w\}$.

According to the above and with the use of Barbalat's Lemma one obtains the asymptotic elimination of the tracking error, that is $lim_{t\rightarrow\infty}e(t) = 0$.

Chapter 3
Main Approaches to Nonlinear Estimation

3.1 Outline

To treat the filtering problem for nonlinear dynamics in financial systems the Extended Kalman Filter is an established approach. However, since this is based on approximate linearization of the system's dynamics and in the truncation of higher order terms in the associated Taylor series expansion, the Unscented Kalman Filter is frequently used in its place. The latter fitler performs state estimation by averaging on state vectors which are selected at iteration of the filter algorithm according to the columns of the estimation error vector covariance matrix. Moreover, to handle the case of non-Gaussian noises in the filtering procedure the particle filter has been proposed. A number of potential state vector values (particles) is updated in time through elitism criteria and out of this set the estimate of the state vector is computed. Moreover, to treat the distributed filtering and state estimation one can apply established methods for decentralized state estimation, such as the Extended Information Filter (EIF) and the Unscented Information Filter (UIF). EIF stands for the distributed implementation of the Extended Kalman Filter while UIF stands for the distributed implementation of the Unscented Kalman Filter. Moreover, to obtain a distributed filtering scheme in this chapter the Derivative-free Extended Information Filter (DEIF) is implemented. This stands for the distributed implementation of a differential flatness theory-based filtering method under the name Derivative-free distributed nonlinear Kalman Filter. The improved performance of DEIF comparing to the EIF and UIF is confirmed both in terms of improved estimation accuracy and in terms of improved speed of computation. Finally, one can note distributed filtering with the use of the distributed Particle filter is proposed. It consists of multiple Particle filters running at distributed computation units while a consensus criterion is used to fuse the local state estimates.

G.G. Rigatos, *State-Space Approaches for Modelling and Control in Financial Engineering*, Intelligent Systems Reference Library 125,
DOI 10.1007/978-3-319-52866-3_3

3.2 Linear State Observers

First, the linear dynamical system of Eq. (3.1) is considered

$$\begin{cases} \dot{x}(t) = Ax(t) + Bu(t) \\ y(t) = Cx(t) \end{cases} \tag{3.1}$$

where $x \in R^{m\times 1}$ is the system's state vector $u \in R^{1\times 1}$ is the control input, and $y \in R^{p\times 1}$ is the system's output. It is assumed that the elements of the state vector x are not completely measurable. In that case the system's state vector can be reconstructed using the sequence of output measurements $y(t)$ and the associated sequence of control inputs $u(t)$. The basic requirement to perform the state vector's reconstruction is the linear system to be observable as defined by the pair of matrices (A, C). An important application of state observers is the design of state estimation-based control schemes. For linear dynamical systems the principle of separation holds: (i) the state feedback controller is designed assuming that the complete state vector of the system is available, (ii) the state observer is designed for those state variables which cannot be measured directly. The concept of state observers is due to Luenberger and includes the Kalman Filter as a special case [120]. Actually, the Kalman Filter is an optimal state observer in the sense that it can compensate in optimal way for the effect that process and measurement noises have on the estimation of the system's state vector.

For the continuous time dynamical system of Eq. (3.1) the state observer is

$$\dot{\hat{x}} = A\hat{x} + Bu + K(y - C\hat{x}) \tag{3.2}$$

From Eq. (3.2) it can be seen that the linear state observer uses the state-space equation of the dynamical system augmented by the additional term $K(y - \hat{y})$ where the signal $y - C\hat{x}$ is called residual (or innovation) and is the difference between the real measurement $y(t)$ and the estimated output $\hat{y}(t)$. Defining the state vector estimation error as $e = x - \hat{x}$, the dynamics of the observer becomes

$$\dot{e}(t) = (A - KC)e(t) \Rightarrow \dot{e}(t) = \hat{A}e(t) \tag{3.3}$$

where $\hat{A} = A - KC$. The main problem in the design of the state observer is to select gain K such that matrix $\hat{A} = A - KC$, which defines the estimation error dynamics, to have eigenvalues strictly in the left complex semi-plane. Two typical methods for the selection of gain K are:

1. All eigenvalues of matrix $\hat{A}$ can be moved to desirable positions at the left complex semi-plane using pole-placement techniques.
2. Gain K can be selected by solving an optimization problem (which finally provides the Kalman Filter). This is expressed as the minimization of a quadratic cost functional (as in the case of Linear Quadratic Regulator - LQR optimal

control) and is performed through the solution of a Riccati equation. In that case the observer's gain K is calculated by $K = PC^T R^{-1}$ considering an optimal control problem for the dual system (A^T, C^T), where the covariance matrix of the estimation error P is found by the solution of the continuous-time Riccati equation of the form

$$\dot{P} = AP + PA^T + Q - PC^T R^{-1} CP \tag{3.4}$$

where matrices Q and R stand for the process and measurement noise covariance matrices, respectively.

3.3 The Continuous-Time Kalman Filter for Linear Models

Next, the continuous-time dynamical system of Eq. (3.5) is assumed [120, 222]:

$$\begin{cases} \dot{x}(t) = Ax(t) + Bu(t) + w(t), \ t \geq t_0 \\ y(t) = Cx(t) + v(t), \ t \geq t_0 \end{cases} \tag{3.5}$$

where again $x \in R^{m \times 1}$ is the system's state vector, and $y \in R^{p \times 1}$ is the system's output. Matrices A, B and C can be time-varying and $w(t)$, $v(t)$ are uncorrelated white Gaussian noises. The covariance matrix of the process noise $w(t)$ is $Q(t)$, while the covariance matrix of the measurement noise is $R(t)$. Then the Kalman Filter is again a linear state observer which is given by

$$\begin{cases} \dot{\hat{x}} = A\hat{x} + Bu + K[y - C\hat{x}], \ \hat{x}(t_0) = 0 \\ K(t) = PC^T R^{-1} \\ \dot{P} = AP + PA^T + Q - PC^T R^{-1} CP \end{cases} \tag{3.6}$$

where $\hat{x}(t)$ is the optimal estimation of the state vector $x(t)$ and $P(t)$ is the covariance matrix of the state vector estimation error with $P(t_0) = P_0$. It can be seen that as in the case of the Luenberger observer, the Kalman Filter consists of the system's state equation plus a corrective term $K[y - C\hat{x}]$. The associated Riccati equation for calculating the covariance matrix $P(t)$ has the same form as Eq. (3.4).

3.4 The Discrete-Time Kalman Filter for Linear Systems

In the discrete-time case the dynamical system is assumed to be expressed in the form of a discrete-time state model:

$$\begin{cases} x(k+1) = \Phi(k)x(k) + L(k)u(k) + w(k) \\ z(k) = Cx(k) + v(k) \end{cases} \tag{3.7}$$

where the state $x(k)$ is a m-vector, $w(k)$ is a m-element process noise vector and Φ is a $m \times m$ real matrix. Moreover the output measurement $z(k)$ is a p-vector, C is an $p \times m$-matrix of real numbers, and $v(k)$ is the measurement noise. It is assumed that the process noise $w(k)$ and the measurement noise $v(k)$ are uncorrelated.

Now the problem of interest is to estimate the state $x(k)$ based on the measurements $z(1), z(2), \ldots, z(k)$. The initial value of the state vector $x(0)$, the initial value of the error covariance matrix $P(0)$ is unknown and an estimation of it is considered, i.e. $\hat{x}(0) =$ a guess of $E[x(0)]$ and $\hat{P}(0) =$ a guess of $Cov[x(0)]$.

For the initialization of matrix P one can set $\hat{P}(0) = \lambda I$, with $\lambda > 0$. The state vector $x(k)$ has to be estimated taking into account $\hat{x}(0)$, $\hat{P}(0)$ and the output measurements $Z = [z(1), z(2), \ldots, z(k)]^T$, i.e. there is a function relationship:

$$\hat{x}(k) = \alpha_n(\hat{x}(0), \hat{P}(0), Z(k)) \tag{3.8}$$

Actually, this is a linear minimum mean squares estimation problem (LMMSE) which is solved recursively, through the function relationship

$$\hat{x}(k+1) = a_{n+1}(\hat{x}(k), z(k+1)) \tag{3.9}$$

The process and output noise are white and their covariance matrices are given by: $E[w(i)w^T(j)] = Q\delta(i-j)$ and $E[v(i)v^T(j)] = R\delta(i-j)$.

Using the above, the discrete-time Kalman Filter can be decomposed into two parts: (i) time update, and (ii) measurement update. The first part employs an estimate of the state vector $x(k)$ made before the output measurement $z(k)$ is available (a priori estimate). The second part estimates $x(k)$ after $z(k)$ has become available (a posteriori estimate).

- When the set of measurements $Z^- = \{z(1), \ldots, z(k-1)\}$ is available, from Z^- an a priori estimation of $x(k)$ is obtained which is denoted by $\hat{x}^-(k) =$ the estimate of $x(k)$ given Z^-.
- When $z(k)$ becomes available, the set of the output measurements becomes $Z = \{z(1), \ldots, z(k)\}$, where $\hat{x}(k) =$ the estimate of $x(k)$ given Z.

The associated estimation errors are defined by

$$\begin{array}{c} e^-(k) = x(k) - \hat{x}^-(k) = \text{the a priori error} \\ e(k) = x(k) - \hat{x}(k) = \text{the a posteriori error} \end{array} \tag{3.10}$$

The estimation error covariance matrices associated with $\hat{x}(k)$ and $\hat{x}^-(k)$ are defined as [104, 120]

$$P^-(k) = Cov[e^-(k)] = E[e^-(k)e^-(k)^T]$$
$$P(k) = Cov[e(k)] = E[e(k)e^T(k)]$$

From the definition of the trace of a matrix, the mean square error of the estimates can be written as

$$MSE(\hat{x}^-(k)) = E[e^-(k)e^-(k)^T] = tr(P^-(k))$$
$$MSE(x(k)) = E[e(k)e^T(k)] = tr(P(k))$$

Finally, the linear Kalman filter equations are

Measurement update: obtain measurement $z(k)$ and compute

$$\begin{aligned} K(k) &= P^-(k)C^T[C{\cdot}P^-(k)C^T + R]^{-1} \\ \hat{x}(k) &= \hat{x}^-(k) + K(k)[z(k) - C\hat{x}^-(k)] \\ P(k) &= P^-(k) - K(k)CP^-(k) \end{aligned} \tag{3.11}$$

Time update: compute

$$\begin{aligned} P^-(k+1) &= \Phi(k)P(k)\Phi^T(k) + Q(k) \\ \hat{x}^-(k+1) &= \Phi(k)\hat{x}(k) + L(k)u(k) \end{aligned} \tag{3.12}$$

3.5 The Extended Kalman Filter for Nonlinear Systems

The following nonlinear time-invariant state model is now considered [198]:

$$\begin{aligned} x(k+1) &= \phi(x(k)) + L(k)u(k) + w(k) \\ z(k) &= \gamma(x(k)) + v(k) \end{aligned} \tag{3.13}$$

where $x{\in}R^{m\times 1}$ is the system's state vector and $z{\in}R^{p\times 1}$ is the system's output, while $w(k)$ and $v(k)$ are uncorrelated, zero-mean, Gaussian zero-mean noise processes with covariance matrices $Q(k)$ and $R(k)$ respectively. The operators $\phi(x)$ and $\gamma(x)$ are $\phi(x) = [\phi_1(x), \phi_2(x), \ldots, \phi_m(x)]^T$, and $\gamma(x) = [\gamma_1(x), \gamma_2(x), \ldots, \gamma_p(x)]^T$, respectively. It is assumed that ϕ and γ are sufficiently smooth in x so that each one has a valid series Taylor expansion. Following a linearization procedure, ϕ is expanded into Taylor series about $\hat{x}$:

$$\phi(x(k)) = \phi(\hat{x}(k)) + J_\phi(\hat{x}(k))[x(k) - \hat{x}(k)] + \cdots \tag{3.14}$$

where $J_\phi(x)$ is the Jacobian of ϕ calculated at $\hat{x}(k)$:

$$J_{\phi}(x) = \frac{\partial \phi}{\partial x}|_{x=\hat{x}(k)} = \begin{pmatrix} \frac{\partial \phi_1}{\partial x_1} & \frac{\partial \phi_1}{\partial x_2} & \cdots & \frac{\partial \phi_1}{\partial x_m} \\ \frac{\partial \phi_2}{\partial x_1} & \frac{\partial \phi_2}{\partial x_2} & \cdots & \frac{\partial \phi_2}{\partial x_m} \\ \vdots & \vdots & \vdots & \vdots \\ \frac{\partial \phi_m}{\partial x_1} & \frac{\partial \phi_m}{\partial x_2} & \cdots & \frac{\partial \phi_m}{\partial x_m} \end{pmatrix} \tag{3.15}$$

Likewise, γ is expanded about $\hat{x}^{-}(k)$

$$\gamma(x(k)) = \gamma(\hat{x}^{-}(k)) + J_{\gamma}[x(k) - \hat{x}^{-}(k)] + \cdots \tag{3.16}$$

where $\hat{x}^{-}(k)$ is the estimation of the state vector $x(k)$ before measurement at the k-th instant to be received and $\hat{x}(k)$ is the updated estimation of the state vector after measurement at the k-th instant has been received. The Jacobian $J_{\gamma}(x)$ is

$$J_{\gamma}(x) = \frac{\partial \gamma}{\partial x}|_{x=\hat{x}^{-}(k)} = \begin{pmatrix} \frac{\partial \gamma_1}{\partial x_1} & \frac{\partial \gamma_1}{\partial x_2} & \cdots & \frac{\partial \gamma_1}{\partial x_m} \\ \frac{\partial \gamma_2}{\partial x_1} & \frac{\partial \gamma_2}{\partial x_2} & \cdots & \frac{\partial \gamma_2}{\partial x_m} \\ \vdots & \vdots & \vdots & \vdots \\ \frac{\partial \gamma_p}{\partial x_1} & \frac{\partial \gamma_p}{\partial x_2} & \cdots & \frac{\partial \gamma_p}{\partial x_m} \end{pmatrix} \tag{3.17}$$

The resulting expressions create first order approximations of ϕ and γ. Thus the linearized version of the system's model is obtained:

$$\begin{aligned} x(k+1) &= \phi(\hat{x}(k)) + J_{\phi}(\hat{x}(k))[x(k) - \hat{x}(k)] + w(k) \\ z(k) &= \gamma(\hat{x}^{-}(k)) + J_{\gamma}(\hat{x}^{-}(k))[x(k) - \hat{x}^{-}(k)] + v(k) \end{aligned} \tag{3.18}$$

Now, the EKF recursion is as follows: First the time update is considered: by $\hat{x}(k)$ the estimation of the state vector at instant k is denoted. Given initial conditions $\hat{x}^{-}(0)$ and $P^{-}(0)$ the recursion proceeds as [198]:

- *Measurement update*. Acquire $z(k)$ and compute:

$$\begin{aligned} K(k) &= P^{-}(k)J_{\gamma}^{T}(\hat{x}^{-}(k))\cdot[J_{\gamma}(\hat{x}^{-}(k))P^{-}(k)J_{\gamma}^{T}(\hat{x}^{-}(k)) + R(k)]^{-1} \\ \hat{x}(k) &= \hat{x}^{-}(k) + K(k)[z(k) - \gamma(\hat{x}^{-}(k))] \\ P(k) &= P^{-}(k) - K(k)J_{\gamma}(\hat{x}^{-}(k))P^{-}(k) \end{aligned} \tag{3.19}$$

- *Time update*. Compute:

$$\begin{aligned} P^{-}(k+1) &= J_{\phi}(\hat{x}(k))P(k)J_{\phi}^{T}(\hat{x}(k)) + Q(k) \\ \hat{x}^{-}(k+1) &= \phi(\hat{x}(k)) + L(k)u(k) \end{aligned} \tag{3.20}$$

The schematic diagram of the EKF loop is given in Fig. 3.1.

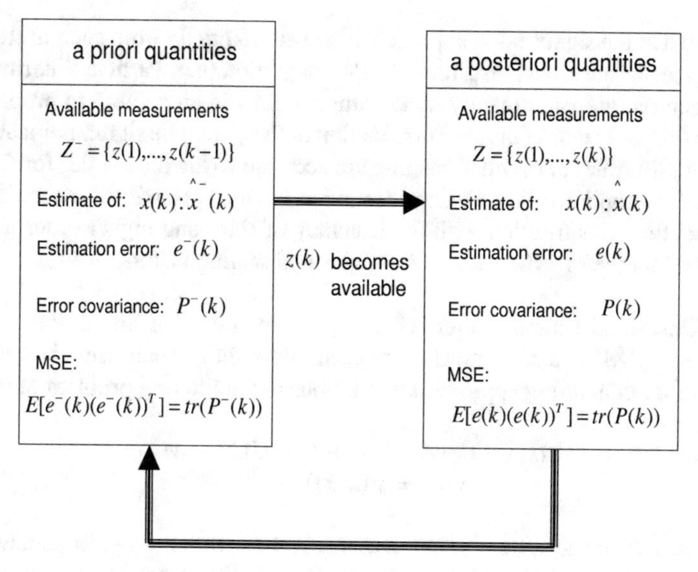

Fig. 3.1 Schematic diagram of the EKF loop

3.6 Sigma-Point Kalman Filters

The Sigma-Point Kalman Filter overcomes the flaws of Extended Kalman Filtering [161]. Unlike EKF no analytical Jacobians of the system equations need to be calculated as in the case of the EKF. This makes the sigma-point approach suitable for application in "black-box" models where analytical expressions of the system dynamics are either not available or not in a form which allows for easy linearization. This is achieved through a different approach for calculating the posterior 1st and 2nd order statistics of a random variable that undergoes a nonlinear transformation. The state distribution is represented again by a Gaussian Random Variable but is now specified using a minimal set of deterministically chosen weighted sample points. The basic sigma-point approach can be described as follows:

1. A set of weighted samples (sigma-points) are deterministically calculated using the mean and square-root decomposition of the covariance matrix of the system's state vector. As a minimal requirement the sigma-point set must completely capture the first and second order moments of the prior random variable. Higher order moments can be captured at the cost of using more sigma-points.
2. The sigma-points are propagated through the true nonlinear function using functional evaluations alone, i.e. no analytical derivatives are used, in order to generate a posterior sigma-point set.
3. The posterior statistics are calculated (approximated) using tractable functions of the propagated sigma-points and weights. Typically, these take on the form of a simple weighted sample mean and covariance calculations of the posterior sigma points.

It is noted that the sigma-point approach differs substantially from general stochastic sampling techniques, such as Monte-Carlo integration (e.g. Particle Filtering methods) which require significantly more sample points in an attempt to propagate an accurate (possibly non-Gaussian) distribution of the state. The sigma-point approach results in posterior approximations that are accurate to the third order for Gaussian inputs for all nonlinearities. For non-Gaussian inputs, approximations are accurate to at least the second-order, with the accuracy of third and higher-order moments determined by the specific choice of weights and scaling factors.

The Unscented Kalman Filter (UKF) is a special case of Sigma-Point Kalman Filters. The UKF is a discrete time filtering algorithm which uses the unscented transform for computing approximate solutions to the filtering problem of the form

$$\begin{aligned} x(k+1) &= \phi(x(k)) + L(k)U(k) + w(k) \\ y(k) &= \gamma(x(k)) + v(k) \end{aligned} \tag{3.21}$$

where $x(k) \in R^n$ is the system's state vector, $y(k) \in R^m$ is the measurement, $w(k) \in R^n$ is a Gaussian process noise $w(k) \sim N(0, Q(k))$, and $v(k) \in R^m$ is a Gaussian measurement noise $v(k) \sim N(0, R(k))$. The mean and covariance of the initial state $x(0)$ are $m(0)$ and $P(0)$, respectively.

Some basic operations performed in the UKF algorithm (*Unscented Transform*) are summarized as follows:

(1) Denoting the current state mean as $\hat{x}$, a set of $2n+1$ sigma points is taken from the columns of the $n \times n$ matrix $\sqrt{(n+\lambda)P_{xx}}$ as follows:

$$\begin{aligned} x^0 &= \hat{x} \\ x^i &= \hat{x} + [\sqrt{(n+\lambda)P_{xx}}]_i, \ i = 1, \ldots, n \\ x^i &= \hat{x} - [\sqrt{(n+\lambda)P_{xx}}]_i, \ i = n+1, \ldots, 2n \end{aligned} \tag{3.22}$$

and the associated weights are computed:

$$\begin{aligned} W_0^{(m)} &= \frac{\lambda}{(n+\lambda)} & W_0^{(c)} &= \frac{\lambda}{(n+\lambda)+(1-\alpha^2+b)} \\ W_i^{(m)} &= \frac{1}{2(n+\lambda)}, \ i = 1, \ldots, 2n & W_i^{(c)} &= \frac{1}{2(n+\lambda)} \end{aligned} \tag{3.23}$$

where $i = 1, 2, \ldots, 2n$ and $\lambda = \alpha^2(n+\kappa) - n$ is a scaling parameter, while α, β and κ are constant parameters. Matrix P_{xx} is the covariance matrix of the state x.

(2) Transform each of the sigma points as

$$z^i = h(x^i) \ i = 0, \ldots, 2n \tag{3.24}$$

(3) Mean and covariance estimates for z can be computed as

$$\hat{z} \simeq \sum_{i=0}^{2n} W_i^{(m)} z^i$$
$$P_{zz} = \sum_{i=0}^{2n} W_i^{(c)} (z^i - \hat{z})(z^i - \hat{z})^T \quad (3.25)$$

(4) The cross-covariance of x and z is estimated as

$$P_{xz} = \simeq \sum_{i=0}^{2n} W_i^{(c)} (x^i - \hat{x})(z^i - \hat{z})^T \quad (3.26)$$

The square root of positive definite matrix P_{xx} means a matrix $A = \sqrt{P_{xx}}$ such that $P_{xx} = AA^T$ and a possible way for calculation is Singular Values Decomposition (SVD).

Next the basic stages of the *Unscented Kalman Filter* are given:

As in the case of the Extended Kalman Filter and the Particle Filter, the Unscented Kalman Filter also consists of a prediction stage (time update) and a correction stage (measurement update) [117, 234].

Time update: Compute the predicted state mean $\hat{x}^-(k)$ and the predicted covariance ${P_{xx}}^-(k)$ as

$$[\hat{x}^-(k), P_{xx}^-(k)] = UT(f_d, \hat{x}(k-1), P_{xx}(k-1))$$
$$P_{xx}^-(k) = P_{xx}(k-1) + Q(k-1) \quad (3.27)$$

Measurement update: Obtain the new output measurement z_k and compute the predicted mean $\hat{z}(k)$ and covariance of the measurement $P_{zz}(k)$, and the cross covariance of the state and measurement $P_{xz}(k)$

$$[\hat{z}(k), P_{zz}(k), P_{xz}(k)] = UT(h_d, \hat{x}^-(k), P_{xx}^-(k))$$
$$P_{zz}(k) = P_{zz}(k) + R(k) \quad (3.28)$$

Then compute the filter gain $K(k)$, the state mean $\hat{x}(k)$ and the covariance $P_{xx}(k)$, conditional to the measurement $y(k)$

$$K(k) = P_{xz}(k) P_{zz}^{-1}(k)$$
$$\hat{x}(k) = \hat{x}^-(k) + K(k)[z(k) - \hat{z}(k)]$$
$$P_{xx}(k) = P_{xx}^-(k) - K(k) P_{zz}(k) K(k)^T \quad (3.29)$$

The filter starts from the initial mean $m(0)$ and covariance $P_{xx}(0)$. The stages of state vector estimation with the use of the unscented filtering algorithm are depicted in Fig. 3.2.

3.7 Particle Filters

3.7.1 The Particle Approximation of Probability Distributions

The functioning of the Particle Filters will be explained next. Particle Filtering is a method for state estimation that is not dependent on the probability density function of the measurements [92, 96]. In the general case, the equations of the optimal filter used for the calculation of the state-vector of a dynamical system do not have an explicit solution. This happens for instance when the process noise and the noise of the output measurement do not follow a Gaussian distribution. In that case approximation through Monte-Carlo methods can be used [5, 36, 65, 257]. A sampling of size N is assumed, i.e. N i.i.d. (independent identically distributed) variables $\xi^1, \xi^2, \ldots, \xi^N$. This sampling follows the p.d.f. $p(x)$ i.e. $\xi^{1:N} \sim p(x)$. Instead of $p(x)$ the function $p(x) \simeq p^N(x) = \frac{1}{N}\sum_{i=1}^N \delta_{\xi^i}(x)$ can be used. It is assumed that all points ξ^i have an equal weighted contribution to the approximation of $p(x)$. A more general approach would be if weight factors were assigned to the points ξ^i, which also satisfy the normality condition $\sum_{i=1}^N w^i = 1$. In the latter case

$$p(x) \simeq p^N(x) = \sum_{i=1}^N w^i \delta_{\xi^i}(x) \tag{3.30}$$

If $p(\xi^i)$ is known then the probability $P(x)$ can be approximated using the discrete values of the p.d.f. $p(\xi^i) = w^i$. If sampling over the p.d.f. $p(x)$ is unavailable, then one can use a p.d.f. $\bar{p}(x)$ with similar support set, i.e. $p(x) = 0 \Rightarrow \bar{p}(x) = 0$. Then it holds $E(\phi(x)) = \int \phi(x) p(x) dx = \int \phi(x) \bar{p}(x) \frac{p(x)}{\bar{p}(x)} dx$. If the N samples of $\bar{p}(x)$

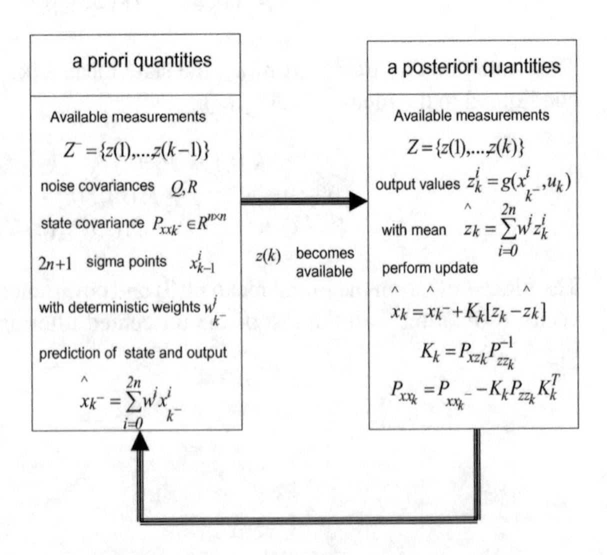

Fig. 3.2 Schematic diagram of the Unscented Kalman Filter loop

are available at the points $\tilde{\xi}^1 \cdots \tilde{\xi}^N$, i.e. $\bar{p}(\tilde{\xi})^i = \delta_{\tilde{\xi}^i}(x)$ and the weight coefficients w^i are defined as $w^i = \frac{p(\tilde{\xi}^i)}{\bar{p}(\tilde{\xi}^i)}$, then it is easily shown that

$$E(\phi(x)) \simeq \sum_{i=1}^{N} w^i \phi(\tilde{\xi}^i), \text{ where } \begin{cases} \tilde{\xi}^{1:N} \sim \bar{p}(x) \\ w^i = p(\tilde{x}^i)/\bar{p}(\tilde{x}^i) \end{cases}. \tag{3.31}$$

The meaning of Eq. (3.31) is as follows: assume that the p.d.f. $p(x)$ is unknown (target distribution), however the p.d.f. $\bar{p}(x)$ (importance law) is available. Then, it is sufficient to sample on $\bar{p}(x)$ and find the associated weight coefficients w^i so as to calculate $E(\phi(x))$.

3.7.2 The Prediction Stage

As in the case of the Kalman Filter or the Extended Kalman Filter the particles filter consists of the measurement update (correction stage) and the time update (prediction stage) [42, 207, 230, 257]. The prediction stage calculates $p(x(k)|Z^-)$ where $Z^- = \{z(1), z(2), \ldots, z(n-1)\}$ according to Eq. (3.30). It holds that:

$$p(x(k-1)|Z^-) = \sum_{i=1}^{N} w_{k-1}^i \delta_{\xi_{k-1}^i}(x(k-1)) \tag{3.32}$$

while from Bayes formula it holds $p(x(k)|Z^-) = \int p(x(k)|x(k-1))p(x(k-1)|Z^-)dx$. Using also Eq. (3.32) one finally obtains

$$\begin{aligned} &p(x(k)|Z^-) = \textstyle\sum_{i=1}^{N} w_{k-1}^i \delta_{\xi_{k^-}^i}(x(k)) \\ &\text{with } \xi_{k^-}^i \sim p(x(k)|x(k-1) = \xi_{k-1}^i) \end{aligned} \tag{3.33}$$

The meaning of Eq. (3.33) is as follows: the state equation of the system is run N times, starting from the N previous values of the state vectors $x(k-1) = \xi_{k-1}^i$

$$\begin{aligned} \hat{x}(k+1) &= \phi(\hat{x}(k)) + L(k)u(k) + w(k) \\ z(k) &= \gamma(\hat{x}(k)) + v(k) \end{aligned} \tag{3.34}$$

Thus estimations of the current value of the state vector $\hat{x}(k)$ are obtained, and consequently the mean value of the state vector will be given from Eq. (3.33). This means that the value of the state vector which is calculated in the prediction stage is the result of the weighted averaging of the state vectors which were calculated after running the state equation, starting from the N previous values of the state vectors ξ_{k-1}^i.

3.7.3 The Correction Stage

The a-posteriori probability density is found using Eq. (3.33). Now, a new position measurement $z(k)$ is obtained and the objective is to calculate the corrected probability density $p(x(k)|Z)$, where $Z = \{z(1), z(2), \ldots, z(k)\}$. From Bayes law it holds that $p(x(k)|Z) = \frac{p(Z|x(k))p(x(k))}{p(Z)}$ which can be also written as

$$p(x(k)|Z) = \frac{p(z(k)|x(k))p(x(k)|Z^-)}{\int p(z(k)|x(k), Z^-)p(x(k)|Z^-)dx} \tag{3.35}$$

Substituting Eq. (3.33) into Eq. (3.35) and after intermediate calculations one finally obtains

$$\begin{array}{c} p(x(k)|Z) = \sum_{i=1}^{N} w_k^i \delta_{\xi_{k^-}^i}(x(k)) \\ \text{where } w_k^i = \frac{w_{k^-}^i p(z(k)|x(k)=\xi_{k^-}^i)}{\sum_{j=1}^{N} w_{k^-}^j p(z(k)|x(k)=\xi_{k^-}^j)} \end{array} \tag{3.36}$$

Eq. (3.36) denotes the corrected value for the state vector. The recursion of the Particle Filter proceeds in a way similar to the update of the Kalman Filter or the Extended Kalman Filter, i.e. [200, 204, 222]:

- *Measurement update*: Acquire $z(k)$ and compute

$$\begin{array}{l} \text{new value of the state vector} \\ p(x(k)|Z) = \sum_{i=1}^{N} w_k^i \delta_{\xi_{k^-}^i}(x(k)) \\ \text{with corrected weights} \\ w_k^i = \frac{w_{k^-}^i p(z(k)|x(k)=\xi_{k^-}^i)}{\sum_{j=1}^{N} w_{k^-}^i p(z(k)|x(k)=\xi_{k^-})^i} \text{ and } \xi_k^i = \xi_{k^-}^i \end{array} \tag{3.37}$$

  Resampling for substitution of the degenerated particles

- *Time update*: compute state vector $x(k+1)$ according to the pdf

$$\begin{array}{c} p(x(k+1)|Z) = \sum_{i=1}^{N} w_k^i \delta_{\xi_k^i}(x(k)) \\ \text{where } \xi_k^i \sim p(x(k+1)|x(k) = \xi_k^i) \end{array} \tag{3.38}$$

The stages of state vector estimation with the use of the Particle Filtering algorithm are depicted in Fig. 3.3.

3.7.4 The Resampling Stage

The algorithm of particle filtering which is described through Eqs. (3.33) and (3.36) has a significant drawback: after a certain number of iterations k, almost all the weights w_k^i become 0. In the ideal case, all the weights should converge to the value $\frac{1}{N}$, i.e. the particles should have the same significance. The criterion used to define

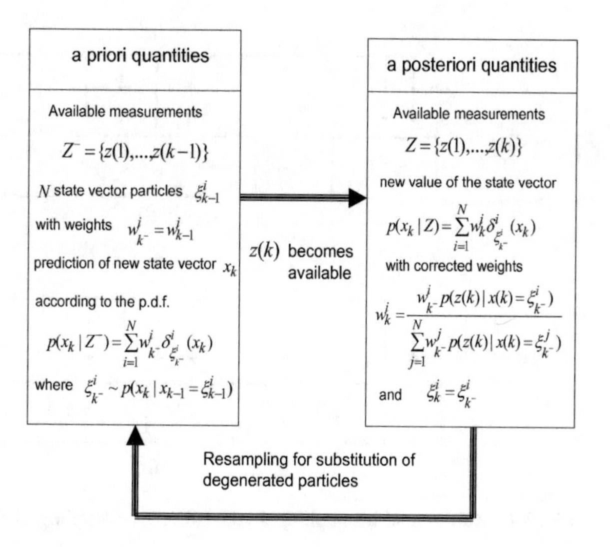

Fig. 3.3 Schematic diagram of the Particle Filter loop

a sufficient number of particles is $N_k^{\mathrm{eff}} = \frac{1}{\sum_{i=1}^{N} w_k^{i\,2}} \in [1, N]$. When N_k^{eff} is close to value N then all particles have almost the same significance. However using the algorithm of Eqs. (3.33) and (3.36) results in $N_k^{\mathrm{eff}} \rightarrow 1$, which means that the particles are degenerated, i.e. they lose their effectiveness. Therefore, it is necessary to modify the algorithm so as to assure that degeneration of the particles will not take place [203, 243, 257, 296].

When N_k^{eff} is small, then most of the particles have weights close to 0 and consequently they have a negligible contribution to the estimation of the state vector. To overcome this drawback, the PF algorithm weakens such particles in favor of particles that have a non-negligible contribution. Therefore, the particles of low weight factors are removed and their place is occupied by replicates of the particles with high weight factors. The total number of particles remains unchanged (equal to N) and therefore this procedure can be viewed as a "resampling" or "redistribution" of the particles set.

The particles resampling presented above maybe slow if not appropriately tuned. There are improved versions of it which substitute the particles of low importance with those of higher importance. A first choice would be to perform a multinomial resampling. N particles are chosen between $\{\xi_k^1, \ldots, \xi_k^N\}$ and the corresponding weights are $w_k^1, \ldots, w_k^N$. The number of times each particle is selected is given by $[j_1, \ldots, j_n]$. Thus a set of N particles is again created, the elements of which are chosen after sampling with the discrete distribution $\sum_{i=1}^{N} w_k^i \delta_{\xi_k^i}(x)$. The particles $\{\xi_k^1, \ldots, \xi_k^N\}$ are chosen according to the probabilities $\{w_k^1, \ldots, w_k^N\}$. The selected particles are assigned with equal weights $\frac{1}{N}$.

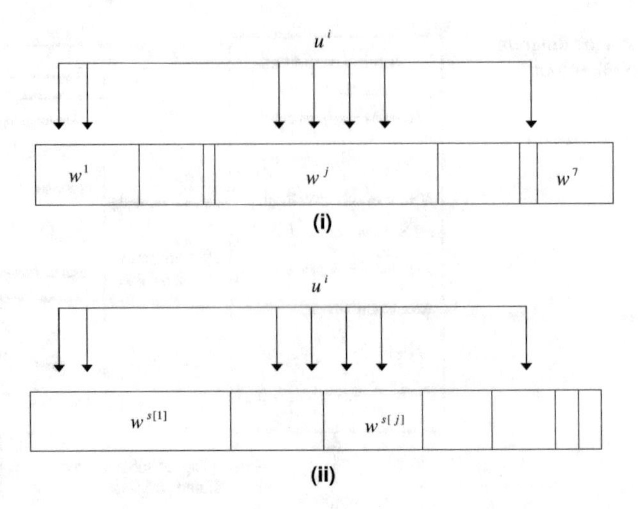

Fig. 3.4 Multinomial resampling: **a** conventional resampling, **b** resampling with sorted weights

3.7.5 Approaches to the Implementation of Resampling

Although sorting of the particles' weights is not necessary for the convergence of the particle filter algorithm, there are variants of the resampling procedure of $(\xi_k^i, w_k^i\ i = 1, \ldots, N)$ which are based on previous sorting in decreasing order of the particles' weights [43]. Sorting of particles' weights gives $w^{s[1]} > w^{s[2]} > \cdots > w^{s[N]}$. A random numbers generator is evoked and the resulting numbers $u^{i:N} \sim U[0, 1]$ fall in the partitions of the interval [0, 1]. The width of these partitions is w^i and thus a redistribution of the particles is generated. For instance, in a wide partition of width w^j will be assigned more particles than in a narrow partition of witdh w^m (see Fig. 3.4).

Two other methods that have been proposed for the implementation of resampling in Particle Filtering are explained in the sequel. These are Kitagawa's approach and the residuals resampling approach [43]. In *Kitagawa's resampling* the speed of the resampling procedure is increased by using less the random numbers generator. The weights are sorted again in decreasing order $w^{s[j]}$ so as to cover the region that corresponds to the interval [0, 1]. Then the random numbers generator is used to produce the variable $u_1 \sim U[0, \frac{1}{N}]$, according to $u_1 \sim U[0, \frac{1}{N}]$ and $u^i = u^1 + \frac{i}{N}$, $i = 2, \ldots, N$. The rest of the variables u^i are produced in a deterministic way (see Fig. 3.5a) [122].

In the *residuals resampling* approach, the redistribution of the residuals is performed as follows: at a first stage particle ξ^i is chosen in a deterministic way $[w^i/N]$ times (with rounding). The residual weights are $\tilde{w}^i = w^i - N[w^i/N]$ and are normalized. Thus, a probability distribution is generated. The rest of the particles are selected according to the multinomial resampling described above (see Fig. 3.5b). The method can be applied if the number $\tilde{N}$ which remains at the second stage is

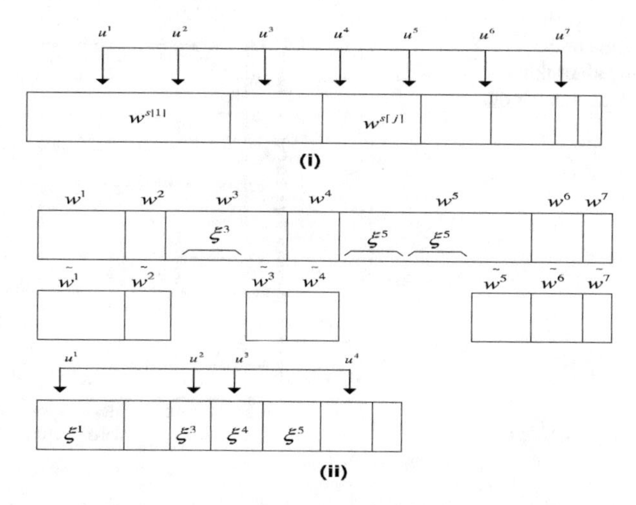

Fig. 3.5 Multinomial resampling: **a** Kitagawa's approach, **b** residuals approach

small, i.e. when $N^{\text{eff}} = 1/\sum_{i=1}^{N} w_i^2$ is small.

Finally, it is mentioned that tuning of the resampling procedure is of importance for succeeding improved performance of the Particle Filter algorithm and convergence to an accurate estimation of the state vector. To tune resampling the following issues are taken into account [200, 257]: (i) control of the diversity of the particles population, (ii) avoidance of particles' impoverishment, i.e. avoidance of absence of particles in the vicinity of the correct state, (iii) selection of the optimal number of particles, (iv) selection of the particles' subsets to be substituted (stratified resampling), and (v) parallel implementation of resampling for improving the speed of the PF algorithm. It is noted that when resampling does not include particles' sorting the computation time of PF scales up linearly with the number of particles [236], as shown in Fig. 3.6. On the other hand when particles' sorting is performed during resampling the computational complexity of PF is $O(Nlog(N))$ [31, 163].

3.8 The Derivative-Free Nonlinear Kalman Filter

3.8.1 Conditions for solving the estimation problem in single-input nonlinear systems

Differential flatness theory will be used for arriving at a filtering method that will have improved performance comparing to the Extended Kalman Filter. It will be shown that through a nonlinear transformation it is possible to design a state estimator for a class of nonlinear systems, which can substitute for the Extended Kalman Filter. The results will be towards derivative-free Kalman Filtering for nonlinear systems.

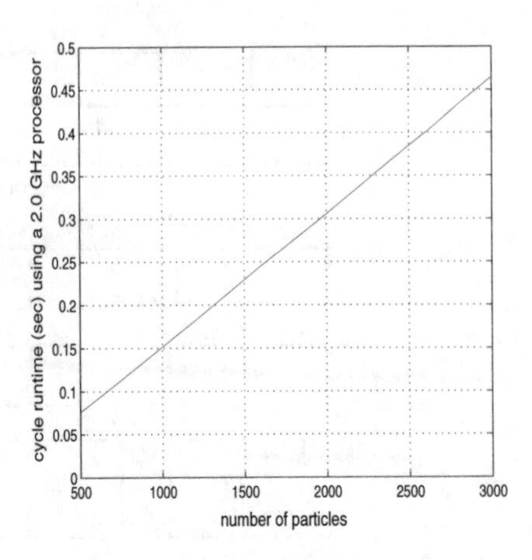

Fig. 3.6 Runtime of the particle filtering algorithm with respect to the number of particles

The following continuous-time nonlinear single-output system is considered [154, 155, 215].

$$\begin{aligned} \dot{x} &= f(x) + q_0(x,u) + \textstyle\sum_{i=1}^{p} \theta_i q_i(x,u), \text{ or} \\ \dot{x} &= f(x) + q_0(x,u) + Q(x,u)\theta \;\; x \in R^n, \;\; u \in R^m, \;\; \theta \in R^p \\ z &= h(x), \;\; z \in R \end{aligned} \tag{3.39}$$

with $q_i : R^n \times R^m \rightarrow R^n, 0 \leq i \leq p, f : R^n \rightarrow R^n, h : R^n \rightarrow R$, smooth functions, $h(x_0) = 0$, $q_0(x, 0) = 0$ for every $x \in R^n$; x is the state vector, $u(x, t) : R^+ \rightarrow R^m$ is the control which is assumed to be known, θ is the parameter vector which is supposed to be constant and y is the scalar output.

The first main assumption on the class of systems considered is the linear dependence on the parameter vector θ. The second main assumption requires that systems of Eq. (3.39) are transformable by a parameter independent state-space change of coordinates in R^n

$$\zeta = T(x), \;\; T(x_0) = 0 \tag{3.40}$$

into the system

$$\begin{aligned} \dot{\zeta} &= A_c\zeta + \psi_0(z,u) + \textstyle\sum_{i=1}^{p} \theta_i \psi_i(z,u) \Rightarrow \\ \dot{\zeta} &= A_c\zeta + \psi_0(z,u) + \Psi(z,u)\theta \\ z &= C_c\zeta \end{aligned} \tag{3.41}$$

with

$$A_c = \begin{pmatrix} 0 & 1 & 0 & \cdots & 0 \\ 0 & 0 & 1 & \cdots & 0 \\ \vdots & \vdots & \vdots & & \vdots \\ 0 & 0 & 0 & \cdots & 0 \end{pmatrix} \tag{3.42}$$

$$C_c = \begin{pmatrix} 1 & 0 & 0 & \cdots & 0 \end{pmatrix} \tag{3.43}$$

and $\psi_i : R \times R^m \rightarrow R^n$ smooth functions for $i = 0, \ldots, p$. The necessary and sufficient conditions for the initial nonlinear system to be transformable into the form of Eq. (3.41) have been given in [154, 155, 215] and are summarized in the following:

(i) rank$\{dh(x), d_{L_f}h(x), \ldots, d_{L_f^{n-1}}h(x)\} = n, \forall x \in R^n$ (which implies local observability). It is noted that $L_f h(x)$ stands for the Lie derivative $L_f h(x) = (\nabla h) f$ and the repeated Lie derivatives are recursively defined as $L_f^0 h = h$ for $i = 0$, $L_f^i h = L_f L_f^{i-1} h = \nabla L_f^{i-1} h f$ for $i = 1, 2, \ldots$.

(ii) $[ad_f^i g, ad_f^j g] = 0,\ 0 \leq i,\ j \leq n-1$. It is noted that $ad_f^i g$ stands for a Lie Bracket which is defined recursively as $ad_f^i g = [f, ad_f^{i-1} g]$ with $ad_f^0 g = g$ and $ad_f g = [f, g] = \nabla g f - \nabla f g$.

(iii) $[q_i, ad_f^i g] = 0,\ 0 \leq i \leq p,\ 0 \leq j \leq n-2\ \forall\ u \in R^m$.

(iv) the vector fields $ad_f^i g,\ 0 \leq i \leq n-1$ are complete, in which g is the vector field satisfying

$$< \begin{pmatrix} dh \\ \vdots \\ d(L_f^{n-1} h) \end{pmatrix}, g >= \begin{pmatrix} 0 \\ \vdots \\ 1 \end{pmatrix} \tag{3.44}$$

where dh has been defined as $dh = \frac{\partial h}{\partial x} = [\frac{\partial h}{\partial x_1}, \frac{\partial h}{\partial x_2}, \ldots, \frac{\partial h}{\partial x_n}]$ Then for every parameter vector θ, the system

$$\begin{aligned} \hat{\zeta} &= A_c \hat{\zeta} + \psi_0(z, u) + \textstyle\sum_{i=1}^{p} \theta_i \psi_i(z, u) + K(z - C_c \hat{\zeta}) \\ \hat{x} &= T^{-1}(\hat{\zeta}) \end{aligned} \tag{3.45}$$

is an asymptotic observer for a suitable choice of K provided that the state $x(t)$ is bounded, with estimation error dynamics

$$\dot{e} = (A_c - K C_c) e = \begin{pmatrix} -k_1 & 1 & 0 & \cdots & 0 \\ -k_2 & 0 & 1 & \cdots & 0 \\ \cdots & \cdots & \cdots & \cdots & \cdots \\ -k_{n-1} & 0 & 0 & \cdots & 1 \\ -k_n & 0 & 0 & \cdots & 0 \end{pmatrix} e \tag{3.46}$$

The eigenvalues of $A_c - K C_c$ can be arbitrarily placed by choosing the vector K, since they coincide with the roots of the polynomial $s^n + k_1 s^{n-1} + \cdots + k_n$.

From, Eq. (3.45) it can be noticed that to obtain estimates of the state vector of the initial nonlinear system, knowledge of the inverse transformation T^{-1} (diffeomorphism) is needed.

3.8.2 *State Estimation with the Derivative-Free Nonlinear Kalman Filter*

Since Eq. (3.45) provides an asympotic observer for the initial nonlinear system of Eq. (3.39) one can consider a case in which the observation error gain matrix K can be provided by the Kalman Filter equations given initially in the continuous-time KF formulation, or in discrete-time form by Eqs. (3.11) and (3.12). The following single-input single-output nonlinear dynamical system is considered

$$x^{(n)} = f(x,t) + g(x,t)u(x,t) \tag{3.47}$$

where $z = x$ is the system's output, and $f(x,t)$, $g(x,t)$ are nonlinear functions. It can be noticed that the system of Eq. (3.47) belongs to the general class of systems of Eq. (3.39). Moreover, Eq. (3.47) is the canonical form one obtains after the application of a differential flatness theory-based transformation. Assuming the transformation $\zeta_i = x^{(i-1)}$, $i = 1,\dots,n$, and $x^{(n)} = f(x,t) + g(x,t)u(x,t) = v(\zeta,t)$, i.e. $\dot{\zeta}_n = v(\zeta,t)$, one obtains the linearized system of the form

$$\begin{aligned}
\dot{\zeta}_1 &= \zeta_2\\
\dot{\zeta}_2 &= \zeta_3\\
&\cdots\cdots\\
\dot{\zeta}_{n-1} &= \zeta_n\\
\dot{\zeta}_n &= v(\zeta,t)
\end{aligned} \tag{3.48}$$

which in turn can be written in state-space equations of the canonical form as

$$\begin{pmatrix} \dot{\zeta}_1 \\ \dot{\zeta}_2 \\ \cdots \\ \dot{\zeta}_{n-1} \\ \dot{\zeta}_n \end{pmatrix} = \begin{pmatrix} 0 & 1 & 0 & \cdots & 0 \\ 0 & 0 & 1 & \cdots & 0 \\ \cdots & \cdots & \cdots & \cdots & \cdots \\ 0 & 0 & 0 & \cdots & 1 \\ 0 & 0 & 0 & \cdots & 0 \end{pmatrix} \begin{pmatrix} \zeta_1 \\ \zeta_2 \\ \cdots \\ \zeta_{n-1} \\ \zeta_n \end{pmatrix} + \begin{pmatrix} 0 \\ 0 \\ \cdots \\ 0 \\ 1 \end{pmatrix} v(\zeta,t) \tag{3.49}$$

$$z = \begin{pmatrix} 1 \, 0 \, 0 \cdots 0 \end{pmatrix} \zeta \tag{3.50}$$

The system of Eqs. (3.49) and (3.50) has been written in the form of Eq. (3.41), which means that Eq. (3.45) is the associated asymptotic observer. Therefore, the observation gain K appearing in Eq. (3.45) can be found using either linear observer design methods (in that case the elements of the observation error gain matrix K have fixed values), or the recursive calculation of the continuous-time Kalman Filter gain described in Sect. 3.3. If the discrete-time Kalman Filter is to be used then one has to apply the recursive formulas of Eqs. (3.11) and (3.12) on the discrete-time equivalent of Eqs. (3.49) and (3.50).

3.8.3 Derivative-Free Kalman Filtering for multivariable Nonlinear Systems

The previous results about transformation of MIMO nonlinear systems into the canonical form will be generalized towards Derivative-free Kalman Filtering for MIMO nonlinear dynamical systems. The proposed method for derivative-free Kalman Filtering for MIMO nonlinear systems will be analyzed through an application example. The following multivariable system is considered:

$$\begin{aligned} \dot{x}_1 &= x_2 \\ \dot{x}_2 &= f_1(x,t) + g_1(x,t)u_1 + \tilde{d}_1 \\ \dot{x}_3 &= x_4 \\ \dot{x}_4 &= f_2(x,t) + g_2(x,t)u_1 + \tilde{d}_2 \end{aligned} \tag{3.51}$$

The previous multi-variable system is considered to have as flat outputs the vector of state variables $y = [x_1, x_3]^T$. Next, considering also the effects of additive disturbances the dynamic model becomes

$$\begin{aligned} \ddot{x}_1 &= f_1(x,t) + g_1(x,t)u + \tilde{d}_1 \\ \ddot{x}_3 &= f_2(x,t) + g_2(x,t)u + \tilde{d}_2 \end{aligned} \tag{3.52}$$

$$\begin{pmatrix} \ddot{x}_1 \\ \ddot{x}_3 \end{pmatrix} = \begin{pmatrix} f_1(x,t) \\ f_2(x,t) \end{pmatrix} + \begin{pmatrix} g_1(x,t) \\ g_2(x,t) \end{pmatrix} u + \begin{pmatrix} \tilde{d}_1 \\ \tilde{d}_2 \end{pmatrix} \tag{3.53}$$

Moreover, considering that the model's structure and parameters are known and that the additive input disturbance $\tilde{d}_i$, $i = 1, 2, \ldots$ are also known, the following feedback linearizing control input is defined

$$u = \begin{pmatrix} g_1(\hat{x},t) \\ g_2(\hat{x},t) \end{pmatrix}^{-1} \{ \begin{pmatrix} \ddot{x}_1^d \\ \ddot{x}_3^d \end{pmatrix} - \begin{pmatrix} f_1(\hat{x},t) \\ f_2(\hat{x},t) \end{pmatrix} - \begin{pmatrix} K_1^T \\ K_2^T \end{pmatrix} e + \begin{pmatrix} u_{c_1} \\ u_{c_2} \end{pmatrix} \} \tag{3.54}$$

where $[u_{c_1} \ u_{c_2}]^T$ is a supervisory control term that is used for the compensation of the model's uncertainties as well as of the external disturbances and $K_i^T = [k_1^i, k_2^i, \ldots, k_{n-1}^i, k_n^i]$ are the rows of the error feedback gain matrix [196, 199]. It is also noted that to perform efficient state estimation under such model uncertainties and external disturbances one can consider results on disturbance observers within a Kalman Filter framework [225].

Finally, the differentially flat nonlinear system is written in the Brunovsky (canonical) form. Considering the state vector $x \in R^{4\times 1}$, with the state variables defined in Eq. (3.52), the following matrices are defined

$$A=\begin{pmatrix}0\,1\,0\,0\\0\,0\,0\,0\\0\,0\,0\,1\\0\,0\,0\,0\end{pmatrix},\quad B=\begin{pmatrix}0\,0\\1\,0\\0\,0\\0\,1\end{pmatrix},\quad C=\begin{pmatrix}1\,0\,0\,0\\0\,0\,1\,0\end{pmatrix} \tag{3.55}$$

Using the matrices of Eq. (3.55) one obtains the Brunovsky form of the MIMO nonlinear model

$$\begin{aligned}\dot{x}&=Ax+Bv\\ y&=Cx\end{aligned} \tag{3.56}$$

where the new input v is given by

$$v=\begin{pmatrix}f_1(x,t)\\f_2(x,t)\end{pmatrix}+\begin{pmatrix}g_1(x,t)\\g_2(x,t)\end{pmatrix}u+\begin{pmatrix}\tilde{d}_1\\\tilde{d}_2\end{pmatrix} \tag{3.57}$$

Prior to performing discrete-time Kalman Filtering for the model of Eq. (3.56) matrices A, B and C of Eq. (3.55) are turned into discrete-time ones using common discretization methods. These discrete-time matrices are denoted as A_d, B_d and C_d respectively.

For the nonlinear system of Eqs. (3.56) and (3.57) state estimation can be performed using the standard Kalman Filter recursion, as described in Eqs. (3.11) and (3.12).

3.9 Distributed Extended Kalman Filtering

3.9.1 *Calculation of Local Extended Kalman Filter Estimations*

Again the discrete-time nonlinear system of Eq. (3.58) is considered.

$$\begin{aligned}x(k+1)&=\phi(x(k))+L(k)u(k)+w(k)\\ z(k)&=\gamma(x(k))+v(k)\end{aligned} \tag{3.58}$$

The Extended Information Filter (EIF) performs fusion of the local state vector estimates which are provided by the local Extended Kalman Filters, using the *Information matrix* and the *Information state vector* [130, 131, 153, 264]. The Information Matrix is the inverse of the state vector covariance matrix, and can be also associated to the Fisher Information matrix [205]. The Information state vector is the product between the Information matrix and the local state vector estimate

$$\begin{aligned}Y(k)&=P^{-1}(k)=I(k)\\ \hat{y}(k)&=P^{-}(k)^{-1}\hat{x}(k)=Y(k)\hat{x}(k)\end{aligned} \tag{3.59}$$

The update equations for the Information Matrix and the Information state vector are given by

$$\begin{aligned} Y(k) &= P^{-}(k)^{-1} + J_{\gamma}^{T}(k)R^{-1}(k)J_{\gamma}(k) \\ &= Y^{-}(k) + I(k) \end{aligned} \tag{3.60}$$

$$\begin{aligned} \hat{y}(k) &= \hat{y}^{-}(k) + J_{\gamma}^{T}R(k)^{-1}[z(k) - \gamma(x(k)) + J_{\gamma}\hat{x}^{-}(k)] \\ &= \hat{y}^{-}(k) + i(k) \end{aligned} \tag{3.61}$$

where

$$\begin{aligned} I(k) &= J_{\gamma}^{T}(k)R(k)^{-1}J_{\gamma}(k) \text{ is the associated information matrix and} \\ i(k) &= J_{\gamma}^{T}R(k)^{-1}[(z(k) - \gamma(x(k))) + J_{\gamma}\hat{x}^{-}(k)] \text{ is the information state contribution} \end{aligned} \tag{3.62}$$

The predicted information state vector and Information matrix are obtained from

$$\begin{aligned} \hat{y}^{-}(k) &= P^{-}(k)^{-1}\hat{x}^{-}(k) \\ Y^{-}(k) = P^{-}(k)^{-1} &= [J_{\phi}(k)P^{-}(k)J_{\phi}(k)^{T} + Q(k)]^{-1} \end{aligned} \tag{3.63}$$

The Extended Information Filter is next formulated for the case that multiple local sensor measurements and local estimates are used to increase the accuracy and reliability of the estimation. It is assumed that an observation vector $z^{i}(k)$ is available for N different measurement sites $i = 1, 2, \ldots, N$ and each site observes a common state according to the local observation model, expressed by

$$z^{i}(k) = \gamma(x(k)) + v^{i}(k),\ i = 1, 2, \ldots, N \tag{3.64}$$

where the local noise vector $v^{i}(k) \sim N(0, R^{i})$ is assumed to be white Gaussian and uncorrelated between measurement sites. The variance of a composite observation noise vector v_{k} is expressed in terms of the block diagonal matrix

$$R(k) = diag[R(k)^{1}, \ldots, R^{N}(k)]^{T} \tag{3.65}$$

The information contribution can be expressed by a linear combination of each local information state contribution i^{i} and the associated information matrix I^{i} at the i-th measurement site

$$\begin{aligned} i(k) &= \sum_{i=1}^{N} J_{\gamma}^{i\,T}(k)R^{i}(k)^{-1}[z^{i}(k) - \gamma^{i}(x(k)) + J_{\gamma}^{i}(k)\hat{x}^{-}(k)] \\ I(k) &= \sum_{i=1}^{N} J_{\gamma}^{i\,T}(k)R^{i}(k)^{-1}J_{\gamma}^{i}(k) \end{aligned} \tag{3.66}$$

Using Eq. (3.66) the update equations for fusing the local state estimates become

$$\begin{aligned} \hat{y}(k) &= \hat{y}^{-}(k) + \sum_{i=1}^{N} J_{\gamma}^{i\,T}(k)R^{i}(k)^{-1}[z^{i}(k) - \gamma^{i}(x(k)) + J_{\gamma}^{i}(k)\hat{x}^{-}(k)] \\ Y(k) &= Y^{-}(k) + \sum_{i=1}^{N} J_{\gamma}^{i\,T}(k)R^{i}(k)^{-1}J_{\gamma}^{i}(k) \end{aligned} \tag{3.67}$$

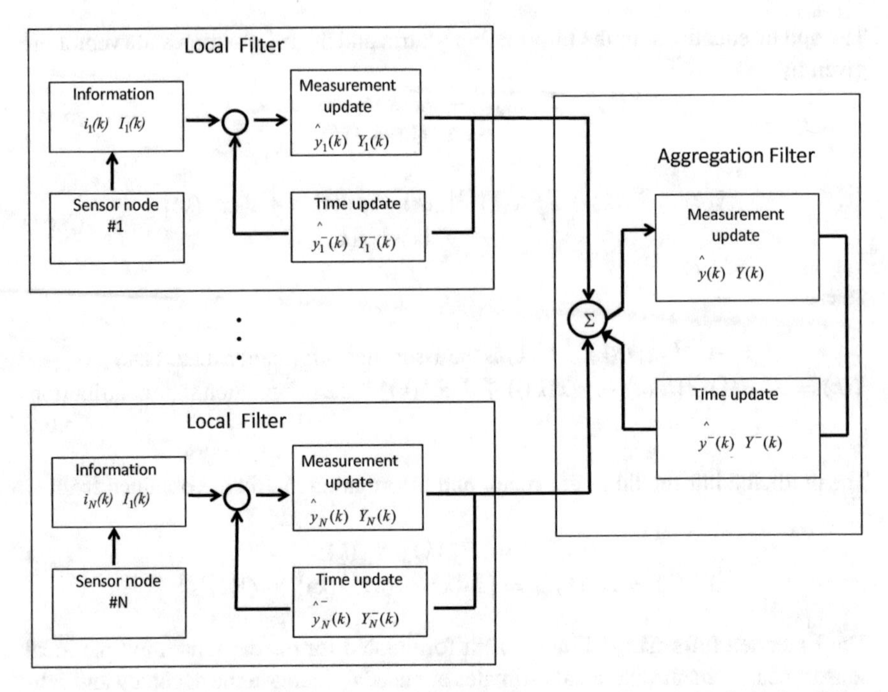

Fig. 3.7 Fusion of the distributed state estimates with the use of the Extended Information Filter

It is noted that in the Extended Information Filter an aggregation (master) fusion filter produces a global estimate by using the measurement site information provided by each local filter.

As in the case of the Extended Kalman Filter the local filters which constitute the Extended information Filter can be written in terms of *time update* and a *measurement update* equation (Figs. 3.7 and 3.8).

Measurement update: Acquire $z(k)$ and compute

$$\begin{gathered} Y(k) = P^{-}(k)^{-1} + J_{\gamma}^{T}(k)R(k)^{-1}J_{\gamma}(k) \\ \text{or } Y(k) = Y^{-}(k) + I(k) \text{ where } I(k) = J_{\gamma}^{T}(k)R^{-1}(k)J_{\gamma}(k) \end{gathered} \tag{3.68}$$

$$\begin{gathered} \hat{y}(k) = \hat{y}^{-}(k) + J_{\gamma}^{T}(k)R(k)^{-1}[z(k) - \gamma(\hat{x}(k)) + J_{\gamma}\hat{x}^{-}(k)] \\ \text{or } \hat{y}(k) = \hat{y}^{-}(k) + i(k) \end{gathered} \tag{3.69}$$

Time update: Compute

$$Y^{-}(k+1) = P^{-}(k+1)^{-1} = [J_{\phi}(k)P(k)J_{\phi}(k)^{T} + Q(k)]^{-1} \tag{3.70}$$

$$y^{-}(k+1) = P^{-}(k+1)^{-1}\hat{x}^{-}(k+1) \tag{3.71}$$

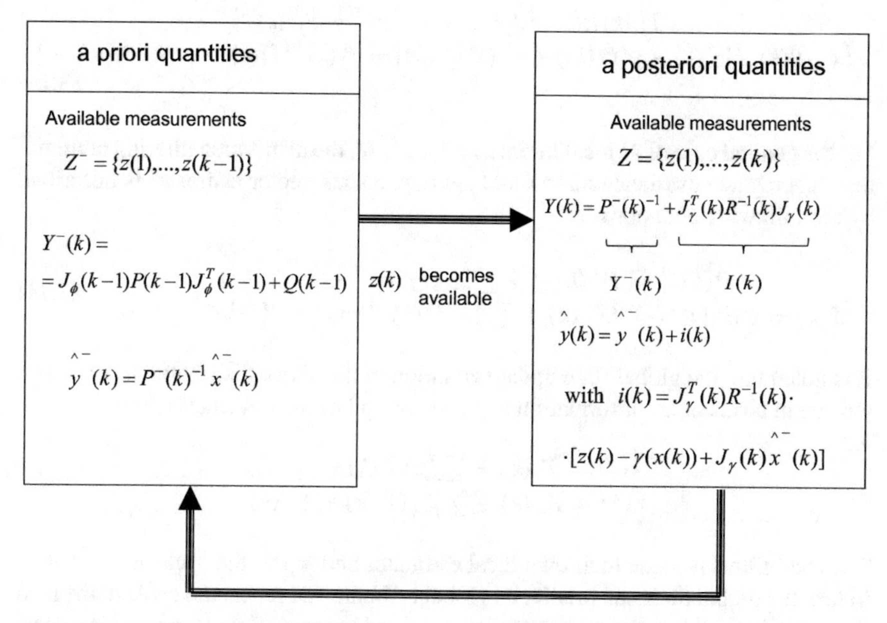

Fig. 3.8 Schematic diagram of the Extended Information Filter loop

3.9.2 Extended Information Filtering for State Estimates Fusion

In the Extended Information Filter each one of the local filters operates independently, processing its own local measurements. It is assumed that there is no sharing of measurements between the local filters and that the aggregation filter (Fig. 3.12) does not have direct access to the raw measurements feeding each local filter. The outputs of the local filters are treated as measurements which are fed into the aggregation fusion filter [130, 131, 264]. Then each local filter is expressed by its respective error covariance and estimate in terms of information contributions given in Eq. (3.63)

$$\begin{aligned} P_i^{-1}(k) &= P_i^-(k)^{-1} + J_\gamma^T R(k)^{-1} J_\gamma(k) \\ \hat{x}_i(k) &= P_i(k)(P_i^-(k)^{-1}\hat{x}_i^-(k)) + J_\gamma^T R(k)^{-1}[z^i(k) - \gamma^i(x(k)) + J_\gamma^i(k)\hat{x}_i^-(k)] \end{aligned} \tag{3.72}$$

It is noted that the local estimates are suboptimal and also conditionally independent given their own measurements. The global estimate and the associated error covariance for the aggregate fusion filter can be rewritten in terms of the computed estimates and covariances from the local filters using the relations

$$
\begin{array}{c}
J_\gamma^T(k)R(k)^{-1}J_\gamma(k) = P_i(k)^{-1} - P_i^-(k)^{-1} \\
J_\gamma^T(k)R(k)^{-1}[z^i(k) - \gamma^i(x(k)) + J_\gamma^i(k)\hat{x}^-(k)] = P_i(k)^{-1}\hat{x}_i(k) - P_i(k)^{-1}\hat{x}_i(k-1)
\end{array}
\tag{3.73}
$$

For the general case of N local filters $i = 1, \ldots, N$, the distributed filtering architecture (aggregate covariance matrix and aggregate state vector estimate) is described by the following equations

$$
\begin{array}{c}
P(k)^{-1} = P^-(k)^{-1} + \sum_{i=1}^N [P_i(k)^{-1} - P_i^-(k)^{-1}] \\
\hat{x}(k) = P(k)[P^-(k)^{-1}\hat{x}^-(k) + \sum_{i=1}^N (P_i(k)^{-1}\hat{x}_i(k) - P_i^-(k)^{-1}\hat{x}_i^-(k))]
\end{array}
\tag{3.74}
$$

It is noted that the global state update equation in the above distributed filter can be written in terms of the information state vector and of the information matrix

$$
\begin{array}{c}
\hat{y}(k) = \hat{y}^-(k) + \sum_{i=1}^N (\hat{y}_i(k) - \hat{y}_i^-(k)) \\
\hat{Y}(k) = \hat{Y}^-(k) + \sum_{i=1}^N (\hat{Y}_i(k) - \hat{Y}_i^-(k))
\end{array}
\tag{3.75}
$$

The local filters provide their own local estimates and repeat the cycle at step $k+1$. In turn the global filter can predict its global estimate and repeat the cycle at the next time step $k+1$ when the new state $\hat{x}(k+1)$ and the new global covariance matrix $P(k+1)$ are calculated. From Eq. (3.74) it can be seen that if a local filter (processing station) fails, then the local covariance matrices and the local state estimates provided by the rest of the filters will enable an accurate computation of the system's state vector.

3.10 Distributed Sigma-Point Kalman Filtering

3.10.1 *Calculation of Local Unscented Kalman Filter Estimations*

It is also possible to perform distributed estimation through the fusion of the estimates provided by local Sigma-Point Kalman Filters. This can be succeeded using the Distributed Sigma-Point Kalman Filter, also known as *Unscented Information Filter* (UIF) [130, 131]. First, the functioning of the local Sigma-Point Kalman Filters will be explained. Each local Sigma-Point Kalman Filter generates an estimation of the state vector by fusing measurements from distributed measurement sites, Sigma-Point Kalman Filtering is proposed [116, 117, 234]. As explained in Sect. 3.6 the Sigma-Point Kalman Filter overcomes some of the flaws of Extended Kalman Filtering. Unlike EKF, in SPKF no analytical Jacobians of the system equations need to be calculated. This makes the sigma-point approach suitable for application in "black-box" models where analytical expressions of the system dynamics are either not available or not in a form which allows for easy linearization. This is achieved through a different approach for calculating the posterior 1st and 2nd order statistics of a random variable that undergoes a nonlinear transformation. The state distribution is represented again by a Gaussian Random Variable but is now specified using a minimal set of deterministically chosen weighted sample points.

The Unscented Information Filter is derived by introducing a linear error propagation based on the unscented transformation into the Extended Information Filter structure. First, an augmented state vector ${x_\alpha}^-(k)$ is considered, along with the process noise vector, and the associated covariance matrix is introduced

$$\hat{x}_\alpha^-(k) = \begin{pmatrix} \hat{x}^-(k) \\ \hat{w}^-(k) \end{pmatrix}, \quad P^{\alpha-}(k) = \begin{pmatrix} P^-(k) & 0 \\ 0 & Q^-(k) \end{pmatrix} \tag{3.76}$$

As in the case of local (lumped) Unscented Kalman Filters, a set of weighted sigma points ${X_\alpha^i}^-(k)$ is generated as

$$\begin{aligned} X_{\alpha,0}^-(k) &= \hat{x}_\alpha^-(k) \\ X_{\alpha,i}^-(k) &= \hat{x}_\alpha^-(k) + [\sqrt{(n_\alpha + \lambda)P_\alpha^-(k-1)}]_i, \ i = 1, \ldots, n \\ X_{\alpha,i}^-(k) &= \hat{x}_\alpha^-(k) + [\sqrt{(n_\alpha + \lambda)P_\alpha^-(k-1)}]_i, \ i = n+1, \ldots, 2n \end{aligned} \tag{3.77}$$

where $\lambda = \alpha^2(n_\alpha + \kappa) - n_\alpha$ is a scaling, while $0 \le \alpha \le 1$ and κ are constant parameters. The corresponding weights for the mean and covariance are defined as in the case of the lumped Unscented Kalman Filter

$$\begin{aligned} W_0^{(m)} &= \frac{\lambda}{n_\alpha+\lambda} & W_0^{(c)} &= \frac{\lambda}{(n_\alpha+\lambda)+(1-\alpha^2+\beta)} \\ W_i^{(m)} &= \frac{1}{2(n_\alpha+\lambda)}, \ i = 1, \ldots, 2n_\alpha & W_i^{(C)} &= \frac{1}{2(n_\alpha+\lambda)}, \ i = 1, \ldots, 2n_\alpha \end{aligned} \tag{3.78}$$

where β is again a constant parameter. The equations of the prediction stage (measurement update) of the information filter, i.e. the calculation of the information matrix and the information state vector of Eq. (3.63) now become

$$\begin{aligned} \hat{y}^-(k) &= Y^-(k)\sum_{i=0}^{2n_\alpha} W_i^m X_i^x(k) \\ Y^-(k) &= P^-(k)^{-1} \end{aligned} \tag{3.79}$$

where X_i^x are the predicted state vectors when using the sigma point vectors X_i^w in the state equation $X_i^x(k+1) = \phi(X_i^{w-}(k)) + L(k)U(k)$. The predicted state covariance matrix is computed as

$$P^-(k) = \sum_{i=0}^{2n_\alpha} W_i^{(c)}[X_i^x(k) - \hat{x}^-(k)][X_i^x(k) - \hat{x}^-(k)]^T \tag{3.80}$$

As noted, the equations of the Extended Information Filter (EIF) are based on the linearized dynamic model of the system and on the inverse of the covariance matrix of the state vector. However, in the equations of the Unscented Kalman Filter (UKF) there is no linearization of the system dynamics, thus the UKF cannot be included directly into the EIF equations. Instead, it is assumed that the nonlinear measurement equation of the system given in Eq. (3.13) can be mapped into a linear function of its statistical mean and covariance, which makes possible to use the information update equations of the EIF. Denoting $Y_i(k) = \gamma(X_i^x(k))$ (i.e. the output of the system calculated through the propagation of the i-th sigma point X^i through the system's nonlinear equation) the observation covariance and its cross-covariance are approximated by

$$\begin{aligned} P_{YY}^-(k) &= E[(z(k) - \hat{z}^-(k))(z(k) - \hat{z}^-(k))^T] \\ &\simeq J_\gamma(k) P^-(k) J_\gamma(k)^T \end{aligned} \tag{3.81}$$

$$\begin{aligned} P_{XY}^-(k) &= E[(x(k) - \hat{x}(k)^-)(z(k) - \hat{z}(k)^-)^T] \\ &\simeq P^-(k) J_\gamma(k)^T \end{aligned} \tag{3.82}$$

where $z(k) = \gamma(x(k))$ and $J_\gamma(k)$ is the Jacobian of the output equation $\gamma(x(k))$. Next, multiplying the predicted covariance and its inverse term on the right side of the information matrix Eq. (3.62) and replacing $P(k)J_\gamma(k)^T$ with $P_{XY}^-(k)$ gives the following representation of the information matrix [130, 131, 264]

$$\begin{aligned} I(k) &= J_\gamma(k)^T R(k)^{-1} J_\gamma(k) \\ &= P^-(k)^{-1} P^-(k) J_\gamma(k)^T R(k)^{-1} J_\gamma^-(k) P^-(k)^T (P^-(k)^{-1})^T \\ &= P^-(k)^{-1} P_{XY}(k) R(k)^{-1} P_{XY}(k)^T (P^-(k)^{-1})^T \end{aligned} \tag{3.83}$$

where $P^-(k)^{-1}$ is calculated according to Eq. (3.80) and the cross-correlation matrix $P_{XY}(k)$ is calculated from

$$P_{XY}^-(k) = \sum_{i=0}^{2n_\alpha} W_i^{(c)} [X_i^x(k) - \hat{x}^-(k)][Y_i(k) - \hat{z}^-(k)]^T \tag{3.84}$$

where $Y_i(k) = \gamma(X_i^x(k))$ and the predicted measurement vector $\hat{z}^-(k)$ is obtained by $\hat{z}^-(k) = \sum_{i=0}^{2n_\alpha} W_i^{(m)} Y_i(k)$. Similarly, the information state vector i_k can be rewritten as

$$\begin{aligned} i(k) &= J_\gamma(k)^T R(k)^{-1} [z(k) - \gamma(x(k)) + J_\gamma(k)^T \hat{x}^-(k)] \\ &= P^-(k)^{-1} P^-(k) J_\gamma(k)^T R(k)^{-1} [z(k) - \gamma(x(k)) + J_\gamma(k)^T (P^-(k))^T (P^-(k)^{-1})^T \hat{x}^-(k)] \\ &= P^-(k)^{-1} P_{XY}^-(k) R(k)^{-1} [z(k) - \gamma(x(k)) + P_{XY}^-(k) (P^-(k)^{-1})^T \hat{x}^-(k)] \end{aligned} \tag{3.85}$$

To complete the analogy to the information contribution equations of the EIF a "measurement" matrix $H^T(k)$ is defined as

$$H(k)^T = P^-(k)^{-1} P_{XY}^-(k) \tag{3.86}$$

In terms of the "measurement" matrix $H(k)$ the information contributions equations are written as

$$\begin{aligned} i(k) &= H^T(k)R(k)^{-1}[z(k) - \gamma(x(k)) + H(k)\hat{x}^-(k)] \\ I(k) &= H^T(k)R(k)^{-1}H(k) \end{aligned} \tag{3.87}$$

The above procedure leads to an implicit linearization in which the nonlinear measurement equation of the system given in Eq. (3.58) is approximated by the statistical error variance and its mean

$$z(k) = \gamma(x(k)) \simeq H(k)x(k) + \bar{u}(k) \tag{3.88}$$

where $\bar{u}(k) = \gamma(\hat{x}^-(k)) - H(k)\hat{x}^-(k)$ is a measurement residual term (3.88).

Next, the local estimations provided by distributed (local) Unscented Kalmans filters will be expressed in terms of the information contributions (information matrix I and information state vector i) of the Unscented Information Filter, which were defined in Eq. (3.87) [130, 131, 264]. It is assumed that the observation vector $\bar{z}_i(k+1)$ is available from N different measurement sites, and that each sensor observes a common state according to the local observation model, expressed by

$$\bar{z}_i(k) = H_i(k)x(k) + \bar{u}_i(k) + v_i(k) \tag{3.89}$$

where the noise vector $v_i(k)$ is taken to be white Gaussian and uncorrelated between sensors. The variance of the composite observation noise vector v_k of all sensors is written in terms of the block diagonal matrix $R(k) = diag[R_1(k)^T, \ldots, R_N(k)^T]^T$. Then one can define the local information matrix $I_i(k)$ and the local information state vector $i_i(k)$ at the i-th sensor, as follows

$$\begin{aligned} i_i(k) &= H_i^T(k)R_i(k)^{-1}[z_i(k) - \gamma^i(x(k)) + H_i(k)\hat{x}^-(k)] \\ I_i(k) &= H_i^T(k)R_i(k)^{-1}H_i(k) \end{aligned} \tag{3.90}$$

Since the information contribution terms have group diagonal structure in terms of the innovation and measurement matrix, the update equations for the multiple state estimation and data fusion are written as a linear combination of the local information contribution terms

$$\begin{aligned} \hat{y}(k) &= \hat{y}^-(k) + \textstyle\sum_{i=1}^N i_i(k) \\ Y(k) &= Y^-(k) + \textstyle\sum_{i=1}^N I_i(k) \end{aligned} \tag{3.91}$$

Then using Eq. (3.79) one can find the mean state vector for the multiple sensor estimation problem.

As in the case of the Unscented Kalman Filter, the Unscented Information Filter running at the i-th measurement processing unit can be written in terms of *measure-*

ment update and *time update* equations:

Measurement update: Acquire measurement $z(k)$ and compute

$$\begin{gathered}Y(k)=P^{-}(k)^{-1}+H^{T}(k)R^{-1}(k)H(k)\\ \text{or } Y(k)=Y^{-}(k)+I(k) \text{ where } I(k)=H^{T}(k)R^{-1}(k)H(k)\end{gathered} \tag{3.92}$$

$$\begin{gathered}\hat{y}(k)=\hat{y}^{-}(k)+H^{T}(k)R^{-1}(k)[z(k)-\gamma(\hat{x}(k))+H(k)\hat{x}^{-}(k)]\\ \text{or } \hat{y}(k)=\hat{y}^{-}(k)+i(k)\end{gathered} \tag{3.93}$$

Time update: Compute

$$\begin{gathered}Y^{-}(k+1)=(P^{-}(k+1))^{-1}\\ \text{where } P^{-}(k+1)=\sum_{i=0}^{2n_{\alpha}}W_i^{(c)}[X_i^{x}(k+1)-\hat{x}^{-}(k+1)][X_i^{x}(k+1)-\hat{x}^{-}(k+1)]^{T}\end{gathered} \tag{3.94}$$

$$\begin{gathered}\hat{y}(k+1)=Y(k+1)\sum_{i=0}^{2n_{\alpha}}W_i^{(m)}X_i^{x}(k+1)\\ \text{where } X_i^{x}(k+1)=\phi(X_i^{w}(k))+L(k)U(k)\end{gathered} \tag{3.95}$$

3.10.2 *Unscented Information Filtering for State Estimates Fusion*

It has been shown that the update of the aggregate state vector of the Unscented Information Filter architecture can be expressed in terms of the local information matrices I_i and of the local information state vectors i_i, which in turn depend on the local covariance matrices P and cross-covariance matrices P_{XY}. Next, it will be shown that the update of the aggregate state vector can be also expressed in terms of the local state vectors $x_i(k)$ and in terms of the local covariance matrices $P_i(k)$ and cross-covariance matrices $P_{XY}^{i}(k)$. It is assumed that the local filters do not have access to each other row measurements and that they are allowed to communicate only their information matrices and their local information state vectors. Thus each local filter is expressed by its respective error covariance and estimate in terms of the local information state contribution i_i and its associated information matrix I_i at the i-th filter site. Then using Eq. (3.79) one obtains (Fig. 3.9)

$$\begin{gathered}P_i(k)^{-1}=P_i^{-}(k)^{-1}+H_i^{T}(k)R_i(k)^{-1}H_i(k)\\ \hat{x}_i=P_i(k)(P_i^{-}(k)\hat{x}_i^{-}(k)+H_i^{T}(k)R_i(k)^{-1}[z_i(k)-\gamma^{i}(x(k))+H_i(k)\hat{x}^{-}(k)])\end{gathered} \tag{3.96}$$

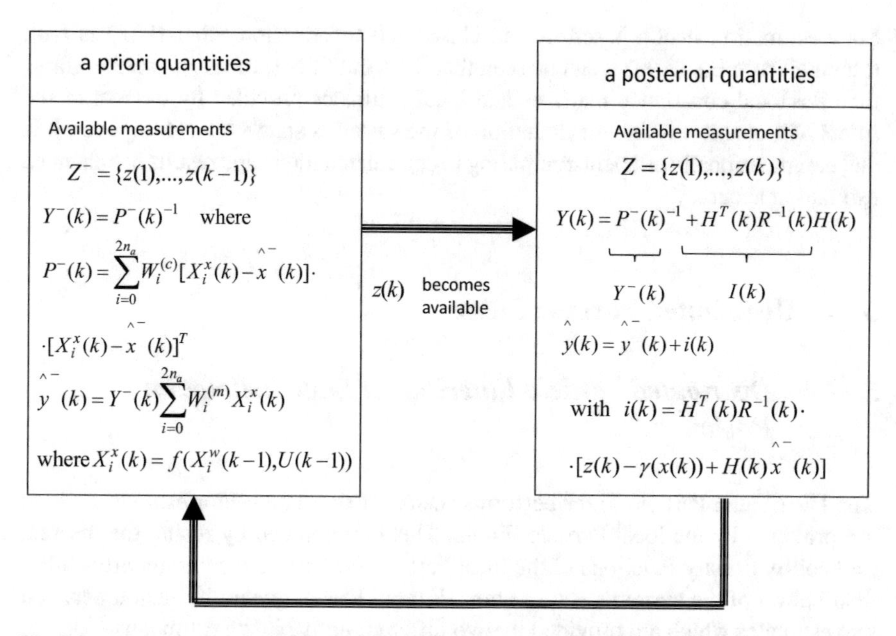

Fig. 3.9 Schematic diagram of the Unscented Information Filter loop

Using Eq. (3.96), each local information state contribution i_i and its associated information matrix I_i at the i-th filter are rewritten in terms of the computed estimates and covariances of the local filters

$$
\begin{array}{c}
H_i^T(k)R_i(k)^{-1}H_i(k) = P_i^{-1}(k) - P_i^-(k)^{-1} \\
H_i^T(k)R_i(k)^{-1}[z_i(k) - \gamma^i(x(k)) + H_i(k)\hat{x}^-(k)] = P_i(k)^{-1}\hat{x}_i(k) - P_i^-(k)^{-1}\hat{x}_i^-(k)
\end{array}
\tag{3.97}
$$

where according to Eq. (3.86) it holds $H_i(k) = P_i^-(k)^{-1}P_{XY,i}^-(k)$. Next, the aggregate estimates of the distributed Unscented Information Filtering are derived for a number of N local filters $i = 1, \ldots, N$ and sensor measurements, first in terms of covariances [130, 131, 264]

$$
\begin{array}{c}
P(k)^{-1} = P^-(k)^{-1} + \sum_{i=1}^N [P_i(k)^{-1} - P_i^-(k)^{-1}] \\
\hat{x}(k) = P(k)[P^-(k)^{-1}\hat{x}^-(k) + \sum_{i=1}^N (P_i(k)^{-1}\hat{x}_i(k) - P_i^-(k)^{-1}\hat{x}_i^-(k))]
\end{array}
\tag{3.98}
$$

and also in terms of the information state vector and of the information state covariance matrix

$$
\begin{array}{c}
\hat{y}(k) = \hat{y}^-(k) + \sum_{i=1}^N (\hat{y}_i(k) - \hat{y}_i^-(k)) \\
Y(k) = Y^-(k) + \sum_{i=1}^N [Y_i(k) - Y_i^-(k)]
\end{array}
\tag{3.99}
$$

State estimation fusion based on the Unscented Information Filter (UIF) is fault tolerant. From Eq. (3.98) it can be seen that if a local filter (processing station) fails, then the local covariance matrices and local estimates provided by the rest of the filters will enable a reliable calculation of the system's state vector. Moreover, it is and computationally efficient comparing to centralized filters and results in enhanced estimation accuracy.

3.11 Distributed Particle Filter

3.11.1 Distributed Particle Filtering for State Estimation Fusion

The Distributed Particle Filter performs fusion of the state vector estimates which are provided by the local Particle Filters. This is succeeded by fusing the discrete probability density functions of the local Particle Filters into a common probability distribution of the system's state vector. Without loss of generality fusion between two estimates which are provided by two different probabilistic estimators (particle filters) is assumed. This amounts to a multiplication and a division operation to remove the common information, and is given by [180, 181]

$$p(x(k)|Z_A \bigcup Z_B) \propto \frac{p(x(k)|Z_A)p(x(k)|Z_B)}{p(x(k)|Z_A \bigcap Z_B)} \tag{3.100}$$

where Z_A is the sequence of measurements associated with the i-th processing unit and Z_B is the sequence of measurements associated with the j-th measurement unit. In the implementation of distributed particle filtering, the following issues arise:

1. Particles from one particle set (which correspond to a local particle filter) do not have the same support (do not cover the same area and points on the samples space) as particles from another particle set (which are associated with another particle filter). Therefore a point-to-point application of Eq. (3.100) is not possible.

2. The transmission of particles representation (i.e. local particle sets and associated weight sets) requires significantly more bandwidth (computing resources) compared to other representations, such as Gaussian mixtures.

Fusion of the estimates provided by the local particle filters (located at different processing units) can be performed through the following stages. First, the discrete particle set of Particle Filter A (Particle Filter B) is transformed into a continuous distribution by placing a Gaussian kernel over each sample (Fig. 3.10) [171]

$$K_h(x) = h^2 K(x) \tag{3.101}$$

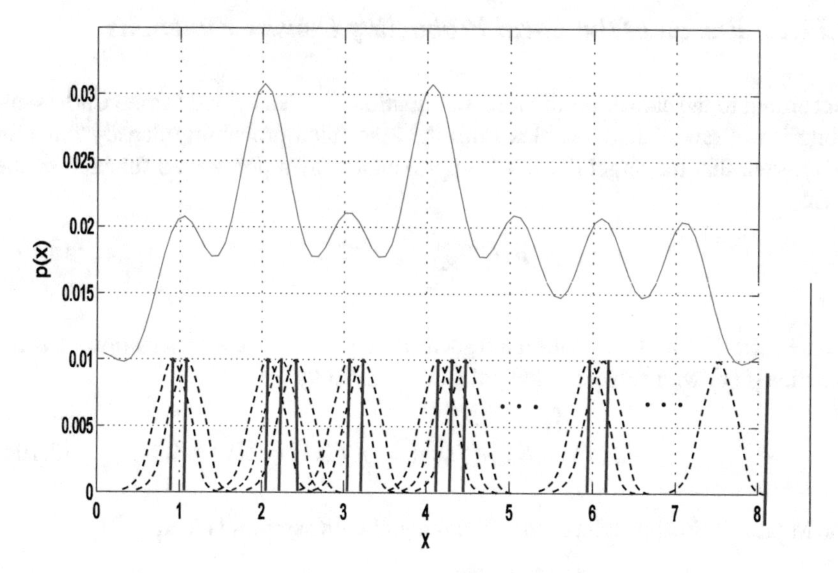

Fig. 3.10 Conversion of the particles discrete probability density function to a continuous distribution, after allocating a Gaussian kernel over each particle

where $K()$ is the rescaled Kernel density and $h > 0$ is the scaling parameter. Then the continuous distribution A (B) is sampled with the other particles set B (A) to obtain the new importance weights, so that the weighted sample corresponds to the numerator of Eq. (3.100) (Fig. 3.11). Such a conversion from discrete particle probability distribution functions $\sum_{i=1}^{N} w_A^{(i)}\delta(x_A^{(i)})$ ($\sum_{i=1}^{N} w_B^{(i)}\delta(x_B^{(i)})$) into continuous distributions is denoted as

$$\sum_{i=1}^{N} w_A^{(i)}\delta(x_A^{(i)}) \rightarrow p_A(x) \quad (\sum_{i=1}^{N} w_B^{(i)}\delta(x_B^{(i)}) \rightarrow p_B(x)) \tag{3.102}$$

The common information appearing in the processing units A and B should not be taken into account in the joint probability distribution which is created after fusing the local probability densities of A and B. This means that in the joint p.d.f. one should sample with importance weights calculated according to Eq. (3.100). The objective is then to create an importance sampling approximation for the joint distribution that will be in accordance to Eq. (3.100). A solution to this can be obtained through Monte Carlo sampling and suitable selection of the so-called "proposal distribution" [180, 181].

3.11.2 Fusion of the Local Probability Density Functions

According to the above, for the joint distribution the idea behind Monte Carlo sampling is to draw N i.i.d samples from the associated probability density function $p(x)$, such that the target density is approximated by a point-mass function of the form

$$p(x) \simeq \sum_{i=1}^{N} w_k^{(i)} \delta(x_k^{(i)}) \tag{3.103}$$

where $\delta(x_k^{(i)})$ is a Dirac delta mass located at $x_k^{(i)}$. Then the expectation of some function $f(x)$ with respect to the pdf $p(x)$ is given by

$$I(f) = E_{p(x)}[f(x)] = \int f(x)p(x)dx \tag{3.104}$$

the Monte-Carlo approximation of the integral with samples is then

$$I_N(f) = \frac{1}{N}\sum_{i=1}^{N} f(x^{(i)}) \tag{3.105}$$

where $x^{(i)} \simeq p(X)$ and $I_N(f) \to I(f)$ for $N \to \infty$. since, the true probability distribution $p(x)$ is hard to sample from, the concept of importance sampling is to select a proposal distribution $\bar{p}(x)$ in place of $p(x)$, with the assumption that $\bar{p}(x)$ includes the support space of $p(x)$. Then the expectation of function $f(x)$, previously given in Eq. (3.104), is now calculated as

$$I(f) = \int f(x)\frac{p(x)}{\bar{p}(x)}\bar{p}(x)dx = \int f(x)w(x)\bar{p}(x)dx \tag{3.106}$$

where $w(x)$ are the importance weights

$$w(x) = \frac{p(x)}{\bar{p}(x)} \tag{3.107}$$

Consequently, the Monte-Carlo estimation of the mean value of function $f(x)$ becomes

$$I_N(f) = \sum_{i=1}^{N} f(x^{(i)})w(x^{(i)}) \tag{3.108}$$

For the division operation, the desired probability distribution is

$$p(x^{(i)}) = \frac{p_A(x^{(i)})}{p_B(x^{(i)})} \tag{3.109}$$

In that case the important weights of the fused probability density functions become

$$w(x^{(i)}) = \frac{p_A(x^{(i)})}{p_B(x^{(i)})\bar{p}(x^{(i)})} \tag{3.110}$$

which is then normalized so that $\sum_{i=1}^{N} w(x^{(i)}) = 1/N$, where N is the number of particles. The next step is to decide what will be the form of the proposal distribution $\bar{p}(x)$. A first option is to take $\bar{p}(x)$ to be a uniform distribution, with a support that covers both of the support sets of the distributions A and B.

$$\bar{p}(x) = U(x) \tag{3.111}$$

Then the sample weights $\bar{p}(x^{(i)})$ are all equal at a constant of value C. Hence the importance weights are

$$w(x^{(i)}) = \frac{p_A(x^{(i)})}{p_B(x^{(i)})C} \tag{3.112}$$

Another suitable proposal distribution that takes more into account the new information received (described as the probability distribution of the second processing unit) is given by

$$\bar{p}(x) = p_B(x) \tag{3.113}$$

and the important weights are then adjusted to be

$$w(x^{(i)}) = \frac{p_A(x^{(i)})}{p_B(x^{(i)})^2} \tag{3.114}$$

3.12 The Derivative-Free Distributed Nonlinear Kalman Filter

3.12.1 Overview

This part of the chapter extends the use of the Derivative-free nonlinear Kalman Filter towards distributed state estimation-based control of nonlinear MIMO systems. Up to now an established approach for performing distributed state estimation has been the Extended Information Fitter [225]. In the Extended Information Filter the local filters do not exchange raw measurements but send to an aggregation filter their local information matrices (local inverse covariance matrices which can be also associated to Fisher Information Matrices) and their associated local information state vectors (products of the local information matrices with the local state vectors) [131, 205]. The Extended Information Filter performs fusion of state estimates from

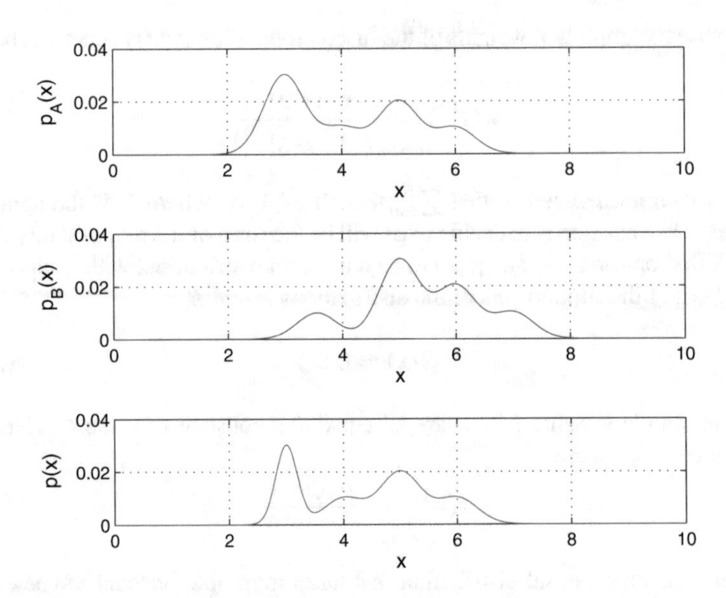

Fig. 3.11 Fusion of the probability density functions produced by the local particle filters

local distributed Extended Kalman Filters which in turn are based on the assumption of linearization of the nonlinear system dynamics by first order Taylor series expansion and truncation of the higher order linearization terms [152, 173, 192]. Moreover, the Extended Kalman Filter requires the computation of Jacobians which in the case of high order nonlinear dynamical systems can be a cumbersome procedure. This approach introduces cumulative errors to the state estimation performed by the local Extended Kalman Filter recursion which is finally transferred to the master filter where the aggregate state estimate of the controlled system is computed. Consequently, these local estimation errors may result in the deterioration of the performance of the associated control loop or even risk its stability [222, 225].

To overcome the aforementioned weaknesses of the Extended Information Filter, a differential flatness theory-based approach to distributed nonlinear Kalman Filtering has been proposed [208, 215]. Distributed filtering is now based on a derivative-free implementation of Kalman Filtering which is shown to be applicable to MIMO nonlinear dynamical systems [211, 222]. In the proposed derivative-free Kalman Filtering method the system is first subjected to a linearization transformation that is based on differential flatness theory [191, 210]. Next state estimation is performed by applying the standard Kalman Filter recursion to the linearized model. Unlike EKF, the proposed method provides estimates of the state vector of the nonlinear system without the need for derivatives and Jacobians calculation. By avoiding linearization approximations, the proposed filtering method improves the accuracy of estimation of the system state variables, and results in smooth control signal variations and in minimization of the tracking error of the associated control loop.

3.12.2 Fusing Estimations from Local Distributed Filters

Again, the discrete-time nonlinear system of Eqs. (3.52) and (3.53) is considered. The Extended Information Filter (EIF) performs fusion of the local state vector estimates which are provided by local Extended Kalman Filters, using the *Information matrix* and the *Information state vector* [131]. The Information Matrix is the inverse of the state vector covariance matrix, and can be also associated to the Fisher Information matrix [205]. The Information state vector is the product between the Information matrix and the local state vector estimate

$$\begin{aligned} Y(k) &= P^{-1}(k) = I(k) \\ \hat{y}(k) &= P^{-}(k)^{-1}\hat{x}(k) = Y(k)\hat{x}(k) \end{aligned} \tag{3.115}$$

The update equation for the Information Matrix and the Information state vector are given by

$$\begin{aligned} Y(k) &= P^{-}(k)^{-1} + J_{\gamma}^{T}(k)R^{-1}(k)J_{\gamma}(k) = Y^{-}(k) + I(k) \\ \hat{y}(k) &= \hat{y}^{-}(k) + J_{\gamma}^{T}R(k)^{-1}[z(k) - \gamma(x(k)) + J_{\gamma}(k)\hat{x}^{-}(k)] = \hat{y}^{-}(k) + i(k), \end{aligned} \tag{3.116}$$

where $I(k) = J_{\gamma}^{T}(k)R(k)^{-1}J_{\gamma}(k)$ is the associated information matrix and, $i(k) = J_{\gamma}^{T}(k)R(k)^{-1}[(z(k) - \gamma(x(k))) + J_{\gamma}\hat{x}^{-}(k)]$ is the information state contribution. The predicted information state vector and Information matrix are obtained from

$$\begin{aligned} \hat{y}^{-}(k) &= P^{-}(k)^{-1}\hat{x}^{-}(k) \\ Y^{-}(k) &= P^{-}(k)^{-1} = [J_{\phi}(k)P^{-}(k)J_{\phi}(k)^{T} + Q(k)]^{-1} \end{aligned} \tag{3.117}$$

It is assumed that an observation vector $z^{i}(k)$ is available for the N different measurement sites (vision nodes) $i = 1, 2, \ldots, N$ and each measurement site observes the system according to the local observation model, expressed by $z^{i}(k) = \gamma(x(k)) + v^{i}(k)$, $i = 1, 2, \ldots, N$, where the local noise vectors $v^{i}(k) \sim N(0, R^{i})$ is assumed to be white Gaussian and uncorrelated. The variance of a composite observation noise vector v_k is expressed in terms of the block diagonal matrix $R(k) = diag[R(k)^{1}, \ldots, R^{N}(k)]^{T}$. The information contribution can be expressed by a linear combination of each local information state contribution i^{i} and the associated information matrix I^{i} at the i-th measurement site

$$\begin{aligned} i(k) &= \textstyle\sum_{i=1}^{N} J_{\gamma}^{i\,T}(k)R^{i}(k)^{-1}[z^{i}(k) - \gamma^{i}(x(k)) + J_{\gamma}^{i}(k)\hat{x}^{-}(k)] \\ I(k) &= \textstyle\sum_{i=1}^{N} J_{\gamma}^{i\,T}(k)R^{i}(k)^{-1}J_{\gamma}^{i}(k). \end{aligned} \tag{3.118}$$

Thus, the update equations for fusing the local state estimates is

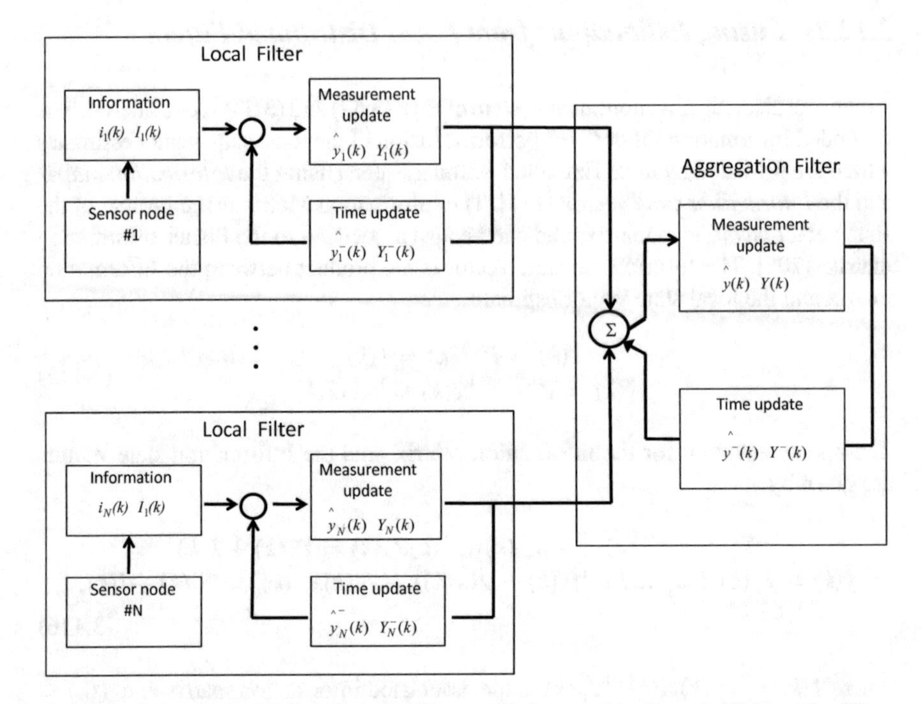

Fig. 3.12 Fusion of the distributed state estimates with the use of the Extended Information Filter

$$\begin{aligned}\hat{y}(k) &= \hat{y}^-(k) + \sum_{i=1}^{N} {J_\gamma^i}^T(k) R^i(k)^{-1}[z^i(k) - \gamma^i(x(k)) + J_\gamma^i(k)\hat{x}^-(k)] \\ Y(k) &= Y^-(k) + \sum_{i=1}^{N} {J_\gamma^i}^T(k) R^i(k)^{-1} J_\gamma^i(k)\end{aligned} \tag{3.119}$$

It is noted that in the Extended Information Filter an aggregation (master) fusion filter produces a global estimate by using the local measurement site and information provided by each local filter. As in the case of the Extended Kalman Filter the local filters which constitute the Extended information Filter can be written in terms of *time update* and *measurement update* equations.

Measurement update: Acquire $z(k)$ and compute

$$\begin{aligned}&Y(k) = P^-(k)^{-1} + J_\gamma^T(k)R(k)^{-1}J_\gamma(k) \\ &\text{or } Y(k) = Y^-(k) + I(k) \\ &\text{where } I(k) = J_\gamma^T(k)R^{-1}(k)J_\gamma(k), \text{ and,} \\ &\hat{y}(k) = \hat{y}^-(k) + J_\gamma^T(k)R(k)^{-1}[z(k) - \gamma(\hat{x}(k)) + J_\gamma\hat{x}^-(k)] \\ &\text{or } \hat{y}(k) = \hat{y}^-(k) + i(k)\end{aligned} \tag{3.120}$$

Time update: Compute

$$\begin{aligned} Y^{-}(k+1) = P^{-}(k+1)^{-1} = [J_{\phi}(k)P(k)J_{\phi}(k)^{T} + Q(k)]^{-1} \\ \text{and } y^{-}(k+1) = P^{-}(k+1)^{-1}\hat{x}^{-}(k+1) \end{aligned} \tag{3.121}$$

If the derivative-free nonlinear Kalman Filter is used in place of the Extended Kalman Filter then in the EIF equations the following matrix substitutions should be performed: $J_{\phi}(k) \rightarrow A_d$, $J_{\gamma}(k) \rightarrow C_d$, where matrices A_d and C_d are the discrete-time equivalents of matrices A_c and C_c which have been defined in Eq. (3.55). Matrices A_d and C_d can be computed using established discretization methods. Moreover, the covariance matrices $P(k)$ and $P^{-}(k)$ are the ones obtained from the linear Kalman Filter update equations given in Sect. 3.4.

3.12.3 Calculation of the Aggregate State Estimation

The outputs of the local filters are treated as measurements which are fed into the aggregation fusion filter (see Fig. 3.12) [131]. Then each local filter is expressed by its respective error covariance and estimate in terms of information contributions and is described by

$$\begin{aligned} P_i^{-1}(k) = P_i^{-}(k)^{-1} + J_{\gamma}^{T}(k)R(k)^{-1}J_{\gamma}(k)\hat{x}_i(k) = \\ = P_i(k)(P_i^{-}(k)^{-1}\hat{x}_i^{-}(k)) + J_{\gamma}^{T}(k)R(k)^{-1}[z^i(k) - \gamma^i(x(k)) + J_{\gamma}^{i}(k)\hat{x}_i^{-}(k)] \end{aligned} \tag{3.122}$$

The global estimate and the associated error covariance for the aggregate fusion filter can be rewritten in terms of the computed estimates and covariances from the local filters using the relations

$$\begin{aligned} J_{\gamma}^{T}(k)R(k)^{-1}J_{\gamma}(k) = P_i(k)^{-1} - P_i^{-}(k)^{-1} \text{ and} \\ J_{\gamma}^{T}(k)R(k)^{-1}[z^i(k) - \gamma^i(x(k)) + J_{\gamma}^{i}(k)\hat{x}^{-}(k)] = \\ = P_i(k)^{-1}\hat{x}_i(k) - P_i(k)^{-1}\hat{x}_i(k-1) \end{aligned} \tag{3.123}$$

For the general case of N local filters $i = 1, \ldots, N$, the distributed filtering architecture is described by the following equations

$$\begin{aligned} P(k)^{-1} = P^{-}(k)^{-1} + \textstyle\sum_{i=1}^{N}[P_i(k)^{-1} - P_i^{-}(k)^{-1}] \\ \hat{x}(k) = P(k)[P^{-}(k)^{-1}\hat{x}^{-}(k) + \textstyle\sum_{i=1}^{N}(P_i(k)^{-1}\hat{x}_i(k) - P_i^{-}(k)^{-1}\hat{x}_i^{-}(k))] \end{aligned} \tag{3.124}$$

The global state update equation in the above distributed filter can be written in terms of the information state vector and of the information matrix, i.e. $\hat{y}(k) = \hat{y}^{-}(k) + \sum_{i=1}^{N}(\hat{y}_i(k) - \hat{y}_i^{-}(k))$, and $\hat{Y}(k) = \hat{Y}^{-}(k) + \sum_{i=1}^{N}(\hat{Y}_i(k) - \hat{Y}_i^{-}(k))$. From Eq. (3.124) it can be seen that if a local filter fails, then the local covariance matrices and the local state estimates provided by the rest of the filters will enable an accurate computation of the target's state vector.

3.12.4 Derivative-Free Extended Information Filtering

As mentioned above, for the system of Eqs. (3.52) and (3.53), state estimation is possible by applying the standard Kalman Filter. The system is first turned into discrete-time form using common discretization methods and then the recursion of the linear Kalman Filter described in Eqs. (3.11) and (3.12) is applied.

If the derivative-free Kalman Filter is used in place of the Extended Kalman Filter then in the EIF equations the following matrix substitutions should be performed: $J_{\phi}(k) \rightarrow A_d$, $J_{\gamma}(k) \rightarrow C_d$, where matrices A_d and C_d are the discrete-time equivalents of matrices A and C which have been defined Eq. (3.55) and which appear also in the measurement and time update of the standard Kalman Filter recursion. Matrices A_d and C_d can be computed using established discretization methods. Moreover, the covariance matrices $P(k)$ and $P^{-}(k)$ are the ones obtained from the linear Kalman Filter update equations given in Sect. 3.4.

Chapter 4
Linearizing Control and Estimation for Nonlinear Dynamics in Financial Systems

4.1 Outline

The previous analysis about nonlinear control and estimation methods will be applied to the problem of stabilization of chaotic finance systems. Chaotic dynamics is often apparent in finance and results in random-like variations of the parameters and variables of markets while also annihilating financial stability conditions [54, 71, 93, 97, 107, 108, 142, 238]. To harness chaotic dynamics and to develop methods that stabilize chaotic finance systems, much work has been done during the last years. One can note results on chaotic finance systems synchronization [40, 41, 263, 304]. Systems' theoretic results have been used in [52, 55, 61, 148, 269, 283] for analyzing the dynamics of chaos in finance. Moreover, methods for feedback control and stabilization of chaotic systems appearing in finance have been given in [4, 51, 270, 288, 303].

This chapter presents an adaptive fuzzy control method for a chaotic finance system that shows interaction between variables such as the interest rate, the investment demand and the price exponent. The method is based on differential flatness theory and on diffeomorphisms (change of state variables) which allow transformation of the initial nonlinear description of the system, into an equivalent linear form. Moreover, the method is implemented only with output feedback thus requiring to monitor only a limited number of state variables in the financial system.

First, it is proven that the dynamic model of the chaotic dynamical system is a differentially flat one. This means that all its state variables and its control inputs can be expressed as differential functions of a primary state variable which is the system's flat output. Moreover, the flat output and its derivatives are differentially independent which means that they are not connected between them with a relation of the type of a differential equation [216, 222, 225, 228]. Next, by applying a change of state variables (diffeomorphism) which is in accordance to differential flatness theory, one arrives at an input-output linearized system. This description is also written in the linear canonical (Brunovsky) state-space form [77, 191, 231]. For

G.G. Rigatos, *State-Space Approaches for Modelling and Control in Financial Engineering*, Intelligent Systems Reference Library 125,
DOI 10.1007/978-3-319-52866-3_4

the latter description the design of a stabilizing state feedback controller becomes possible.

Since there is no knowledge about the financial system's dynamics and the control method is a model-free one, the unknown parts of the dynamics are identified in real-time with the use of neurofuzzy approximators. The information obtained about the system's dynamics is used for the computation of the control input, and thus an indirect adaptive control scheme is established. The update of the approximators' weights is based on a gradient-type algorithm [10, 198, 205]. The learning rate of the neurofuzzy approximators is obtained from the requirement the first derivative of the system's Lyapunov function to be always negative. The computation of the control signal requires also the solution of two algebraic Riccati equations. Lyapunov stability analysis proves that the control loop satisfies the H-infinity tracking performance ciretion and this signifies elevated robustness against model uncertainty and external perturbations. Moreover, under moderate conditions global asymptotic stability is proven.

4.2 Dynamic Model of the Chaotic Finance System

The considered macroeconomics model is derived after using accumulated knowledge about the interaction between parameters such as the interest rate, the investments demand and the price exponent (this indirectly expresses the inflation rate) [148, 288]. Thus one has:

(i) The change of the interest rate in time is proportional to the difference between investments demand and savings. Moreover, it is proportional to the price exponent (interest rate) which implies an adjustment to consumption goods' prices. The above can be written in the form of the differential equation:

$$\dot{x} = f_1(y - SV)x + f_2 z \tag{4.1}$$

where y is the investments demand, SV is the amount of savings and f_1, f_2 are constants.

(ii) The change of the investment demand is proportional to the benefit from the rate of investments, while (a) it is inhibited in a proportional manner by the investments demand itself, (b) it is inhibited in an exponential (square) manner by the value of the interest rate. The previous are expressed through the following relation:

$$\dot{y} = f_2(BEN - \alpha y - \beta x^2) \tag{4.2}$$

where BEN is the benefit rate of investments, f_2, α and β are constants.

(iii) The price exponent expresses a contradiction (discrepancy) between supply and demand in a commercial market. The price exponent is an indication of the inflation rate. The change of the price exponent is inhibited in a proportional manner by the value of the inflation rate itself. It is also inhibited in a proportional manner by the value of the interest rate. The previous are expressed through the following relation

$$\dot{z} = -f_4 z - f_5 x \tag{4.3}$$

where f_4 and f_5 are constants.

4.2.1 State-Space Model of the Chaotic Financial System

The following state variables notation is used next: $x_1 = x$, $x_2 = y$ and $x_3 = z$. Moreover, the coefficient of the previous equations are denoted as a, b and c. Thus, the dynamics of the chaotic finance system is now given by [288]

$$\begin{aligned} \dot{x}_1 &= x_3 + (x_2 - a)x_1 \\ \dot{x}_2 &= 1 - bx_2 - x_1^2 \\ \dot{x}_3 &= -x_1 - cx_3 \end{aligned} \tag{4.4}$$

As previously noted, in state vector $x = [x_1, x_2, x_3]^T$, x_1 is the interest rate, x_2 is the investment demand and x_3 is the price exponent. Moreover, a is the savings amount, b is the cost per investment, and c is the elasticity of demand. The dynamics of the financial system is complemented with the inclusion of control inputs [288]

$$\begin{aligned} \dot{x}_1 &= x_3 + (x_2 - a)x_1 \\ \dot{x}_2 &= 1 - bx_2 - x_1^2 + u \\ \dot{x}_3 &= -x_1 - cx_3 \end{aligned} \tag{4.5}$$

The financial system is also written in the state-space form:

$$\dot{x} = f(x) + g(x)u \tag{4.6}$$

where

$$f(x) = \begin{pmatrix} x_3 + (x_2 - a)x_1 \\ 1 - bx_2 - x_1^2 \\ -x_1 - cx_3 \end{pmatrix} \quad g(x) = \begin{pmatrix} 0 \\ 1 \\ 0 \end{pmatrix} \tag{4.7}$$

4.2.2 Chaotic Dynamics of the Finance System

The finance system exhibits chaotic dynamics. This means that in steady state it has a behavior that can be neither characterized as a stable equilibrium nor as a periodic or almost periodic oscillation. As time advances, the behavior of the system changes in a random-like manner and this depends on its initial conditions. Although the system is deterministic, it exhibits randomness in the way it evolves in time. By selecting the parameters' values to be $a = 0.9$, $b = 0.2$, $c = 1.2$ and the initial condition to be $x_0 = [1, 3, 2]$ one arrives at a chaotic behavior for the finance system as depicted in Figs. 4.1 and 4.2.

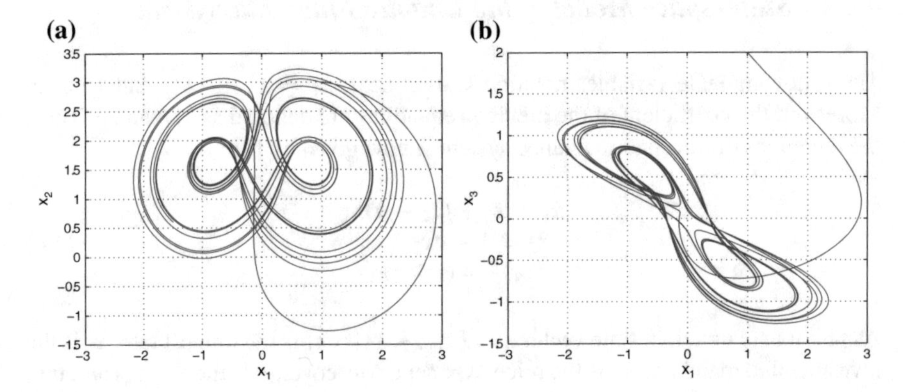

Fig. 4.1 Chaotic dynamics of the finance system: **a** phase diagram of state variables x_1 and x_2, **b** phase diagram of state variables x_1 and x_3

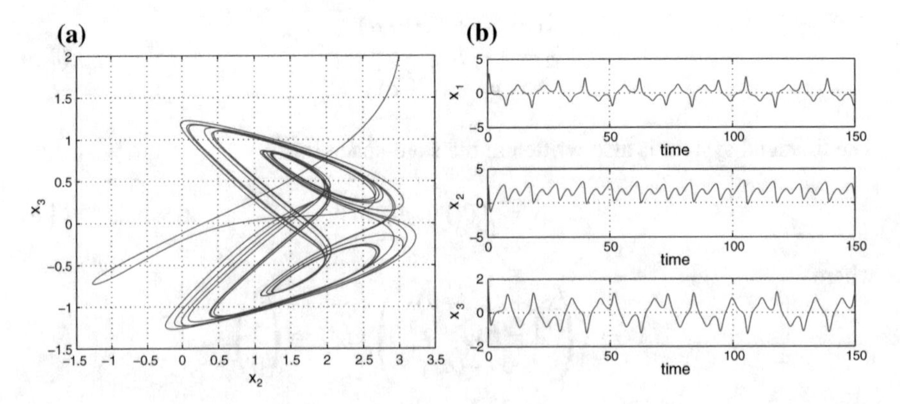

Fig. 4.2 Chaotic dynamics of the finance system: **a** phase diagram of state variables x_2 and x_3, **b** variation in time of state variables x_1, x_3 and x_3

4.3 Overview of Differential Flatness Theory

4.3.1 Conditions for Applying the Differential Flatness Theory

Differential flatness theory is currently a main direction in the analysis of nonlinear dynamical systems [222, 225, 231]. To conclude if a dynamical system is differentially flat, the following should be examined: (i) the existence of the so-called flat output, i.e. a new variable which is expressed as a function of the system's state variables. It should hold that the flat output and its derivatives should not be coupled in the form of an ordinary differential equation, (ii) the components of the system (i.e. state variables and control input) should be expressed as functions of the flat output and its derivatives [77, 134, 191, 222].

First, the generic class of nonlinear systems $\dot{x} = f(x, u)$ (including multivariable systems) is considered. Such systems can be transformed to the form of an affine in-the-input system by adding an integrator to each input [34, 134]

$$\dot{x} = f(x) + \sum_{i=1}^{m} g_i(x)u_i \tag{4.8}$$

The following definitions are now used [215]:

(i) Lie derivative: $L_f h(x)$ stands for the Lie derivative $L_f h(x) = (\nabla h)f$ and the repeated Lie derivatives are recursively defined as $L_f^0 h = h$ for $i = 0$, $L_f^i h = L_f L_f^{i-1} h = \nabla L_f^{i-1} h f$ for $i = 1, 2, \ldots$.
(ii) Lie Bracket: $ad_f^i g$ stands for a Lie Bracket which is defined recursively as $ad_f^i g = [f, ad_f^{i-1} g$ with $ad_f^0 g = g$ and $ad_f g = [f, g] = \nabla g f - \nabla f g$.

If the system of Eq. (4.8) can be linearized by a diffeomorphism $z = \phi(x)$ and a static state feedback $u = \alpha(x) + \beta(x)v$ into the following form

$$\begin{array}{c} \dot{z}_{i,j} = z_{i+1,j} \text{ for } 1 \le j \le m \text{ and } 1 \le i \le v_j - 1 \\ \dot{z}_{v_{i,j}} = v_j \end{array} \tag{4.9}$$

with $\sum_{j=1}^{m} v_j = n$, then $y_j = z_{1,j}$ for $1 \le j \le m$ are the 0-flat outputs which can be written as functions of only the elements of the state vector x. To define conditions for transforming the system of Eq. (4.8) into the canonical form described in Eq. (4.9) the following theorem holds [34]

Theorem *For nonlinear systems described by Eq. (4.8) the following variables are defined: (i)* $G_0 = span[g_1, \ldots, g_m]$*, (ii)* $G_1 = span[g_1, \ldots, g_m, ad_f g_1, \ldots, ad_f g_m]$*, … (k)* $G_k = span\{ad_f^j g_i \text{ for } 0 \le j \le k,\ 1 \le i \le m\}$*. Then, the linearization problem for the system of Eq. (4.8) can be solved if and only if: (1). The dimension of*

G_i, $i = 1, \ldots, k$ is constant for $x \in X \subseteq R^n$ and for $1 \le i \le n-1$, (2). The dimension of G_{n-1} if of order n, (3). The distribution G_k is involutive for each $1 \le k \le n-2$.

4.3.2 Transformation of Nonlinear Systems into Canonical Forms

It is assumed now that after defining the flat outputs of the initial nonlinear system and after expressing the system state variables and control inputs as functions of the flat output and of the associated derivatives, the system can be transformed in the Brunovsky canonical form:

$$\begin{array}{ll} \dot{x}_1 = x_2 & \\ \dot{x}_2 = x_3 & \\ \cdots & \\ \dot{x}_{r_1-1} = x_{r_1} & \\ \dot{x}_{r_1} = f_1(x) + \sum_{j=1}^{p} g_{1_j}(x)u_j + d_1 & y_1 = x_1 \\ & y_2 = x_2 \\ & \cdots \\ \dot{x}_{r_1+1} = x_{r_1+2} & \\ & y_p = x_{n-r_p+1} \\ \dot{x}_{r_1+2} = x_{r_1+3} & \\ \cdots & \\ \dot{x}_{p-1} = x_p & \\ \dot{x}_p = f_p(x) + \sum_{j=1}^{p} g_{p_j}(x)u_j + d_p & \end{array} \qquad (4.10)$$

where $x = [x_1, \ldots, x_n]^T$ is the state vector of the transformed system (according to the differential flatness formulation), $u = [u_1, \ldots, u_p]^T$ is the set of control inputs, $y = [y_1, \ldots, y_p]^T$ is the output vector, f_i are the drift functions and $g_{i,j}$, $i, j = 1, 2, \ldots, p$ are smooth functions corresponding to the control input gains, while d_j is a variable associated to external disturbances. It holds that $r_1 + r_2 + \cdots + r_p = n$. Having written the initial nonlinear system into the canonical (Brunovsky) form it holds

$$y_i^{(r_i)} = f_i(x) + \sum_{j=1}^{p} g_{ij}(x)u_j + d_j \qquad (4.11)$$

Next the following vectors and matrices can be defined: $f(x) = [f_1(x), \ldots, f_n(x)]^T$, $g(x) = [g_1(x), \ldots, g_n(x)]^T$, with $g_i(x) = [g_{1i}(x), \ldots, g_{pi}(x)]^T$, $A = diag[A_1, \ldots, A_p]$, $B = diag[B_1, \ldots, B_p]$, $C^T = diag[C_1, \ldots, C_p]$, $d = [d_1, \ldots, d_p]^T$, where matrix A has the multivariable systems canonical form, i.e. with block-diagonal elements

$$A_i = \begin{pmatrix} 0 & 1 & \cdots & 0 \\ 0 & 0 & \cdots & 0 \\ \vdots & \vdots & \cdots & \vdots \\ 0 & 0 & \cdots & 1 \\ 0 & 0 & \cdots & 0 \end{pmatrix}_{r_i \times r_i}$$
$$B_i^T = \begin{pmatrix} 0 & 0 & \cdots & 0 & 1 \end{pmatrix}_{1 \times r_i}$$
$$C_i = \begin{pmatrix} 1 & 0 & \cdots & 0 & 0 \end{pmatrix}_{1 \times r_i} \tag{4.12}$$

Thus, Eq. (4.11) can be written in state-space form

$$\begin{aligned} \dot{x} &= Ax + Bv + B\tilde{d} \\ y &= Cx \end{aligned} \tag{4.13}$$

where the control input is written as $v = f(x) + g(x)u$. The system of Eqs. (4.12) and (4.13) is in controller and observer canonical form.

4.4 Flatness-Based Control of the Chaotic Finance Dynamics

4.4.1 Differential Flatness of the Chaotic Finance System

The state-space model of the chaotic finance system, complemented by the application of an external control input, is

$$\begin{aligned} \dot{x}_1 &= x_3 + (x_2 - a)x_1 \\ \dot{x}_2 &= 1 - bx_2 - x_1^2 + u \\ \dot{x}_3 &= -x_1 - cx_3 \end{aligned} \tag{4.14}$$

The flat output of the system is taken to be the state variable $y = x_3$. From the third row of Eq. (4.14) one has

$$x_1 = -\dot{x}_3 - cx_3 \Rightarrow x_1 = f_1(y, \dot{y}) \tag{4.15}$$

From the first row of Eq. (4.14) one has

$$x_2 = \frac{\dot{x}_1 - x_3}{x_1} + a \Rightarrow x_2 = f_2(y, \dot{y}, \ddot{y}) \tag{4.16}$$

From the second row of Eq. (4.14) one has

$$u = \dot{x}_2 - 1 + bx_2 + x_1^2 \Rightarrow u = f_3(y, \dot{y}, \ddot{y}, y^{(3)}) \tag{4.17}$$

Since all state variables and the control input can be written as differential functions of the flat output, it is confirmed that the system is differentially flat.

4.4.2 Design of a Stabilizing Feedback Controller

By deriving twice the third row of Eq. (4.14) with respect to time, and by substituting the time derivatives $\dot{x}_i,\ i=1,2,3$ again in accordance to the rows of Eq. (4.14) one has

$$\begin{aligned} x_3^{(3)} &= (x_1+cx_3)(1-bx_2-x_1^2)x_1 + x_2[x_3+(x_2-a)x_1]+ \\ & a[x_3+(x_2-a)x_1] + c[x_3+(x_2-a)x_1] + c^2[-x_1-cx_3] - x_1u \end{aligned} \tag{4.18}$$

or equivalently

$$x_3^{(3)} = f(x) + g(x)u \tag{4.19}$$

or in the form

$$y^{(3)} = f(y,\dot{y},\ddot{y}) + g(y,\dot{y},\ddot{y})u \tag{4.20}$$

where

$$\begin{aligned} f(y,\dot{y},\ddot{y}) &= (x_1+cx_3)(1-bx_2-x_1^2)x_1 + x_2[x_3+(x_2-a)x_1]+ \\ & a[x_3+(x_2-a)x_1] + c[x_3+(x_2-a)x_1] + c^2[-x_1-cx_3] \end{aligned} \tag{4.21}$$

$$g(y,\dot{y},\ddot{y}) = -x_1 \tag{4.22}$$

By defining the transformed control input $v = f(y,\dot{y},\ddot{y}) + g(y,\dot{y},\ddot{y})u$ one has that

$$y^{(3)} = v \tag{4.23}$$

For the linearized description of the finance system given in Eq. (4.23), and using the notation $z_1 = y$, $z_2 = \dot{y}$ and $z_3 = \ddot{y}$, and $v = f(y,\dot{y},\ddot{y}) + g(y,\dot{y},\ddot{y})u$ one arrives also at the state-space description

$$\begin{pmatrix} \dot{z}_1 \\ \dot{z}_2 \\ \dot{z}_3 \end{pmatrix} = \begin{pmatrix} 0 & 1 & 0 \\ 0 & 0 & 1 \\ 0 & 0 & 0 \end{pmatrix} \begin{pmatrix} z_1 \\ z_2 \\ z_3 \end{pmatrix} + \begin{pmatrix} 0 \\ 0 \\ 1 \end{pmatrix} v \tag{4.24}$$

$$z^{meas} = (1\ 0\ 0) \begin{pmatrix} z_1 \\ z_2 \\ z_3 \end{pmatrix} \tag{4.25}$$

and the stabilizing feedback control input is given by

$$v = y_d^{(3)} - k_1(\ddot{y} - \ddot{y}_d) - k_2(\dot{y} - \dot{y}_d) - k_3(y - y_d) \tag{4.26}$$

and the control input that is actually applied to the financial system is

$$u = g^{-1}(y, \dot{y}, \ddot{y})[v - f(y, \dot{y}, \ddot{y})] \tag{4.27}$$

The previous control signal results in the tracking error dynamics of the form

$$e^{(3)}(t) + k_1\ddot{e}(t) + k_2\dot{e}(t) + k_3 e(t) = 0 \tag{4.28}$$

By selecting the feedback gains k_i, $i = 1, 2, 3$ such that the characteristic polynomial of Eq. (4.28) to be a Hurwitz one, it assured that $lim_{t\to\infty}e(t) = 0$.

4.5 Adaptive Fuzzy Control of the Chaotic Finance System Using Output Feedback

4.5.1 Problem Statement

Adaptive fuzzy control aims at solving the control problem of the chaotic finance system in case that its dynamics is unknown and the state vector is not completely measurable. It has been shown that after applying the differential flatness theory-based transformation, the following non-linear SISO system is obtained:

$$x^{(n)} = f(x, t) + g(x, t)u + \tilde{d} \tag{4.29}$$

where $f(x, t)$, $g(x, t)$ are unknown nonlinear functions and $\tilde{d}$ is an unknown additive disturbance. The objective is to force the system's output $y = x$ to follow a given bounded reference signal x_d. In the presence of non-gaussian disturbances w, successful tracking of the reference signal is denoted by the H_∞ criterion [222]

$$\int_0^T e^T Qedt \leq \rho^2 \int_0^T w^T wdt \tag{4.30}$$

where ρ is the attenuation level and corresponds to the maximum singular value of the transfer function $G(s)$ of the linearized equivalent of Eq. (4.29).

4.5.2 Transformation of Tracking into a Regulation Problem

The H_∞ approach to nonlinear systems control consists of the following steps: (i) linearization is applied: (ii) the unknown system dynamics is approximated by neural of fuzzy estimators, (iii) an H_∞ control term, is employed to compensate for estimation errors and external disturbances. If the state vector is not measurable, this can be reconstructed with the use of an observer.

For measurable state vector x, desirable state vector x_m and uncertain functions $f(x,t)$ and $g(x,t)$ an appropriate control law for (4.29) would be

$$u = \frac{1}{\hat{g}(x,t)}[x_m^{(n)} - \hat{f}(x,t) + K^T e + u_c] \tag{4.31}$$

where, $\hat{f}$ and $\hat{g}$ are the approximations of the unknown parts of the system dynamics f and g respectively, and which can be given by the outputs of suitably trained neuro-fuzzy networks. The term u_c denotes a supervisory controller which compensates for the approximation error $w = [f(x,t) - \hat{f}(x,t)] + [g(x,t) - \hat{g}(x,t)]u$, as well as for the additive disturbance $\tilde{d}$. Moreover the vectors $K^T = [k_n, k_{n-1}, \ldots, k_1]$, and $e^T = [e, \dot{e}, \ddot{e}, \ldots, e^{(n-1)}]^T$ are chosen such that the polynomial $e^{(n)} + k_1 e^{(n-1)} + k_2 e^{(n-2)} + \cdots + k_n e$ is Hurwitz. The substitution of the control law of Eq. (4.31) in Eq. (4.29) results into

$$\begin{aligned}
x^{(n)} &= f(x,t) + g(x,t)\tfrac{1}{\hat{g}(x,t)}[x_m^{(n)} - \hat{f}(x,t) - K^T e + u_c] + \tilde{d} \Rightarrow \\
x^{(n)} &= f(x,t) + \{\hat{g}(x,t) + [g(x,t) - \hat{g}(x,t)]\}\tfrac{1}{\hat{g}(x,t)}[x_m^{(n)} - \hat{f}(x,t) - K^T e + u_c] + \tilde{d} \Rightarrow \\
x^{(n)} &= f(x,t) + \{\tfrac{\hat{g}(x,t)}{\hat{g}(x,t)}[x_m^{(n)} - \hat{f}(x,t) - K^T e + u_c] + [g(x,t) - \hat{g}(x,t)]u + \tilde{d} \Rightarrow \\
x^{(n)} &= f(x,t) + x_m^{(n)} - \hat{f}(x,t) - K^T e + u_c + [g(x,t) - \hat{g}(x,t)]u + u_c + \tilde{d} \Rightarrow \\
x^{(n)} - x_m^{(n)} &= -K^T e + [f(x,t) - \hat{f}(x,t)] + [g(x,t) - \hat{g}(x,t)]u + u_c + \tilde{d} \Rightarrow \\
x^{(n)} &= -K^T e + u_c + [f(x,t) - \hat{f}(x,t)] + [g(x,t) - \hat{g}(x,t)]u + \tilde{d}
\end{aligned} \tag{4.32}$$

The above relation can be written in a state-equations form. The state vector is taken to be $e^T = [e, \dot{e}, \ldots, e^{(n-1)}]$, which yields

$$\dot{e} = Ae - BK^T e + Bu_c + B\{[f(x,t) - \hat{f}(x,t)] + [g(x,t) - \hat{g}(x,t)]u + \tilde{d}\} \tag{4.33}$$

or equivalently

$$\dot{e} = (A - BK^T)e + Bu_c + B\{[f(x,t) - \hat{f}(x,t)] + [g(x,t) - \hat{g}(x,t)]u + \tilde{d}\} \tag{4.34}$$

$$e_1 = C^T e$$

where

$$A = \begin{pmatrix} 0 & 1 & 0 & \cdots\cdots & 0 \\ 0 & 0 & 1 & \cdots\cdots & 0 \\ \cdots & \cdots & \cdots & \cdots\cdots & \cdots \\ \cdots & \cdots & \cdots & \cdots\cdots & \cdots \\ 0 & 0 & 0 & \cdots\cdots & 1 \\ 0 & 0 & 0 & \cdots\cdots & 0 \end{pmatrix} \tag{4.35}$$

$$B^T = \left(0, 0, \ldots, 0, 1\right),\ C^T = \left(1, 0, \ldots, 0, 0\right)$$

$$K^T = \left(k_0, k_1, \ldots, k_{n-2}, k_{n-1}\right)$$

where e_1 denotes the output error $e_1 = x - x_m$. Equation (4.34) describes a regulation problem.

4.5.3 Estimation of the State Vector

The control of the system described by Eq. (4.29) becomes more complicated when the state vector x is not directly measurable and has to be reconstructed through a state observer. The following definitions are used

- error of the state vector $e = x - x_m$
- error of the estimated state vector $\hat{e} = \hat{x} - x_m$
- observation error $\tilde{e} = e - \hat{e} = (x - x_m) - (\hat{x} - x_m)$

When an observer is used to reconstruct the state vector, the control law of Eq. (4.31) is written as

$$u = \frac{1}{\hat{g}(\hat{x}, t)}[x_m^{(n)} - \hat{f}(\hat{x}, t) + K^T e + u_c] \tag{4.36}$$

Applying Eq. (4.36) to the nonlinear system described by Eq. (4.29), after some operations results into

$$\begin{aligned} x^{(n)} = x_m^{(n)} - K^T\hat{e} + u_c + [f(x,t) - \hat{f}(\hat{x},t)]+ \\ [g(x,t) - \hat{g}(\hat{x},t)]u + \tilde{d} \end{aligned}$$

It holds $e = x - x_m \Rightarrow x^{(n)} = e^{(n)} + x_m^{(n)}$. Substituting $x^{(n)}$ in the above equation gives

$$\begin{aligned} e^{(n)} + x_m^{(n)} = x_m^{(n)} - K^T\hat{e} + u_c + [f(x,t) - \hat{f}(\hat{x},t)]+ \\ +[g(x,t) - \hat{g}(\hat{x},t)]u + \tilde{d} \Rightarrow \end{aligned} \tag{4.37}$$

$$\begin{aligned} \dot{e} = Ae - BK^T\hat{e} + Bu_c + B\{[f(x,t) - \hat{f}(\hat{x},t)]+ \\ +[g(x,t) - \hat{g}(\hat{x},t)]u + \tilde{d}\} \end{aligned} \tag{4.38}$$

$$e_1 = C^T e \tag{4.39}$$

where $e = [e, \dot{e}, \ddot{e}, \ldots, e^{(n-1)}]^T$, and $\hat{e} = [\hat{e}, \dot{\hat{e}}, \ddot{\hat{e}}, \ldots, \hat{e}^{(n-1)}]^T$.

The state observer is designed according to Eqs. (4.38) and (4.39) and is given by [222]:

$$\dot{\hat{e}} = A\hat{e} - BK^T\hat{e} + K_o[e_1 - C^T\hat{e}] \tag{4.40}$$

$$\hat{e}_1 = C^T\hat{e} \tag{4.41}$$

The observation gain $K_o = [k_{o_0}, k_{o_1}, \ldots, k_{o_{n-2}}, k_{o_{n-1}}]^T$ is selected so as to assure the convergence of the observer.

4.5.4 The Additional Control Term u_c

The additional term u_c which appeared in Eq. (4.31) is also introduced in the observer-based control to compensate for:

- The external disturbances $\tilde{d}$
- The state vector estimation error $\tilde{e} = e - \hat{e} = x - \hat{x}$
- The approximation error of the nonlinear functions $f(x,t)$ and $g(x,t)$, denoted as $w = [f(x,t) - \hat{f}(\hat{x},t)] + [g(x,t) - \hat{g}(\hat{x},t)]u$

The control signal u_c consists of 2 terms, namely:

- the H_∞ control term, $u_a = -\frac{1}{r}B^T P\tilde{e}$ for the compensation of d and w
- the control term u_b for the compensation of the observation error $\tilde{e}$

4.5.5 Dynamics of the Observation Error

The observation error is defined as $\tilde{e} = e - \hat{e} = x - \hat{x}$. Subtracting Eq. (4.40) from Eq. (4.38) as well as Eq. (4.41) from Eq. (4.39) one gets

$$\begin{aligned} \dot{e} - \dot{\hat{e}} &= A(e - \hat{e}) + Bu_c + B\{[f(x,t) - \hat{f}(\hat{x},t)] + \\ &+ [g(x,t) - \hat{g}(\hat{x},t)]u + \tilde{d}\} - K_o C^T(e - \hat{e}) \\ e_1 - \hat{e}_1 &= C^T(e - \hat{e}) \end{aligned}$$

i.e.

$$\begin{aligned} \dot{\tilde{e}} &= A\tilde{e} + Bu_c + B\{[f(x,t) - \hat{f}(\hat{x},t)] + \\ &+ [g(x,t) - \hat{g}(\hat{x},t)]u + \tilde{d}\} - K_o C^T\tilde{e} \\ \tilde{e}_1 &= C^T\tilde{e} \end{aligned}$$

which can be written as

$$\dot{\tilde{e}} = (A - K_o C^T)\tilde{e} + Bu_c + B\{[f(x,t) - \hat{f}(\hat{x},t)] + [g(x,t) - \hat{g}(\hat{x},t)]u + \tilde{d}\} \quad (4.42)$$

$$\tilde{e}_1 = C\tilde{e} \quad (4.43)$$

4.5.6 Approximation of Unknown Nonlinear Dynamics

Neurofuzzy networks can been trained on-line to approximate parts of the dynamic equation of non-linear systems, or to compensate for external disturbances. The approximation of functions $f(x,t)$ and $g(x,t)$ of Eq. (4.29) can be carried out with

Takagi-Sugeno neuro-fuzzy networks of zero or first order (Fig. 4.3). These consist of rules of the form:

R^l : IF $\hat{x}$ is A_1^l AND $\dot{\hat{x}}$ is A_2^l AND $\cdots$ AND $\hat{x}^{(n-1)}$ is A_n^l THEN $\bar{y}^l = \sum_{i=1}^n w_i^l \hat{x}_i + b^l, \ \ l = 1, 2, \ldots, L$

The output of the neuro-fuzzy model is calculated by taking the average of the consequent part of the rules

$$\hat{y} = \frac{\sum_{l=1}^L \bar{y}^l \prod_{i=1}^n \mu_{A_i^l}(\hat{x}_i)}{\sum_{l=1}^L \prod_{i=1}^n \mu_{A_i^l}(\hat{x}_i)} \tag{4.44}$$

where $\mu_{A_i^l}$ is the membership function of x_i in the fuzzy set A_i^l. The training of the neuro-fuzzy networks is carried out with 1st order gradient algorithms, in pattern mode, i.e. by processing only one data pair (x_i, y_i) at every time step i. The estimation of $f(x, t)$ and $g(x, t)$ can be written as

$$\hat{f}(\hat{x}|\theta_f) = \theta_f^T \phi(\hat{x}) \hat{g}(\hat{x}|\theta_g) = \theta_g^T \phi(\hat{x}) \tag{4.45}$$

where $\phi(\hat{x})$ are kernel functions with elements $\phi^l(\hat{x}) = \frac{\prod_{i=1}^n \mu_{A_i^l}(\hat{x}_i)}{\sum_{l=1}^L \prod_{i=1}^n \mu_{A_i^l}(\hat{x}_i)} \ \ l = 1, 2, \ldots, L$. It is assumed that that the weights θ_f and θ_g vary in the bounded areas M_{θ_f} and M_{θ_g} which are defined as

$$\begin{aligned} M_{\theta_f} &= \{\theta_f \in R^h : ||\theta_f|| \leq m_{\theta_f}\} \\ M_{\theta_g} &= \{\theta_g \in R^h : ||\theta_g|| \leq m_{\theta_g}\} \end{aligned} \tag{4.46}$$

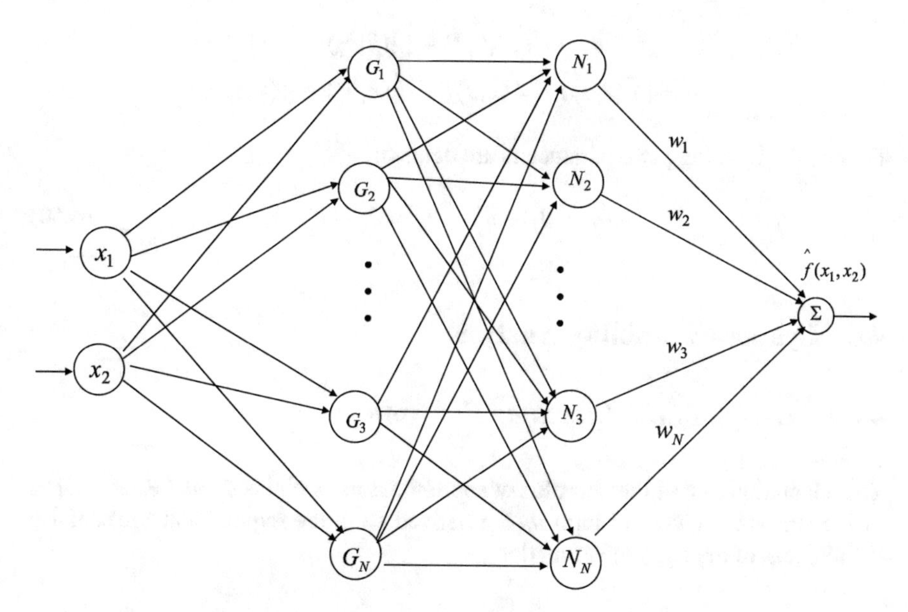

Fig. 4.3 Neuro-fuzzy approximator: G_i Gaussian basis function, N_i: normalization unit

with m_{θ_f} and m_{θ_g} positive constants. The values of θ_f and θ_g for which optimal approximation is succeeded are:

$$\begin{aligned}\theta_f^* &= arg\ min_{\theta_f \in M_{\theta_f}}[sup_{x\in U_x, \hat{x}\in U_{\hat{x}}}|f(x) - \hat{f}(\hat{x}|\theta_f)|]\\ \theta_g^* &= arg\ min_{\theta_g \in M_{\theta_g}}[sup_{x\in U_x, \hat{x}\in U_{\hat{x}}}|g(x) - \hat{g}(\hat{x}|\theta_g)|]\end{aligned} \tag{4.47}$$

The variation ranges of x and $\hat{x}$ are the compact sets

$$\begin{aligned}U_x &= \{x \in R^n : ||x|| \le m_x < \infty\},\\ U_{\hat{x}} &= \{\hat{x} \in R^n : ||\hat{x}|| \le m_{\hat{x}} < \infty\}\end{aligned} \tag{4.48}$$

The approximation error of $f(x,t)$ and $g(x,t)$ is given by

$$\begin{aligned}&w = [\hat{f}(\hat{x}|\theta_f^*) - f(x,t)] + [\hat{g}(\hat{x}|\theta_g^*) - g(x,t)]u \Rightarrow\\ &w = \{[\hat{f}(\hat{x}|\theta_f^*) - f(x|\theta_f^*)] + [f(x|\theta_f^*) - f(x,t)]\}+\\ &\{[\hat{g}(\hat{x}|\theta_g^*) - g(\hat{x}|\theta_g^*)] + [g(\hat{x}|\theta_g^*)g(x,t)]\}u\end{aligned} \tag{4.49}$$

where

- $\hat{f}(\hat{x}|\theta_f^*)$ is the approximation of f for the best estimation θ_f^* of the weights' vector θ_f.
- $\hat{g}(\hat{x}|\theta_g^*)$ is the approximation of g for the best estimation θ_g^* of the weights' vector θ_g.

The approximation error w can be decomposed into w_a and w_b, where

$$\begin{aligned}w_a &= [\hat{f}(\hat{x}|\theta_f) - \hat{f}(\hat{x}|\theta_f^*)] + [\hat{g}(\hat{x}|\theta_g) - \hat{g}(\hat{x}|\theta_g^*)]u\\ w_b &= [\hat{f}(\hat{x}|\theta_f^*) - f(x,t)] + [\hat{g}(\hat{x}|\theta_g^*) - g(x,t)]u\end{aligned}$$

Finally, the following two parameters are defined:

$$\tilde{\theta}_f = \theta_f - \theta_f^*,\quad \tilde{\theta}_g = \theta_g - \theta_g^* \tag{4.50}$$

4.6 Lyapunov Stability Analysis

4.6.1 Design of the Lyapunov Function

The adaptation law of the neurofuzzy approximators' weights θ_f and θ_g as well as of the supervisory control term u_c are derived from the requirement for negative definiteness of the Lyapunov function

$$V = \frac{1}{2}\hat{e}^T P_1 \hat{e} + \frac{1}{2}\tilde{e}^T P_2 \tilde{e} + \frac{1}{2\gamma_1}\tilde{\theta}_f^T \tilde{\theta}_f + \frac{1}{2\gamma_2}\tilde{\theta}_g^T \tilde{\theta}_g \tag{4.51}$$

The selection of the Lyapunov function is based on the following principle of indirect adaptive control $\hat{e}$: $\lim_{t\to\infty}\hat{x}(t) = x_d(t)$ and $\tilde{e}$: $\lim_{t\to\infty}\hat{x}(t) = x(t)$. This yields $\lim_{t\to\infty} x(t) = x_d(t)$. Substituting Eqs. (4.38), (4.39) and (4.42), (4.43) into Eq. (4.51) and differentiating results into

$$\dot{V} = \frac{1}{2}\dot{\hat{e}}^T P_1 \hat{e} + \frac{1}{2}\hat{e}^T P_1 \dot{\hat{e}} + \frac{1}{2}\dot{\tilde{e}}^T P_2 \tilde{e} + \frac{1}{2}\tilde{e}^T P_2 \dot{\tilde{e}} + \frac{1}{\gamma_1}\tilde{\theta}_f^T \dot{\tilde{\theta}}_f + \frac{1}{\gamma_2}\tilde{\theta}_g^T \dot{\tilde{\theta}}_g \tag{4.52}$$

which in turn gives

$$\begin{aligned}\dot{V} = \tfrac{1}{2}\{(A - BK^T)\hat{e} + K_o C^T \tilde{e}\}^T P_1 \hat{e} + \tfrac{1}{2}\hat{e}^T P_1\{(A - BK^T)\hat{e} + K_o C^T \tilde{e}\} + \\ +\tfrac{1}{2}\{(A - K_o C^T)\tilde{e} + Bu_c + Bd + Bw\}^T P_2 \tilde{e} + \tfrac{1}{2}\tilde{e}^T P_2\{(A - K_o C^T)\tilde{e} + Bu_c + Bd + Bw\} + \\ +\tfrac{1}{\gamma_1}\tilde{\theta}_f^T \dot{\tilde{\theta}}_f + \tfrac{1}{\gamma_2}\tilde{\theta}_g^T \dot{\tilde{\theta}}_g\end{aligned} \tag{4.53}$$

or, equivalently

$$\begin{aligned}\dot{V} = \tfrac{1}{2}\{\hat{e}^T (A - BK^T)^T + \tilde{e}^T C K_o^T\} P_1 \hat{e} + \tfrac{1}{2}\hat{e}^T P_1\{(A - BK^T)\hat{e} + K_o C^T \tilde{e}\} + \\ +\tfrac{1}{2}\{\tilde{e}^T (A - K_o C^T)^T + B^T u_c + B^T w + B^T d\} P_2 \tilde{e} + \\ \tfrac{1}{2}\tilde{e}^T P_2\{(A - K_o C^T)\tilde{e} + Bu_c + Bw + Bd\} + \tfrac{1}{\gamma_1}\tilde{\theta}_f^T \dot{\tilde{\theta}}_f + \tfrac{1}{\gamma_2}\tilde{\theta}_g^T \dot{\tilde{\theta}}_g\end{aligned} \tag{4.54}$$

$$\begin{aligned}\dot{V} = \tfrac{1}{2}\hat{e}^T (A - BK^T)^T P_1 \hat{e} + \tfrac{1}{2}\tilde{e}^T C K_o^T P_1 \hat{e} + +\tfrac{1}{2}\hat{e}^T P_1 (A - BK^T)\hat{e} + \tfrac{1}{2}\hat{e}^T P_1 K_o C^T \tilde{e} + \\ +\tfrac{1}{2}\tilde{e}^T (A - K_o C^T)^T P_2 \tilde{e} + \tfrac{1}{2}B^T P_2 \tilde{e}(u_c + w + d) + \\ \tfrac{1}{2}\tilde{e}^T P_2 (A - K_o C^T)\tilde{e} + \tfrac{1}{2}\tilde{e}^T P_2 B(u_c + w + d) + \tfrac{1}{\gamma_1}\tilde{\theta}_f^T \dot{\tilde{\theta}}_f + \tfrac{1}{\gamma_2}\tilde{\theta}_g^T \dot{\tilde{\theta}}_g\end{aligned} \tag{4.55}$$

Assumption 1 For given positive definite matrices Q_1 and Q_2 there exist positive definite matrices P_1 and P_2, which are the solution of the following Riccati equations [222]

$$(A - BK^T)^T P_1 + P_1 (A - BK^T) + Q_1 = 0 \tag{4.56}$$

$$\begin{aligned}(A - K_o C^T)^T P_2 + P_2 (A - K_o C^T) - \\ -P_2 B(\tfrac{2}{r} - \tfrac{1}{\rho^2}) B^T P_2 + Q_2 = 0\end{aligned} \tag{4.57}$$

The conditions given in Eqs. (4.56)–(4.57) are related to the requirement that the systems described by Eqs. (4.40), (4.41) and Eqs. (4.42), (4.43) are strictly positive real. Substituting Eqs. (4.56)–(4.57) into $\dot{V}$ yields

$$\begin{array}{c}\dot{V} = \frac{1}{2}\hat{e}^T\{(A-BK^T)^T P_1 + P_1(A-BK^T)\}\hat{e} + \tilde{e}^T C K_o^T P_1\hat{e} + \\ +\frac{1}{2}\tilde{e}^T\{(A-K_oC^T)^T P_2 + P_2(A-K_oC^T)\}\tilde{e} + B^T P_2\tilde{e}(u_c+w+d) + \\ +\frac{1}{\gamma_1}\tilde{\theta}_f^T\dot{\tilde{\theta}}_f + \frac{1}{\gamma_2}\tilde{\theta}_g^T\dot{\tilde{\theta}}_g\end{array} \tag{4.58}$$

which is also written as

$$\begin{array}{c}\dot{V} = -\frac{1}{2}\hat{e}^T Q_1\hat{e} + \tilde{e}^T C K_o^T P_1\hat{e} - \frac{1}{2}\tilde{e}^T\{Q_2 - P_2B(\frac{2}{r}-\frac{1}{\rho^2})B^T P_2\}\tilde{e} + B^T P_2\tilde{e}(u_c+w+d) + \\ +\frac{1}{\gamma_1}\tilde{\theta}_f^T\dot{\tilde{\theta}}_f + \frac{1}{\gamma_2}\tilde{\theta}_g^T\dot{\tilde{\theta}}_g\end{array} \tag{4.59}$$

The supervisory control u_c is decomposed in two terms, u_a and u_b

$$u_a = -\frac{1}{r}p_{1n}\tilde{e}_1 = -\frac{1}{r}\tilde{e}^T P_2 B + \frac{1}{r}(p_{2n}\tilde{e}_2 + \cdots + p_{nn}\tilde{e}_n) = -\frac{1}{r}\tilde{e}^T P_2 B + \Delta u_a \tag{4.60}$$

where p_{1n} stands for the last (n-th) element of the first row of matrix P_2, and

$$u_b = -[(P_2B)^T(P_2B)]^{-1}(P_2B)^T C K_o^T P_1\hat{e} \tag{4.61}$$

- u_a is an H_∞ control used for the compensation of the approximation error w and the additive disturbance $\tilde{d}$. Its first component $-\frac{1}{r}\tilde{e}^T P_2 B$ has been chosen so as to compensate for the term $\frac{1}{r}\tilde{e}^T P_2 B B^T P_2\tilde{e}$, which appears in Eq. (4.59). By subtracting the second component $-\frac{1}{r}(p_{2n}\tilde{e}_2 + \cdots + p_{nn}\tilde{e}_n)$ one has that $u_a = -\frac{1}{r}p_{1n}\tilde{e}_1$, which means that u_a is computed based on the feedback the measurable variable $\tilde{e}_1$. Equation (4.60) is finally rewritten as $u_a = -\frac{1}{r}\tilde{e}^T P_2 B + \Delta u_a$.
- u_b is a control used for the compensation of the observation error the control term u_b has been chosen so as to satisfy the condition $\tilde{e}^T P_2 B u_b = -\tilde{e}^T C K_o^T P_1\hat{e}$.

The control scheme is depicted in Fig. 4.4. Substituting Eqs. (4.60) and (4.61) in $\dot{V}$, one gets

$$\begin{array}{c}\dot{V} = -\frac{1}{2}\hat{e}^T Q_1\hat{e} + \tilde{e}^T C K_o^T P_1\hat{e} - \frac{1}{2}\tilde{e}^T Q_2\tilde{e} + \frac{1}{r}\tilde{e}^T P_2 B B^T P_2\tilde{e} - \\ -\frac{1}{2\rho^2}\tilde{e}^T P_2 B B^T P_2\tilde{e} + \tilde{e}^T P_2 B u_b - \frac{1}{r}\tilde{e}^T P_2 B B^T P_2\tilde{e} + B^T P_2\tilde{e}(w+d+\Delta u_a) + \\ +\frac{1}{\gamma_1}\tilde{\theta}_f^T\dot{\tilde{\theta}}_f + \frac{1}{\gamma_2}\tilde{\theta}_g^T\dot{\tilde{\theta}}_g\end{array} \tag{4.62}$$

or equivalently,

$$\begin{array}{c}\dot{V} = -\frac{1}{2}\hat{e}^T Q_1\hat{e} - \frac{1}{2}\tilde{e}^T Q_2\tilde{e} - \frac{1}{2\rho^2}\tilde{e}^T P_2 B B^T P_2\tilde{e} + B^T P_2\tilde{e}(w+d+\Delta u_a) + \\ +\frac{1}{\gamma_1}\tilde{\theta}_f^T\dot{\tilde{\theta}}_f + \frac{1}{\gamma_2}\tilde{\theta}_g^T\dot{\tilde{\theta}}_g\end{array} \tag{4.63}$$

It holds that $\dot{\tilde{\theta}}_f = \dot{\theta}_f - \dot{\theta}_f^* = \dot{\theta}_f$ and $\dot{\tilde{\theta}}_g = \dot{\theta}_g - \dot{\theta}_g^* = \dot{\theta}_g$. The following weight adaptation laws are considered:

$$\dot{\theta}_f = \{\begin{matrix}-\gamma_1\tilde{e}^T P_2 B\phi(\hat{x}) \ if \ ||\theta_f|| < m_{\theta_f} \\ 0 \ ||\theta_f|| \geq m_{\theta_f}\end{matrix} \tag{4.64}$$

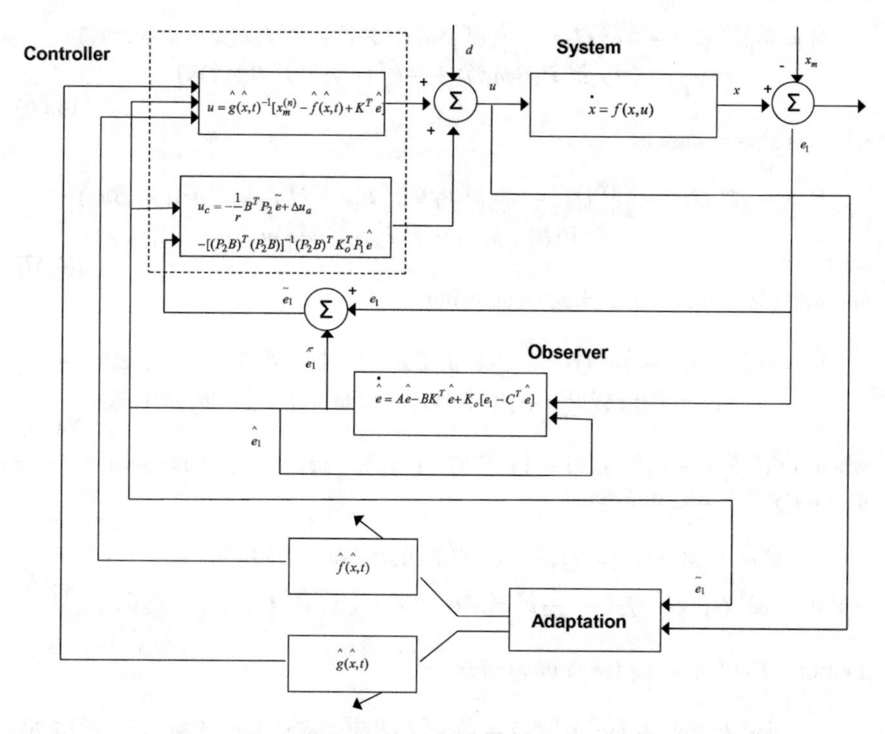

Fig. 4.4 The proposed H_∞ control scheme

$$\dot{\theta}_g = \{ \begin{matrix} -\gamma_2 \tilde{e}^T P_2 B \phi(\hat{x}) u_c \ \ if \ ||\theta_g|| < m_{\theta_g} \\ 0 \ \ ||\theta_g|| \geq m_{\theta_g} \end{matrix} \tag{4.65}$$

To set $\dot{\theta}_f$ and $\dot{\theta}_g$ equal to 0, when $||\theta_f \geq m_{\theta_f}||$, and $||\theta_g \geq m_{\theta_g}||$ the projection operator is employed [225]:

$$P\{\gamma_1 \tilde{e}^T P_2 B \phi(\hat{x})\} = -\gamma_1 \tilde{e}^T P_2 B \phi(\hat{x}) + \\ +\gamma_1 \tilde{e}^T P_2 B \frac{\theta_f \theta_f^T}{||\theta_f||^2} \phi(\hat{x})$$

$$P\{\gamma_1 \tilde{e}^T P_2 B \phi(\hat{x}) u_c\} = -\gamma_1 \tilde{e}^T P_2 B \phi(\hat{x}) u_c + \\ +\gamma_1 \tilde{e}^T P_2 B \frac{\theta_f \theta_f^T}{||\theta_f||^2} \phi(\hat{x}) u_c$$

The update of θ_f stems from a gradient algorithm on the cost function $\frac{1}{2}(f - \hat{f})^2$ [198–205]. The update of θ_g is also of the gradient type, while u_c implicitly tunes the adaptation gain γ_2. Substituting Eqs. (4.64) and (4.65) in $\dot{V}$ gives

$$\begin{array}{c}\dot{V}=-\frac{1}{2}\hat{e}^T Q_1\hat{e}-\frac{1}{2}\tilde{e}^T Q_2\tilde{e}-\frac{1}{2\rho^2}\tilde{e}^T P_2 B B^T P_2\tilde{e}+B^T P_2\tilde{e}(w+d+\Delta u_a)+\\ +\frac{1}{\gamma_1}\tilde{\theta}_f^T(-\gamma_1\tilde{e}^T P_2 B\phi(\hat{x}))+\frac{1}{\gamma_2}\tilde{\theta}_g^T(-\gamma_2\tilde{e}^T P_2 B\phi(\hat{x})u)\end{array} \tag{4.66}$$

which is also written as

$$\begin{array}{c}\dot{V}=-\frac{1}{2}\hat{e}^T Q_1\hat{e}-\frac{1}{2}\tilde{e}^T Q_2\tilde{e}-\frac{1}{2\rho^2}\tilde{e}^T P_2 B B^T P_2\tilde{e}+\tilde{e}^T P_2 B(w+d+\Delta u_a)-\\ -\tilde{e}^T P_2 B\tilde{\theta}_f^T\phi(\hat{x})-\tilde{e}^T P_2 B\tilde{\theta}_g^T\phi(\hat{x})u\end{array} \tag{4.67}$$

and using Eqs. (4.45) and (4.50) results into

$$\begin{array}{c}\dot{V}=-\frac{1}{2}\hat{e}^T Q_1\hat{e}-\frac{1}{2}\tilde{e}^T Q_2\tilde{e}-\frac{1}{2\rho^2}\tilde{e}^T P_2 B B^T P_2\tilde{e}+\tilde{e}^T P_2 B(w+d+\Delta u_a)-\\ -\tilde{e}^T P_2 B\{[\hat{f}(\hat{x}|\theta_f)+\hat{g}(\hat{x}|\theta_f)u]-[\hat{f}(\hat{x}|\theta_f^*)+\hat{g}(\hat{x}|\theta_g^*)u]\}\end{array} \tag{4.68}$$

where $[\hat{f}(\hat{x}|\theta_f)+\hat{g}(\hat{x}|\theta_f)u]-[\hat{f}(\hat{x}|\theta_f^*)+\hat{g}(\hat{x}|\theta_g^*)u]=w_a$. Thus setting $w_1=w+w_a+d+\Delta u_a$ one gets

$$\begin{array}{c}\dot{V}=-\frac{1}{2}\hat{e}^T Q_1\hat{e}\frac{1}{2}\tilde{e}^T Q_2\tilde{e}-\frac{1}{2\rho^2}\tilde{e}^T P_2 B B^T P_2\tilde{e}+B^T P_2\tilde{e}w_1\Rightarrow\\ \dot{V}=-\frac{1}{2}\hat{e}^T Q_1\hat{e}\frac{1}{2}\tilde{e}^T Q_2\tilde{e}-\frac{1}{2\rho^2}\tilde{e}^T P_2 B B^T P_2\tilde{e}+\frac{1}{2}w_1^T B^T P_2\tilde{e}+\frac{1}{2}\tilde{e}^T P_2 B w_1\end{array} \tag{4.69}$$

Lemma *The following inequality holds*

$$\frac{1}{2}\tilde{e}^T P_2 B w_1+\frac{1}{2}w_1^T B^T P_2\tilde{e}-\frac{1}{2\rho^2}\tilde{e}^T P_2 B B^T P_2\tilde{e}\leq\frac{1}{2}\rho^2 w_1^T w_1 \tag{4.70}$$

Proof The binomial $(\rho a-\frac{1}{\rho}b)^2\geq 0$ is considered. Expanding the left part of the above inequality one gets

$$\begin{array}{c}\rho^2 a^2+\frac{1}{\rho^2}b^2-2ab\geq 0\Rightarrow\frac{1}{2}\rho^2 a^2+\frac{1}{2\rho^2}b^2-ab\geq 0\\ \Rightarrow ab-\frac{1}{2\rho^2}b^2\leq\frac{1}{2}\rho^2 a^2\Rightarrow\frac{1}{2}ab+\frac{1}{2}ab-\frac{1}{2\rho^2}b^2\leq\frac{1}{2}\rho^2 a^2\end{array} \tag{4.71}$$

The following substitutions are carried out: $a=w_1$ and $b=\tilde{e}^T P_2 B$ and the previous relation becomes

$$\begin{array}{c}\frac{1}{2}w_1^T B^T P_2\tilde{e}+\frac{1}{2}\tilde{e}^T P_2 B w_1-\frac{1}{2\rho^2}\tilde{e}^T P_2 B B^T P_2\tilde{e}\\ \leq\frac{1}{2}\rho^2 w_1^T w_1\end{array} \tag{4.72}$$

The above inequality is used in $\dot{V}$, and the right part of the associated inequality is enforced

$$\dot{V}\leq-\frac{1}{2}\hat{e}^T Q_1\hat{e}-\frac{1}{2}\tilde{e}^T Q_2\tilde{e}+\frac{1}{2}\rho^2 w_1^T w_1 \tag{4.73}$$

Thus, Eq. (4.73) can be written as

$$\dot{V}\leq-\frac{1}{2}E^T Q E+\frac{1}{2}\rho^2 w_1^T w_1 \tag{4.74}$$

where

$$E = \begin{pmatrix} \hat{e} \\ \tilde{e} \end{pmatrix}, \quad Q = \begin{pmatrix} Q_1 & 0 \\ 0 & Q_2 \end{pmatrix} = diag[Q_1, Q_2] \tag{4.75}$$

Hence, the H_∞ performance criterion is derived. For ρ sufficiently small Eq. (4.73) will be true and the H_∞ tracking criterion will be satisfied. In that case, the integration of $\dot{V}$ from 0 to T gives

$$\begin{aligned} \int_0^T \dot{V}(t)dt \leq -\tfrac{1}{2}\int_0^T ||E||^2 dt + \tfrac{1}{2}\rho^2 \int_0^T ||w_1||^2 dt \Rightarrow \\ 2V(T) - 2V(0) \leq -\int_0^T ||E||_Q^2 dt + \rho^2 \int_0^T ||w_1||^2 dt \Rightarrow \\ 2V(T) + \int_0^T ||E||_Q^2 dt \leq 2V(0) + \rho^2 \int_0^T ||w_1||^2 dt \end{aligned}$$

It is assumed that there exists a positive constant $M_w > 0$ such that $\int_0^\infty ||w_1||^2 dt \leq M_w$. Therefore for the integral $\int_0^T ||E||_Q^2 dt$ one gets

$$\int_0^\infty ||E||_Q^2 dt \leq 2V(0) + \rho^2 M_w \tag{4.76}$$

Thus, the integral $\int_0^\infty ||E||_Q^2 dt$ is bounded and according to Barbalat's Lemma

$$\lim_{t\to\infty} E(t) = 0 \Rightarrow \begin{matrix} \lim_{t\to\infty} \hat{e}(t) = 0 \\ \lim_{t\to\infty} \tilde{e}(t) = 0 \end{matrix}$$

Therefore $\lim_{t\to\infty} e(t) = 0$.

4.6.2 The Role of Riccati Equation Coefficients in H_∞ Control Robustness

The linear system of Eqs. (4.42) and (4.43) is considered again

$$\begin{aligned} \dot{\tilde{e}} = (A - K_o C^T)\tilde{e} + Bu_c + B\{[f(x,t) - \hat{f}(\hat{x},t)] + [g(x,t) - \hat{g}(\hat{x},t)]u + \tilde{d}\} \\ e_1 = C^T \tilde{e} \end{aligned}$$

The aim of H_∞ control is to eliminate the impact of the modelling errors $w = [f(x,t) - \hat{f}(\hat{x},t)] + [g(x,t) - \hat{g}(\hat{x},t)]u$ and the external disturbances $\tilde{d}$ which are not white noise signals. This implies the minimization of the quadratic cost function [222]:

$$J(t) = \tfrac{1}{2}\int_0^T [\tilde{e}^T(t)\tilde{e}(t) + r u_c^T(t) u_c(t) - \rho^2 (w + \tilde{d})^T (w + \tilde{d})]dt, \quad r, \rho > 0 \tag{4.77}$$

The weight r determines how much the control signal should be penalized and the weight ρ determines how much the disturbances influence should be rewarded in the sense of a mini-max differential game. The control input u_c has been defined as the sum of the terms described in Eqs. (4.60) and (4.61).

The parameter ρ in Eq. (4.77), is an indication of the closed-loop system robustness. If the values of $\rho > 0$ are excessively decreased with respect to r, then the solution of the Riccati equation is no longer a positive definite matrix. Consequently there is a lower bound ρ_{min} of ρ for which the H_∞ control problem has a solution. The acceptable values of ρ lie in the interval $[\rho_{min}, \infty)$. If ρ_{min} is found and used in the design of the H_∞ controller, then the closed-loop system will have increased robustness. Unlike this, if a value $\rho > \rho_{min}$ is used, then an admissible stabilizing H_∞ controller will be derived but it will be a suboptimal one. The Hamiltonian matrix

$$H = \begin{pmatrix} A - K_o C^T & -\left(\frac{2}{r} - \frac{1}{\rho^2}\right) B B^T \\ -Q & -(A - K_o C^T)^T \end{pmatrix} \tag{4.78}$$

provides a criterion for the existence of a solution of the Riccati equation (4.57). A necessary condition for the solution of the algebraic Riccati equation to be a positive semi-definite symmetric matrix is that H has no imaginary eigenvalues [222].

4.7 Simulation Tests

The proposed adaptive fuzzy control method has been applied to the problem of stabilization of the dynamics of the chaotic finance system defined in Eq. (4.5) and its

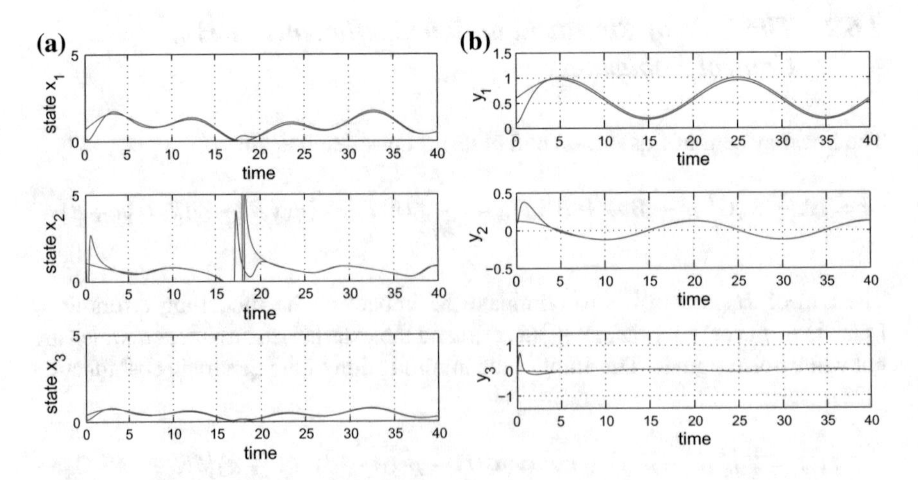

Fig. 4.5 Tracking of reference setpoint 1 (*red line*): **a** state variables x_1, x_2, and x_3 (*blue line*), **b** flat output $y_1 = x_3$ and its time derivatives $y_2 = \dot{y}_1$, $y_3 = \ddot{y}_1$ (*blue lines*)

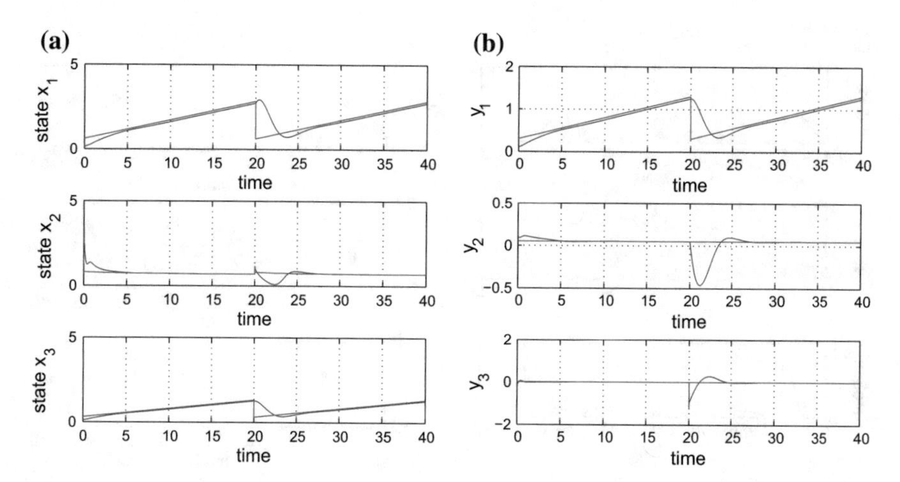

Fig. 4.6 Tracking of reference setpoint 2 (*red line*): **a** state variables x_1, x_2, and x_3 (*blue line*), **b** flat output $y_1 = x_3$ and its time derivatives $y_2 = \dot{y}_1$, $y_3 = \ddot{y}_1$ (*blue lines*)

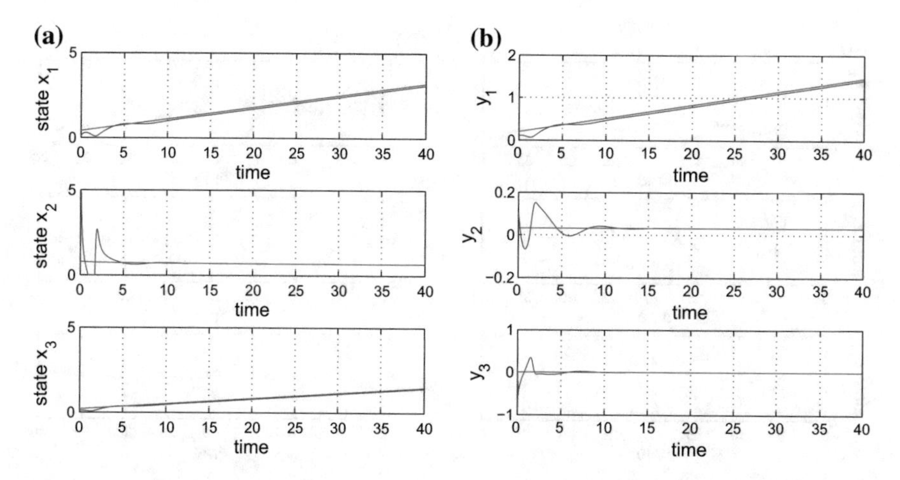

Fig. 4.7 Tracking of reference setpoint 3 (*red line*): **a** state variables x_1, x_2, and x_3 (*blue line*), **b** flat output $y_1 = x_3$ and its time derivatives $y_2 = \dot{y}_1$, $y_3 = \ddot{y}_1$ (*blue lines*)

performance has been checked through simulation experiments in the case of tracking of several reference trajectories. The presented results are depicted in Figs. 4.5, 4.6, 4.7, 4.8 and 4.9. It has been confirmed that all state variables converged fast to the reference trajectories and that the tracking error was minimized. Moreover, the control inputs computed by the nonlinear H-infinity controller varied smoothly.

The estimation of the unknown dynamics of the system with the use of neuro-fuzzy approximators has been explained in Sect. 4.5.6. The approximators' inputs were the system's state variables x_1: interest rate, x_2: investment demand and x_3: price exponent. Knowing that there are $i = 3$ state variables for the chaotic finance

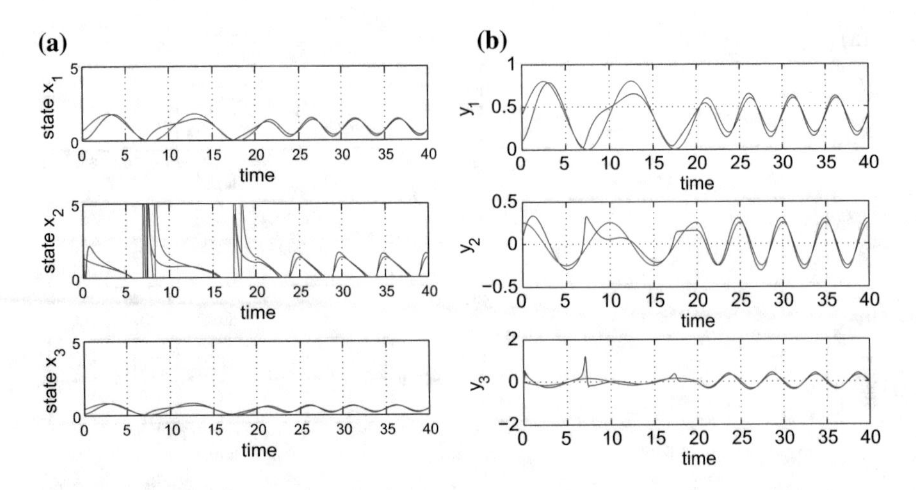

Fig. 4.8 Tracking of reference setpoint 4 (*red line*): **a** state variables x_1, x_2, and x_3 (*blue line*), **b** flat output $y_1 = x_3$ and its time derivatives $y_2 = \dot{y}_1$, $y_3 = \ddot{y}_1$ (*blue lines*)

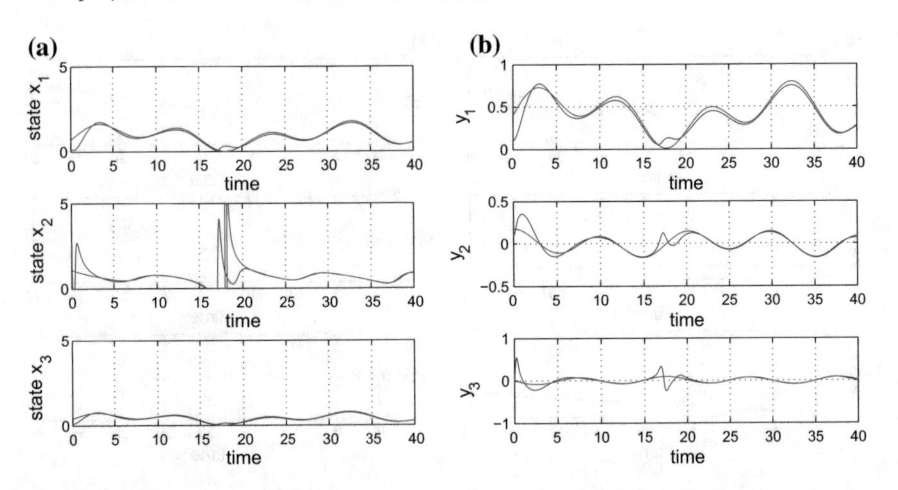

Fig. 4.9 Tracking of reference setpoint 5 (*red line*): **a** state variables x_1, x_2, and x_3 (*blue line*), **b** flat output $y_1 = x_3$ and its time derivatives $y_2 = \dot{y}_1$, $y_3 = \ddot{y}_1$ (*blue lines*)

system and that each such variable comprises $n = 3$ fuzzy sets, the total number of rules in the fuzzy rule base should be $n^m = 3^3 = 27$. The aggregate output of the neuro-fuzzy approximator (rule-base) for function $f(x)$ is given by Eq. (4.44). The centers $c_i^{(l)}$, $i = 1, \ldots, 3$ and the variances $v^{(l)}$ of each rule are summarized in Table 4.1. Similar is the structure of the neuro-fuzzy approximator for function $g(x)$.

The control loop was based on simultaneous estimation of the unknown chaotic finance system's dynamics (this was performed with the use of neuro-fuzzy approximators) and of the nonmeasurable elements of the chaotic finance system's state

Table 4.1 Parameters of the fuzzy rule base

Rule	$c_1^{(l)}$	$c_2^{(l)}$	$c_3^{(l)}$	$v^{(l)}$
$R^{(1)}$	−1.0	−1.0	−1.0	3
$R^{(2)}$	−1.0	−1.0	0.0	3
$R^{(3)}$	−1.0	−1.0	1.0	3
$R^{(4)}$	−1.0	0.0	−1.0	3
$R^{(5)}$	−1.0	0.0	0.0	3
$R^{(6)}$	−1.0	0.0	1.0	3
...	...	...	...	...
...	...	...	...	...
$R^{(27)}$	1.0	1.0	1.0	3

vector, that is of the interest rate x_1 and of the investment demand x_2 (this was performed with the use of the state observer).

The proposed adaptive fuzzy control method is a model-free one. This means that it needs no prior knowledge about the finance system's dynamic model. Moreover it is implemented only with the use of output feedback. The control scheme provides simultaneously solution to three optimization problems: (i) minimization of the state vector's tracking error, that is of the distance of the state vector from the reference trajectories, (ii) minimization of the state-vector's estimation error, that is the state vector computed by the state observer finally converges to the real value of the system's state vector, (iii) minimization of the estimation error for the unknown dynamics of the financial system being identified with the use of neuro-fuzzy approximators. The control method is applicable if there exists a solution for the two Riccati equations given in Eqs. (4.56) and (4.57).

Chapter 5
Nonlinear Optimal Control and Filtering for Financial Systems

5.1 Outline

Based on the previous results on control and estimation of nonlinear dynamical systems, another nonlinear control method will be developed for the stabilization of chaotic dynamics that may appear in macroeconomics. As already noted, several financial systems may exhibit chaotic dynamics [54, 71, 93, 97, 107, 108, 142]. Thus the study of chaotic financial systems and the development of methods trying to annihilate such chaotic dynamics have attracted much research interest during the last years [40, 41, 263, 304]. One can note results on the dynamics analysis and stability features of chaotic finance systems [52, 61, 148, 269, 283]. Moreover, there are noteworthy results on feedback control and stabilization of chaotic systems in finance [51, 270, 288, 303]. In this chapter chaotic dynamics of a financial system was considered, in which the state-space model expressed interactions between state variables such as the interest rate, the investment demand, the price exponent and the profit margin.

A nonlinear optimal controller is developed for the chaotic finance system, according to the concept of H-infinity control theory [216, 221, 222, 225, 228]. The controller makes use of an approximately linearized model of the system's dynamics, that is obtained after Taylor series expansion and the computation of the associated Jacobian matrices [10, 198, 205]. Since truncation of higher order terms is performed in the Taylor series expansion, a modelling error is introduced which is proven to be compensated by the robustness of the control method. For the linearized equivalent model of the system an H-infinity feedback controller is designed. The controller implements the solution of a nonlinear optimal control problem for the financial system under model uncertainties and external perturbations. It expresses a mini-max differential game taking place between the control input (trying to minimize a quadratic cost functional of the system) and model uncertainty and perturbation inputs (trying to maximize this cost functional).

G.G. Rigatos, *State-Space Approaches for Modelling and Control in Financial Engineering*, Intelligent Systems Reference Library 125,
DOI 10.1007/978-3-319-52866-3_5

The computation of the feedback control gain requires the solution of an algebraic Riccati equation at each iteration of the control algorithm. Through Lyapunov stability analysis it is proven that the control loop satisfies an H-infinity performance criterion which signifies improved robustness to the approximate linearization errors as well as to external perturbations [66, 145, 259]. Under moderate conditions which are related to the selection of the attenuation level coefficient (a parameter appearing in the algebraic Riccati equation), the global asymptotic stability of the control scheme is also assured. Furthermore, to estimate the state vector of the chaotic finance system with measurements from a limited number of state variables, the use of an H-infinity Kalman Filter is proposed [88, 244].

5.2 Chaotic Dynamics in a Macroeconomics Model

5.2.1 Dynamic Model of the Chaotic Finance System

An extension of the macroeconomic model examined in the previous chapter is considered here. The specific macroeconomics model is derived after using accumulated knowledge about the interaction between parameters such as the interest rate, the investments demand, the price exponent (this indirectly expresses the inflation rate) and the average profit margin [148, 288]. Thus one has:

(i) The change of the interest rate in time is proportional to the difference between investments demand and savings. Moreover, it is proportional to the price exponent (interest rate) which implies an adjustment to consumption goods' prices. The above can written in the form of the differential equation:

$$\dot{x} = f_1(y - SV)x + f_2 z \tag{5.1}$$

where y is the investments demand, SV is the amount of savings and f_1, f_2 are constants.
(ii) The change of the investment demand is proportional to the benefit from the rate of investments, while (a) it is inhibited in a proportional manner by the investments demand itself, (b) it is inhibited in an exponential (square) manner by the value of the interest rate. The previous are expressed through the following relation:

$$\dot{y} = f_2(BEN - \alpha y - \beta x^2) \tag{5.2}$$

where BEN is the benefit rate of investments, f_2, α and β are constants.
(iii) The price exponent expresses a contradiction (discrepancy) between supply and demand in a commercial market. The price exponent is an indication of the inflation rate. The change of the price exponent is inhibited in a proportional manner by the value of the inflation rate itself. It is also inhibited in a proportional manner by the value of the interest rate. The previous are expressed through the following relation

$$\dot{z} = -f_4 z - f_5 x \tag{5.3}$$

where f_4 and f_5 are constants.
(iv) The rate of change of the average profit margin is inhibited jointly by the interest rate and by the value of investment demand. Moreover, it is inhibited in a proportional manner by the value of the average profit margin itself. The above are expressed through the relation

$$\dot{w} = -f_6 xy - f_7 w \tag{5.4}$$

where again f_6 and f_7 are constants. The previous model has been confirmed to exhibit chaotic dynamics, that is its phase diagram is affected in a severe and unpredicted manner by initial conditions.

5.2.2 State-Space Model of the Chaotic Financial System

The following state variables notation is used next: $x_1 = x$, $x_2 = y$, $x_3 = z$ and $x_4 = w$. Moreover, the coefficient of the previous equations are denoted as a,b,c.d and q. The model is complemented by considering the effects of state variable x_4 in the first row of the state-space description. Thus, the dynamics of the chaotic finance system is now given by [288]

$$\begin{aligned}\dot{x}_1 &= x_3 + (x_2 - a)x_1 + x_4\\ \dot{x}_2 &= 1 - bx_2 - x_1^2\\ \dot{x}_3 &= -x_1 - cx_3\\ \dot{x}_4 &= -dx_1x_2 - qx_4\end{aligned} \tag{5.5}$$

As previously noted, in state vector $x = [x_1, x_2, x_3, x_4]^T$, x_1 is the interest rate, x_2 is the investment demand, x_3 is the price exponent, and x_4 is the average profit margin. Moreover, a is the savings amount, b is the cost per investment, and c is the elasticity of demand. The dynamics of the financial system is complemented with the inclusion of control inputs [288]

$$\begin{aligned}\dot{x}_1 &= x_3 + (x_2 - a)x_1 + x_4 + u_1\\ \dot{x}_2 &= 1 - bx_2 - x_1^2 + u_2\\ \dot{x}_3 &= -x_1 - cx_3\\ \dot{x}_4 &= -dx_1x_2 - kx_4 + u_3\end{aligned} \tag{5.6}$$

The financial system is also written in the state-space form:

$$\dot{x} = f(x) + g(x)u \tag{5.7}$$

where

$$f(x) = \begin{pmatrix} x_3 + (x_2 - a)x_1 + x_4 \\ 1 - bx_2 - x_1^2 \\ -x_1 - cx_3 \\ -dx_1x_2 - kx_4 \end{pmatrix} \quad g(x) = \begin{pmatrix} 1\,0\,0 \\ 0\,1\,0 \\ 0\,0\,0 \\ 0\,0\,1 \end{pmatrix} \tag{5.8}$$

5.2.3 Chaotic Dynamics of the Finance System

The fixed points (equilibria) of the finance system are computed from the following condition [288]:

$$\begin{array}{c} x_3 + (x_2 - a)x_1 + x_4 = 0 \\ 1 - bx_2 - x_1^2 = 0 \\ -x_1 - cx_3 = 0 \\ -dx_1x_2 - ex_4 = 0 \end{array} \tag{5.9}$$

When the parameters of the model satisfy the following condition

$$\frac{eb+dc+abce-ce}{c(d-e)} > 0 \tag{5.10}$$

then the system exhibits three fixed points (equilibria) [288]

$$\begin{array}{c} P_0 = (0, \frac{1}{b}, 0, 0) \\ P_1 = (\theta, \frac{e+ace}{c(e-d)}, -\frac{\theta}{c}, \frac{d\theta(1+ac)}{cd-ce}) \\ P_2 = (-\theta, \frac{e+ace}{c(e-d)}, \frac{\theta}{c}, \frac{d\theta(1+ac)}{cd-ce}) \end{array} \tag{5.11}$$

where $\theta = \sqrt{\frac{eb_abce}{c(d-e)} + 1}$. For the uncontrolled system, local stability features are confirmed after computing the associated Jacobian matrices and their eigenvalues at the equilibria. This procedure shows that the aforementioned equilibria are unstable saddle points because the Jacobian matrices have both negative and positive real eigenvalues.

The finance system exhibits chaotic dynamics. This means that in steady state it has a behavior that can be neither characterized as a stable equilibrium nor as a periodic or almost periodic oscillation. As time advances, the behavior of the system changes in a random-like manner and this depends on its initial conditions. Although the system is deterministic, it exhibits randomness in the way it evolves in time. By selecting the parameters' values to be $a = 0.9$, $b = 0.2$, $c = 1.5$, $d = 0.2$ and $e = 0.17$ and the initial condition to be $x_0 = [1, 2, 0.5, 0.5]$ one arrives at a chaotic behavior for the finance system as depicted in Fig. 5.1.

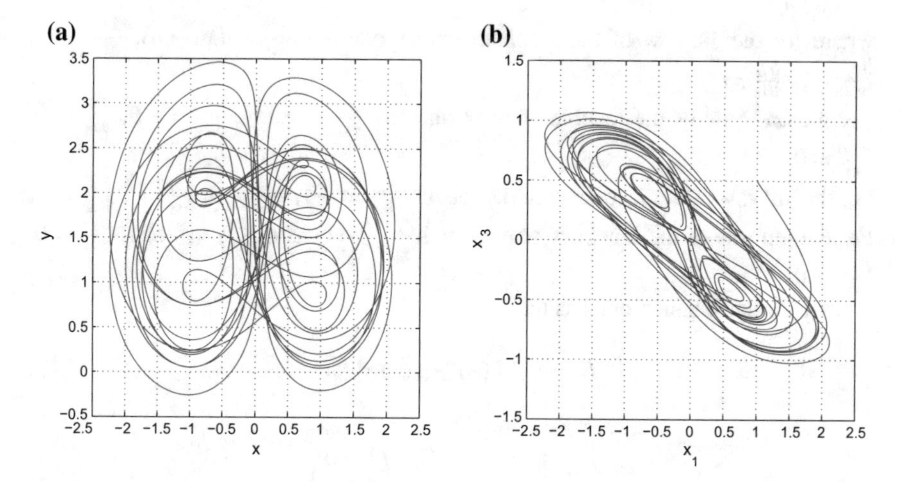

Fig. 5.1 Chaotic dynamics of the finance system: **a** phase diagram of state variables x_1 and x_2, **b** phase diagram of state variables x_1 and x_3

5.3 Design of an H-Infinity Nonlinear Feedback Controller

5.3.1 Approximate Linearization of the Chaotic Finance System

The system's dynamic model undergoes linearization round its present operating point (x^*, u^*), where x^* is the present value of the finance system's state vector and u^* is the last value of the control input vector that was exerted on it. Thus one arrives at the approximately linearized description of the system:

$$\dot{x} = Ax + Bu + \tilde{d} \tag{5.12}$$

where $\tilde{d}$ is the linearization error due to truncation of higher-order terms in the Taylor series expansion and

$$A = \nabla_x[f(x) + g(x)u] \mid_{x^*,u^*} \tag{5.13}$$

that is

$$A = \begin{pmatrix} \frac{\partial f_1}{\partial x_1} & \frac{\partial f_1}{\partial x_2} & \frac{\partial f_1}{\partial x_3} & \frac{\partial f_1}{\partial x_4} \\ \frac{\partial f_2}{\partial x_1} & \frac{\partial f_2}{\partial x_2} & \frac{\partial f_2}{\partial x_3} & \frac{\partial f_2}{\partial x_4} \\ \frac{\partial f_3}{\partial x_1} & \frac{\partial f_3}{\partial x_2} & \frac{\partial f_3}{\partial x_3} & \frac{\partial f_3}{\partial x_4} \\ \frac{\partial f_4}{\partial x_1} & \frac{\partial f_4}{\partial x_2} & \frac{\partial f_4}{\partial x_3} & \frac{\partial f_4}{\partial x_4} \end{pmatrix} \tag{5.14}$$

where for the 1st row of the Jacobian matrix one has $\frac{\partial f_1}{\partial x_1}=(x_2-u)$, $\frac{\partial f_1}{\partial x_2}=x_1$, $\frac{\partial f_1}{\partial x_3}=1$, $\frac{\partial f_1}{\partial x_4}=1$.
For the 2nd row of the Jacobian matrix one has $\frac{\partial f_2}{\partial x_1}=-2x_1$, $\frac{\partial f_2}{\partial x_2}=-b$, $\frac{\partial f_2}{\partial x_3}=0$, $\frac{\partial f_2}{\partial x_4}=0$.
For the 3rd row of the Jacobian matrix one has $\frac{\partial f_3}{\partial x_1}=-1$, $\frac{\partial f_3}{\partial x_2}=0$, $\frac{\partial f_3}{\partial x_3}=0$ $\frac{\partial f_3}{\partial x_4}=0$.
For the 4th row of the Jacobian matrix one has $\frac{\partial f_4}{\partial x_1}=-dx_2$, $\frac{\partial f_4}{\partial x_2}=-dx_1$, $\frac{\partial f_4}{\partial x_3}=0$, $\frac{\partial f_4}{\partial x_4}=0$.

In a similar manner, one has that

$$B=\nabla_u[f(x)+g(x)u]\mid_{x^*,u^*} \tag{5.15}$$

$$B=g(x)\Rightarrow B=\begin{pmatrix}1&0&0\\0&1&0\\0&0&0\\0&0&1\end{pmatrix} \tag{5.16}$$

5.3.2 *Equivalent Linearized Dynamics of the Chaotic Finance System*

After linearization round its current operating point, the chaotic finance system's dynamic model is written as

$$\dot{x}=Ax+Bu+d_1 \tag{5.17}$$

Parameter d_1 stands for the linearization error in the chaotic finance system's dynamic model appearing in Eq. (5.17). The reference setpoints for the chaotic finance system's state vector are denoted by $\mathbf{x_d}=[x_1^d,\dots,x_4^d]$. Tracking of this trajectory is succeeded after applying the control input u^*. At every time instant the control input u^* is assumed to differ from the control input u appearing in Eq. (5.17) by an amount equal to Δu, that is $u^*=u+\Delta u$

$$\dot{x}_d=Ax_d+Bu^*+d_2 \tag{5.18}$$

The dynamics of the controlled system described in Eq. (5.17) can be also written as

$$\dot{x}=Ax+Bu+Bu^*-Bu^*+d_1 \tag{5.19}$$

and by denoting $d_3=-Bu^*+d_1$ as an aggregate disturbance term one obtains

$$\dot{x}=Ax+Bu+Bu^*+d_3 \tag{5.20}$$

By subtracting Eq. (5.18) from Eq. (5.20) one has

$$\dot{x} - \dot{x}_d = A(x - x_d) + Bu + d_3 - d_2 \tag{5.21}$$

By denoting the tracking error as $e = x - x_d$ and the aggregate disturbance term as $\tilde{d} = d_3 - d_2$, the tracking error dynamics becomes

$$\dot{e} = Ae + Bu + \tilde{d} \tag{5.22}$$

The above linearized form of the chaotic finance system's model can be efficiently controlled after applying an H-infinity feedback control scheme.

5.3.3 The Nonlinear H-Infinity Control

The initial nonlinear model of the chaotic finance system is in the form

$$\dot{x} = f(x, u) \;\; x \in R^n, \;\; u \in R^m \tag{5.23}$$

Linearization of the chaotic finance system is performed at each iteration of the control algorithm round its present operating point $(x^*, u^*) = (x(t), u(t - T_s))$. The linearized equivalent of the system is described by

$$\dot{x} = Ax + Bu + L\tilde{d} \;\; x{\in}R^n, \;\; u{\in}R^m, \;\; \tilde{d}{\in}R^q \tag{5.24}$$

where, as previously explained, matrices A and B are obtained from the computation of the Jacobians

$$A = \begin{pmatrix} \frac{\partial f_1}{\partial x_1} & \frac{\partial f_1}{\partial x_2} & \cdots & \frac{\partial f_1}{\partial x_n} \\ \frac{\partial f_2}{\partial x_1} & \frac{\partial f_2}{\partial x_2} & \cdots & \frac{\partial f_2}{\partial x_n} \\ \cdots & \cdots & \cdots & \cdots \\ \frac{\partial f_n}{\partial x_1} & \frac{\partial f_n}{\partial x_2} & \cdots & \frac{\partial f_n}{\partial x_n} \end{pmatrix} |_{(x^*,u^*)} \;\; B = \begin{pmatrix} \frac{\partial f_1}{\partial u_1} & \frac{\partial f_1}{\partial u_2} & \cdots & \frac{\partial f_1}{\partial u_m} \\ \frac{\partial f_2}{\partial u_1} & \frac{\partial f_2}{\partial u_2} & \cdots & \frac{\partial f_2}{\partial u_m} \\ \cdots & \cdots & \cdots & \cdots \\ \frac{\partial f_n}{\partial u_1} & \frac{\partial f_n}{\partial u_2} & \cdots & \frac{\partial f_n}{\partial u_m} \end{pmatrix} |_{(x^*,u^*)} \tag{5.25}$$

and vector $\tilde{d}$ denotes disturbance terms due to linearization errors. The problem of disturbance rejection for the linearized model that is described by

$$\begin{aligned} \dot{x} &= Ax + Bu + L\tilde{d} \\ y &= Cx \end{aligned} \tag{5.26}$$

where $x \in R^n, u \in R^m, \tilde{d} \in R^q$ and $y \in R^p$, cannot be handled efficiently if the classical LQR control scheme is applied. This is because of the existence of the perturbation term $\tilde{d}$. The disturbance term $\tilde{d}$ apart from modeling (parametric) uncertainty and external perturbation terms can also represent noise terms of any distribution.

In the H_∞ control approach, a feedback control scheme is designed for trajectory tracking by the system's state vector and simultaneous disturbance rejection, considering that the disturbance affects the system in the worst possible manner. The disturbances' effects are incorporated in the following quadratic cost function:

$$J(t) = \frac{1}{2}\int_0^T [y^T(t)y(t) + ru^T(t)u(t) - \rho^2\tilde{d}^T(t)\tilde{d}(t)]dt, \quad r, \rho > 0 \tag{5.27}$$

The significance of the negative sign in the cost function's term that is associated with the perturbation variable $\tilde{d}(t)$ is that the disturbance tries to maximize the cost function $J(t)$ while the control signal $u(t)$ tries to minimize it. The physical meaning of the relation given above is that the control signal and the disturbances compete to each other within a mini-max differential game. This problem of mini-max optimization can be written as

$$min_u max_{\tilde{d}} J(u, \tilde{d}) \tag{5.28}$$

The objective of the optimization procedure is to compute a control signal $u(t)$ which can compensate for the worst possible disturbance, that is externally imposed to the system. However, the solution to the mini-max optimization problem is directly related to the value of the parameter ρ. This means that there is an upper bound in the disturbances magnitude that can be annihilated by the control signal.

5.3.4 Computation of the Feedback Control Gains

For the linearized system given by Eq. (5.26) the cost function of Eq. (5.27) is defined, where the coefficient r determines the penalization of the control input and the weight coefficient ρ determines the reward of the disturbances' effects.

It is assumed that (i) The energy that is transferred from the disturbances signal $\tilde{d}(t)$ is bounded, that is $\int_0^\infty \tilde{d}^T(t)\tilde{d}(t)dt < \infty$, (ii) matrices $[A, B]$ and $[A, L]$ are stabilizable, (iii) matrix $[A, C]$ is detectable. Then, the optimal feedback control law is given by

$$u(t) = -Kx(t) \tag{5.29}$$

with

$$K = \frac{1}{r}B^T P \tag{5.30}$$

where P is a positive semi-definite symmetric matrix which is obtained from the solution of the Riccati equation

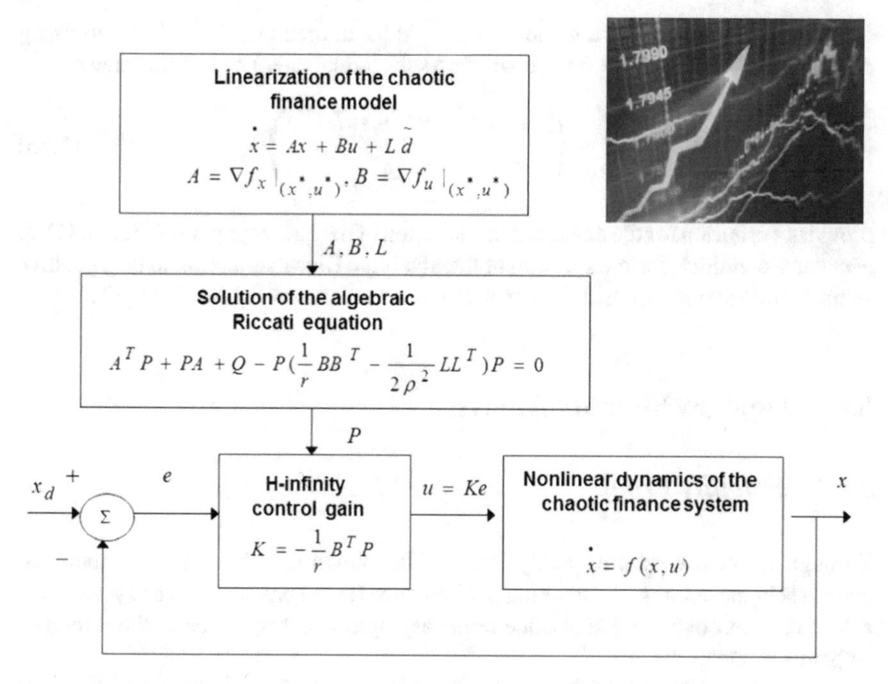

Fig. 5.2 Diagram of the control scheme for stabilization of the chaotic finance system

$$A^T P + PA + Q - P\left(\tfrac{1}{r} BB^T - \tfrac{1}{2\rho^2} LL^T\right) P = 0 \tag{5.31}$$

where Q is also a positive definite symmetric matrix. The worst case disturbance is given by

$$\tilde{d}(t) = \tfrac{1}{\rho^2} L^T P x(t) \tag{5.32}$$

The diagram of the considered control loop is depicted in Fig. 5.2.

5.3.5 *The Role of Riccati Equation Coefficients in H_∞ Control Robustness*

The parameter ρ in Eq. (5.27), is an indication of the closed-loop system robustness [259–265]. If the values of $\rho > 0$ are excessively decreased with respect to r, then the solution of the Riccati equation is no longer a positive definite matrix. Consequently there is a lower bound ρ_{min} of ρ for which the H_∞ control problem has a solution. The acceptable values of ρ lie in the interval $[\rho_{min}, \infty)$. If ρ_{min} is found and used in the design of the H_∞ controller, then the closed-loop system will have increased robustness. Unlike this, if a value $\rho > \rho_{min}$ is used, then an admissible stabilizing H_∞ controller will be derived but it will be a suboptimal one. The Hamiltonian matrix

$$H = \begin{pmatrix} A & -(\frac{1}{r}BB^T - \frac{1}{\rho^2}LL^T) \\ -Q & -A^T \end{pmatrix} \tag{5.33}$$

provides a criterion for the existence of a solution of the Riccati equation Eq. (5.31). A necessary condition for the solution of the algebraic Riccati equation to be a positive semi-definite symmetric matrix is that H has no imaginary eigenvalues [225].

5.4 Lyapunov Stability Analysis

5.4.1 *Stability Proof*

Through Lyapunov stability analysis it will be shown that the proposed nonlinear control scheme assures H_∞ tracking performance for the chaotic finance system, and that in case of bounded disturbance terms asymptotic convergence to the reference setpoints is succeeded.

The tracking error dynamics for the chaotic finance system is written in the form

$$\dot{e} = Ae + Bu + L\tilde{d} \tag{5.34}$$

where in the finance system's case $L = I \in R^2$ with I being the identity matrix. Variable $\tilde{d}$ denotes model uncertainties and external disturbances of the chaotic finance system's model. The following Lyapunov equation is considered

$$V = \frac{1}{2}e^T P e \tag{5.35}$$

where $e = x - x_d$ is the tracking error. By differentiating with respect to time one obtains

$$\begin{aligned} \dot{V} &= \tfrac{1}{2}\dot{e}^T P e + \tfrac{1}{2}e^T P \dot{e} \Rightarrow \\ \dot{V} &= \tfrac{1}{2}[Ae + Bu + L\tilde{d}]^T P e + \tfrac{1}{2}e^T P[Ae + Bu + L\tilde{d}] \Rightarrow \end{aligned} \tag{5.36}$$

$$\begin{aligned} \dot{V} = \tfrac{1}{2}[e^T A^T + u^T B^T + \tilde{d}^T L^T]Pe + \\ + \tfrac{1}{2}e^T P[Ae + Bu + L\tilde{d}] \Rightarrow \end{aligned} \tag{5.37}$$

$$\begin{aligned} \dot{V} = \tfrac{1}{2}e^T A^T Pe + \tfrac{1}{2}u^T B^T Pe + \tfrac{1}{2}\tilde{d}^T L^T Pe + \\ \tfrac{1}{2}e^T PAe + \tfrac{1}{2}e^T PBu + \tfrac{1}{2}e^T PL\tilde{d} \end{aligned} \tag{5.38}$$

The previous equation is rewritten as

$$\dot{V} = \frac{1}{2}e^T(A^T P + PA)e + \left(\frac{1}{2}u^T B^T Pe + \frac{1}{2}e^T PBu\right) + \\ + \left(\frac{1}{2}\tilde{d}^T L^T Pe + \frac{1}{2}e^T PL\tilde{d}\right) \tag{5.39}$$

Assumption: For given positive definite matrix Q and coefficients r and ρ there exists a positive definite matrix P, which is the solution of the following matrix equation

$$A^T P + PA = -Q + P\left(\frac{2}{r}BB^T - \frac{1}{\rho^2}LL^T\right)P \tag{5.40}$$

Moreover, the following feedback control law is applied to the system

$$u = -\frac{1}{r}B^T Pe \tag{5.41}$$

By substituting Eqs. (5.40) and (5.41) one obtains

$$\dot{V} = \frac{1}{2}e^T[-Q + P\left(\frac{2}{r}BB^T - \frac{1}{\rho^2}LL^T\right)P]e + \\ + e^T PB\left(-\frac{1}{r}B^T Pe\right) + e^T PL\tilde{d} \Rightarrow \tag{5.42}$$

$$\dot{V} = -\frac{1}{2}e^T Qe + \frac{1}{r}PBB^T Pe - \frac{1}{2\rho^2}e^T PLL^T Pe \\ - \frac{1}{r}e^T PBB^T Pe + e^T PL\tilde{d} \tag{5.43}$$

which after intermediate operations gives

$$\dot{V} = -\frac{1}{2}e^T Qe - \frac{1}{2\rho^2}e^T PLL^T Pe + e^T PL\tilde{d} \tag{5.44}$$

or, equivalently

$$\dot{V} = -\frac{1}{2}e^T Qe - \frac{1}{2\rho^2}e^T PLL^T Pe + \\ + \frac{1}{2}e^T PL\tilde{d} + \frac{1}{2}\tilde{d}^T L^T Pe \tag{5.45}$$

Lemma: The following inequality holds

$$\frac{1}{2}e^T PL\tilde{d} + \frac{1}{2}\tilde{d}L^T Pe - \frac{1}{2\rho^2}e^T PLL^T Pe \leq \frac{1}{2}\rho^2\tilde{d}^T\tilde{d} \tag{5.46}$$

Proof: The binomial $(\rho\alpha - \frac{1}{\rho}b)^2$ is considered. Expanding the left part of the above inequality one gets

$$\rho^2 a^2 + \frac{1}{\rho^2}b^2 - 2ab \geq 0 \Rightarrow \frac{1}{2}\rho^2 a^2 + \frac{1}{2\rho^2}b^2 - ab \geq 0 \Rightarrow \\ ab - \frac{1}{2\rho^2}b^2 \leq \frac{1}{2}\rho^2 a^2 \Rightarrow \frac{1}{2}ab + \frac{1}{2}ab - \frac{1}{2\rho^2}b^2 \leq \frac{1}{2}\rho^2 a^2 \tag{5.47}$$

The following substitutions are carried out: $a = \tilde{d}$ and $b = e^T PL$ and the previous relation becomes

$$\tfrac{1}{2}\tilde{d}^T L^T Pe + \tfrac{1}{2}e^T PL\tilde{d} - \tfrac{1}{2\rho^2}e^T PLL^T Pe \leq \tfrac{1}{2}\rho^2\tilde{d}^T\tilde{d} \tag{5.48}$$

Equation (5.48) is substituted in Eq. (5.45) and the inequality is enforced, thus giving

$$\dot{V} \leq -\tfrac{1}{2}e^T Qe + \tfrac{1}{2}\rho^2\tilde{d}^T\tilde{d} \tag{5.49}$$

Equation (5.49) shows that the H_∞ tracking performance criterion is satisfied. The integration of $\dot{V}$ from 0 to T gives

$$\begin{aligned} &\int_0^T \dot{V}(t)dt \leq -\tfrac{1}{2}\int_0^T ||e||_Q^2 dt + \tfrac{1}{2}\rho^2\int_0^T ||\tilde{d}||^2 dt \Rightarrow \\ &2V(T) + \int_0^T ||e||_Q^2 dt \leq 2V(0) + \rho^2\int_0^T ||\tilde{d}||^2 dt \end{aligned} \tag{5.50}$$

Moreover, if there exists a positive constant $M_d > 0$ such that

$$\int_0^\infty ||\tilde{d}||^2 dt \leq M_d \tag{5.51}$$

then one gets

$$\int_0^\infty ||e||_Q^2 dt \leq 2V(0) + \rho^2 M_d \tag{5.52}$$

Thus, the integral $\int_0^\infty ||e||_Q^2 dt$ is bounded. Moreover, $V(T)$ is bounded and from the definition of the Lyapunov function V in Eq. (5.35) it becomes clear that $e(t)$ will be also bounded since $e(t) \in \Omega_e = \{e|e^T Pe \leq 2V(0) + \rho^2 M_d\}$.

According to the above and with the use of Barbalat's Lemma one obtains $lim_{t\to\infty} e(t) = 0$.

5.4.2 Robust State Estimation with the Use of the H_∞ Kalman Filter

The control loop has to be implemented by measuring and processing only a small number of state variables. To reconstruct the missing information about the state vector of the financial system it is proposed to use a filtering scheme and based on it to apply state estimation-based control [222]. The recursion of the H_∞ Kalman Filter, for the model of the chaotic finance system, can be formulated in terms of a *measurement update* and a *time update* part

Measurement update:

$$\begin{aligned} D(k) &= [I - \theta W(k)P^{-}(k) + C^{T}(k)R(k)^{-1}C(k)P^{-}(k)]^{-1} \\ K(k) &= P^{-}(k)D(k)C^{T}(k)R(k)^{-1} \\ \hat{x}(k) &= \hat{x}^{-}(k) + K(k)[y(k) - C\hat{x}^{-}(k)] \end{aligned} \tag{5.53}$$

Time update:

$$\begin{aligned} \hat{x}^{-}(k+1) &= A(k)x(k) + B(k)u(k) \\ P^{-}(k+1) &= A(k)P^{-}(k)D(k)A^{T}(k) + Q(k) \end{aligned} \tag{5.54}$$

where it is assumed that parameter θ is sufficiently small to assure that the covariance matrix $P^{-}(k) - \theta W(k) + C^{T}(k)R(k)^{-1}C(k)$ will be positive definite. When $\theta = 0$ the H_{∞} Kalman Filter becomes equivalent to the standard Kalman Filter. One can measure only a part of the state vector of the chaotic finance system, and estimate through filtering the rest of the state vector elements. The filtering method can be also used for forecasting the variations of the chaotic finance system.

5.5 Simulation Tests

The proposed nonlinear H-infinity control method has been applied to the problem of stabilization of the dynamics of the chaotic finance system defined in Eq. (5.6) and its performance has been checked through simulation experiments in the case

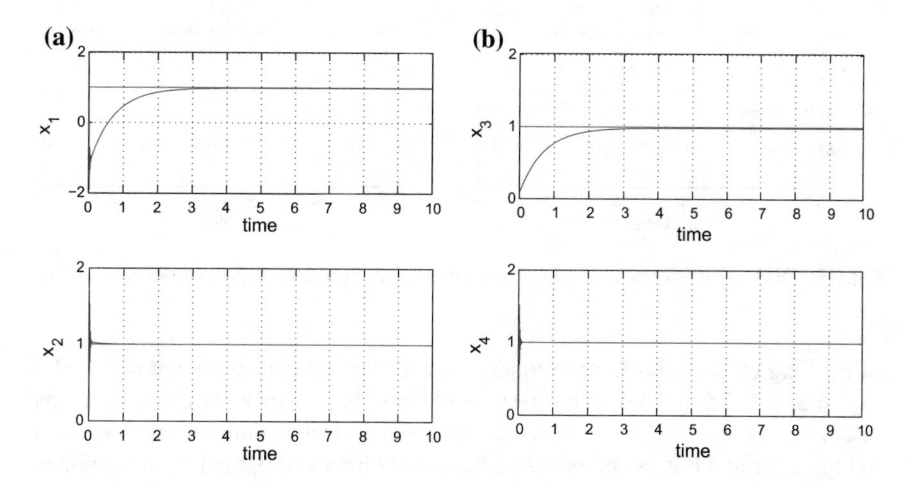

Fig. 5.3 Tracking of reference setpoint 1: **a** state variables x_1 and x_2, **b** state variables x_3 and x_4

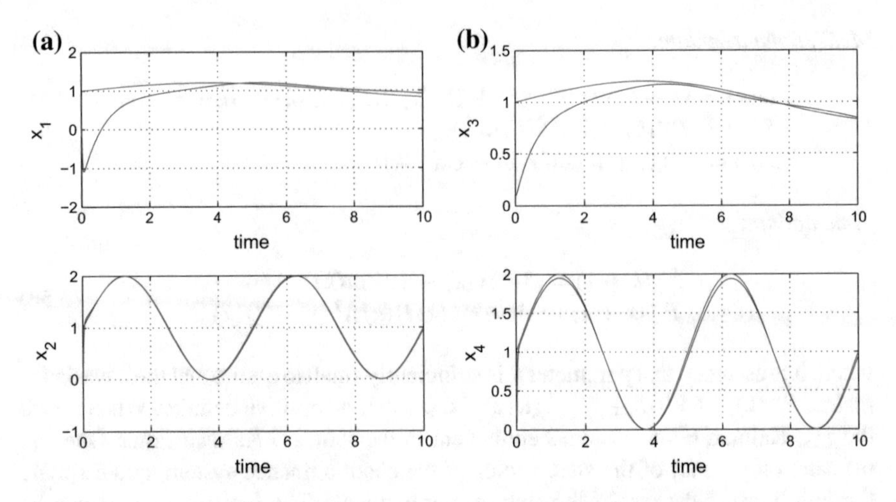

Fig. 5.4 Tracking of reference setpoint 2: **a** state variables x_1 and x_2, **b** state variables x_3 and x_4

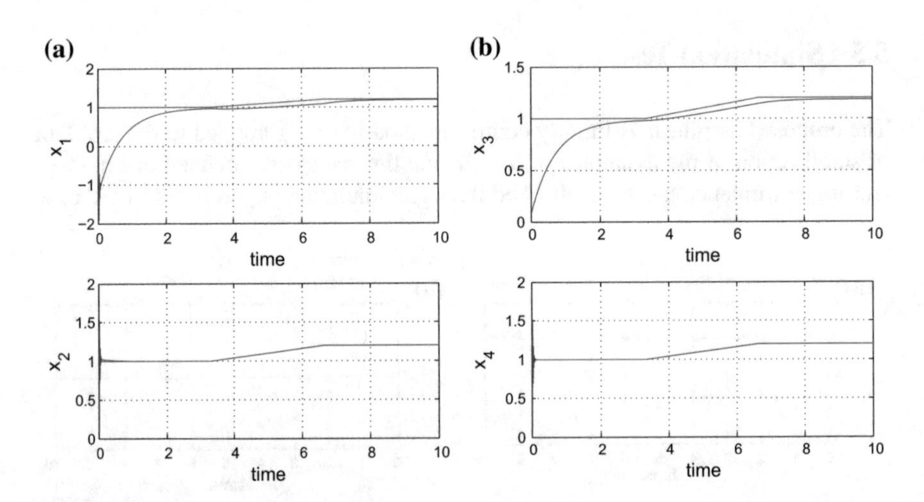

Fig. 5.5 Tracking of reference setpoint 3: **a** state variables x_1 and x_2, **b** state variables x_3 and x_4

of tracking of several reference trajectories. The presented results are depicted in Figs. 5.3, 5.4, 5.5 and 5.6. It has been confirmed that all state variables converged fast to the reference trajectories and that the tracking error was minimized. Moreover, in Figs. 5.7 and 5.8 it can be seen that the control inputs computed by the nonlinear H-infinity controller varied smoothly.

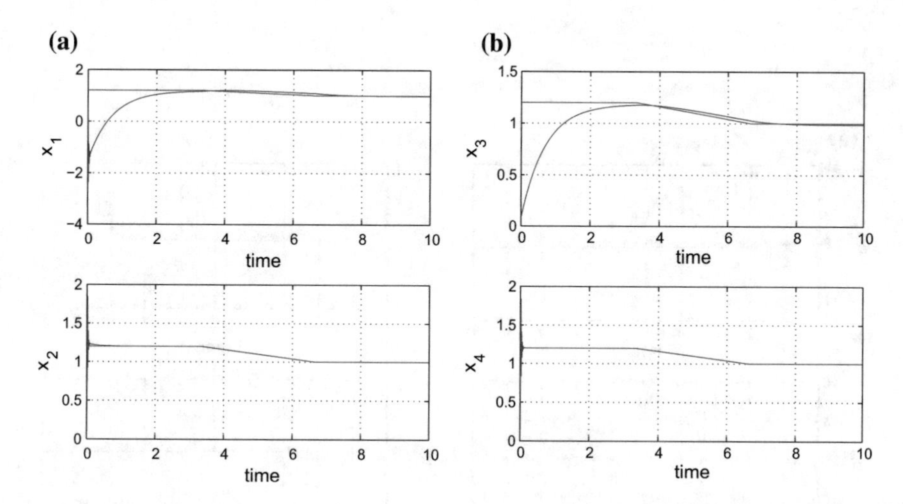

Fig. 5.6 Tracking of reference setpoint 4: **a** state variables x_1 and x_2, **b** state variables x_3 and x_4

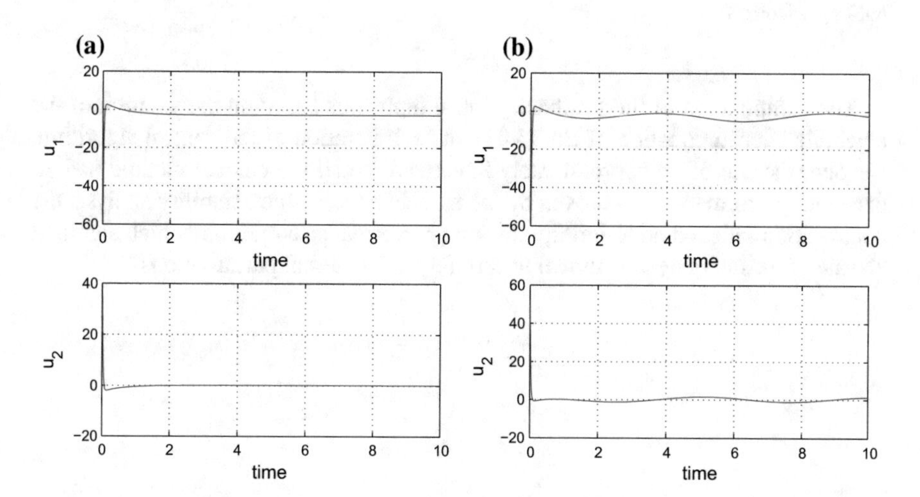

Fig. 5.7 **a** control inputs u_1 and u_2 when tracking setpoint 1, **b** control inputs u_1 and u_2 when tracking setpoint 2

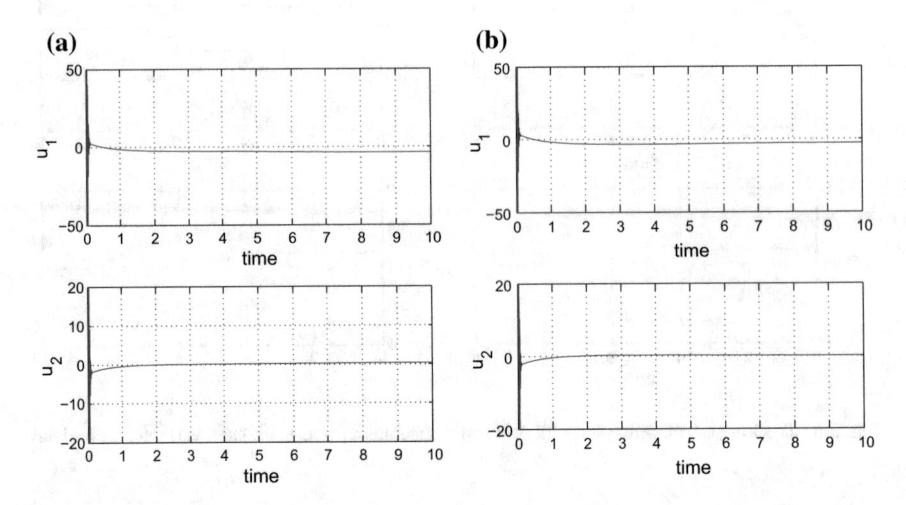

Fig. 5.8 **a** control inputs u_1 and u_2 when tracking setpoint 3, **b** control inputs u_1 and u_2 when tracking setpoint 4

The computation of the feedback control input was based on the solution of the algebraic Riccati equation of Eq. (5.31), at each iteration of the control algorithm. Despite, the use of an approximately linearized model the control method had an excellent performance. As proven by the associated Lyapunov stability analysis the control loop satisfied an H-infinity tracking performance condition, which signified elevated robustness against model uncertainty and external perturbations.

Chapter 6
Kalman Filtering Approach for Detection of Option Mispricing in the Black–Scholes PDE

6.1 Outline

Up to now the dynamics of finance systems has been considered to be described by nonlinear differential equations. In the present chapter finance systems dynamics in the form of partial differential equations (PDEs) will be examined. The Black–Scholes model of option pricing is considered to be a major tool in financial engineering since it explains the changes of the options value as a function of time and of the security index. The Black–Scholes model is actually a diffusion partial differential equation and its numerical solution gives a precise view of how the value of options varies [102, 189]. The chapter proposes the new Kalman Filtering method, under the name of Derivative-free nonlinear Kalman Filtering for estimation of option prices, based on the use of Black–Scholes model, without depending on initial conditions [216, 222, 225].

Using the method for numerical solution of the PDE through discretization the initial partial differential equation is decomposed into a set of nonlinear ordinary differential equations with respect to time [216]. Next, each one of the local models associated with the ordinary differential equations is transformed into a model of the linear canonical (Brunovsky) form through a change of coordinates (diffeomorphism) which is based on differential flatness theory. This transformation provides an extended model of the nonlinear PDE for which state estimation is possible by application of the standard Kalman Filter recursion [34, 77, 134, 169, 231]. Unlike other nonlinear estimation methods (e.g. Extended Kalman Filter) the application of the standard Kalman Filter recursion to the linearized equivalent of the Black–Scholes PDE model does need the computation of Jacobian matrices and partial derivatives [216, 222, 225].

The application of the proposed filtering method is computationally efficient and fast. Through simulation experiments it was confirmed that the derivative-free nonlinear Kalman Filter approximated with accuracy the two-dimensional model of the Black–Scholes PDE and consequently enabled to estimate successfully variations of

G.G. Rigatos, *State-Space Approaches for Modelling and Control in Financial Engineering*, Intelligent Systems Reference Library 125,
DOI 10.1007/978-3-319-52866-3_6

option prices (despite no knowledge of initial conditions). Based on these estimates, validation of the Black–Scholes PDE model can be performed. Thus one can find out if the parameters appearing in the model, such as interest rate and volatility are precise enough, to capture the real dynamics of the financial system [39, 58, 73, 74, 137]. If it is observed that the accuracy of the model is lost then a retuning of its parameters can be carried out.

6.2 Option Pricing Modeling with the Use of the Black–Scholes PDE

6.2.1 Option Pricing Modeling with the Use of Stochastic Differential Equations

A financial derivative is called a European option if it gives the right for pay-off according to a function $H : [0, \infty) \rightarrow R$ which has as principle variable the security index S_t, while there is also a predefined expiration date $T \in [0, \infty)$. If the payoff can take place before the expiration date then one has an American option.

The value function of a European option is denoted as $V(t, S_t)$, where t varies in the interval $t \in [0, T]$, while the associated payoff function is defined as H and the maturity time is $T \in [0, \infty)$. The value function $V : [0, T) \times [0, \infty] \rightarrow R$ for a European option, can be differentiated with respect to time while it is also twice differentiable with respect to the security index S.

The option price model assumes that the security price $S = \{S_t, t \in [0, T]\}$ with initial value equal to S_0 follows a geometric Brownian motion

$$dS_t = \alpha_t S_t dt + \sigma_t S_t dW_t \tag{6.1}$$

where $\alpha = \{\alpha_t, \ t \in [0, T]\}$ is the appreciation rate which is a positive variable, and $\sigma = \{\sigma_t, \ t \in [0, T]\}$ is the volatility parameter. Variable W denoted a Wiener process, $W = \{W_t, \ t \in [0, T]\}$.

There is also a second stochastic process in the model which is given by

$$dB_t = r_t B_t dt \tag{6.2}$$

where B_t corresponds to the domestic savings parameter $B = \{B_t, \ t \in [0, T]\}$ with initial value $B_0 = 1$. Variable r_t is the interest rate $r = \{r_t, \ t \in [0, T]\}$. The domestic savings account is also called *locally riskless asset* when there is no noise term in the SDE model given in Eq. (6.2).

6.2.2 *The Black–Scholes PDE*

The Black–Scholes partial differential equation is given by

$$\frac{\partial V(t,S)}{\partial t} + r_t S\frac{\partial V(t,S)}{\partial S} + \frac{1}{2}\sigma_t^2 S^2 \frac{\partial V^2(t,S)}{\partial S^2} - r_t V(t,S) = 0 \quad (6.3)$$

for $t \in [0,T)$ and $S \in (0,\infty)$ where r is the interest rate and σ_t is the volatility, while the associated terminal condition is

$$V(T,S) = H(S) \quad (6.4)$$

6.2.3 *Solution of the Black–Scholes PDE*

The solution of the Black–Scholes PDE in a closed form became possible after a change of variables that transformed it into an equivalent heat diffusion equation. In this solution, an option is considered with underlying security $S = \{S_t,\ t \in [0,T]\}$ which follows the stochastic differential equation

$$dS_t = \alpha_t S_t dt + \sigma_t S_t dW \quad (6.5)$$

Moreover, it is considered that the payoff function is given by

$$H(S) = (S-K)^+ \quad (6.6)$$

for $S \in (0,\infty)$ with strike price equal to K and final payoff (maturity date) equal to T. The notation a^+ means $a^+ = max(0,a)$. The closed-form solution of the Black–Scholes PDE of Eq. (6.3) is the value function of the option $c_{T,K}(t,S_t)$, at time instant t. In particular it holds

$$C_{T,K}(t,S_t) = S_t N(d_1(t)) - K\frac{B_t}{B_T}N(d_2(t)) \quad (6.7)$$

with

$$d_1(t) = \frac{ln(\frac{S_t}{K}) + \int_t^T (r_s + \frac{1}{2}\sigma_s^2)ds}{\sqrt{\int_t^T \sigma_s^2 ds}} \quad (6.8)$$

and

$$d_2(t) = d_1(t) - \sqrt{\int_t^T \sigma_s^2 ds} \quad (6.9)$$

for $t \in [0,T)$. Variable B_t denotes the domestic savings account. Variable S_t denotes the security variable. Variable r denotes the interest rate. Variable σ denotes volatility. Variable $N()$ denotes the cumulative normal distribution with density $N(x)=\frac{1}{\sqrt{2\pi}}exp\left\{-\frac{x^2}{2}\right\}$.

For small values of S_t, the functions $d_1(t)$, $d_2(t)$, as well as functions $N(d_1(t))$, $N(d_2(t))$ take small values. For larger values of S_t, functions $d_1(t)$ and $d_2(t)$ take also larger values, while functions $N(d_1(t))$ and $N(d_2(t))$ converge to 1.

6.2.4 Sensitivities of the European Call Option

The pricing function $C_{T,K}$ for the European call option depends on the following variables, which are (i) the security price S_t, (ii) the time left to maturity T_t, (iii) the volatility σ for the underlying security, (iv) the interest rate r, and (v) the strike price K.

A Taylor series expansion is performed to the Black–Scholes equation. This gives

$$\begin{array}{c}V(t+h,S_{t+h})-V(t,S_t)=\frac{\partial V(t,S_t)}{\partial S}(S_{t+h}-S_t)+\frac{1}{2}\frac{\partial^2 V(t,S_t)}{\partial S^2}(S_{t+h}-S_t)^2+\\ +\frac{\partial V(t,S_t)}{\partial t}h+\frac{\partial V(t,S_t)}{\partial \sigma}(\sigma_{t+h}-\sigma_t)+\frac{\partial V(t,S_t)}{\partial r}(r_{t+h}-r_t)\end{array} \tag{6.10}$$

or, equivalently

$$\begin{array}{c}V(t+h,S_{t+h})-V(t,S_t)=\Delta(S_{t+h}-S_t)+\frac{1}{2}\Gamma(S_{t+h}-S_t)^2-\\ -\Theta h+V(\sigma_{t+h}-\sigma_t)+\rho(r_{t+h}-r_t)\end{array} \tag{6.11}$$

where

$$\begin{array}{c}\Delta=\frac{\partial V(t,S_t)}{\partial S}\quad \Gamma=\frac{\partial^2 V(t,S_t)}{\partial S^2}\quad \Theta=\frac{\partial V(t,S_t)}{\partial t}\\ V=\frac{\partial V(t,S_t)}{\partial \sigma}\quad \rho=\frac{\partial V(t,S_t)}{\partial r}\end{array} \tag{6.12}$$

for $t\in[0,T)$. Variables Δ, Γ, Θ, V and ρ denote the associated partial derivatives of the option's value function, which are also called *sensitivities* or *greeks*.

6.2.5 Nonlinearities in the Black–Scholes PDE

Nonlinearities are introduced in the Black–Scholes PDE through the volatility variable σ, which is considered to be a nonlinear function of the second order partial derivative of the option's price V_{ss}. Indicative nonlinearities in the Black–Scholes PDE are as follows:

1. *Leland's model*: the volatility takes the form $\sigma^2(t, S, V_{ss}) = \sigma_0^2(1 + A{\cdot}sign(V_{ss}))$ where $A = \sqrt{\frac{2}{\pi}}\frac{K}{\sigma_0\sqrt{\delta t}}$ is Leland's number, σ_0 is the volatility coefficient, K stands for the round-trip cost and δt is the time between adjustment of the portfolios.

2. *Soner's and Barles model*: the volatility takes the form $\sigma^2(t, S, V_{ss}) = \sigma_0^2(1 + \psi(e^{r(T-\tau)}a^2S^2V_{ss}))$ where the differential relation for function $\psi(x)$ is given by $\psi'(x) = \frac{\psi(x)+1}{2\sqrt{x\psi(x)}+x}$ for $x{\neq}0$ and $\psi(0) = 0$.

3. *UV model*: the volatility takes the form $\sigma^2(t, S, V_{ss}) = \sigma^2_{max}$ if $V_{ss}{\leq}0$ and $\sigma^2(t, S, V_{ss}) = \sigma^2_{min}$ if $V_{ss} > 0$. The volatility is considered to be unknown but varying between σ_{max} and σ_{min}.

4. *Frey's illiquidity model*: the volatility takes the form $\sigma^2(t, S, V_{ss}) = \frac{\sigma_0^2}{1-\rho S V_{ss}{}^2}$ where ρ is a parameter denoting the liquidity of the market.

6.2.6 Derivative Pricing

The computation of an option's value using the Black–Scholes model is usually performed through numerical solution of the associated partial differential equation, and after applying the finite differences method. In general it is considered that the change of the financial derivative $X = \{X_t, t \in [0, T]\}$ takes place according to the following stochastic differential equation:

$$dX_t = \alpha(t, X_t)dt + b(t, X_t)d\tilde{W}_t \tag{6.13}$$

for $t \in [0, T]$ with $X_0{\in}R$. In particular, a European option with security index $X = S$ is considered with payoff function $V(T, S_T) = h(S_T)$, where T is the payoff (maturity) time. Assume that V is the variable that expresses the value of the option with respect to the security index S and the time t. To describe this dynamics, a modified form of the Black–Scholes PDE is used, that is

$$\frac{\partial V(t,S)}{\partial t} + a(S,t)\frac{\partial V(t,S)}{\partial S} + \frac{1}{2}(b(S,t))^2\frac{\partial^2 V(t,S)}{\partial S^2} = 0 \tag{6.14}$$

for $t \in (0, T)$ and $S \in [0, \infty)$, with terminal condition $V(T, S) = H(S) = max(S_T - K, 0)$. Initial and boundary conditions can be also formulated as follows: $V(S, 0)$ for $S{\in}[0, S_{max}]$, $V(0, t) = 0$ for $t \in [0, T]$ and $V(S_{max}, T) = S_{max} - Ke^{-\int_{T-t}^{T} r(s)ds}$. The maximum value of the security index S_{max} is usually taken to be $S_{max} = 4K$ where K is the strike price. Variable $r(s)$ stands for the interest rate.

6.3 Estimation of Nonlinear Diffusion Dynamics

6.3.1 Filtering in Distributed Parameter Systems

Filtering in distributed parameter systems, such as the Black–Scholes PDE is such a case is a complicated problem that has been little explored up to now. A reason for this is that state estimation methods used for residual generation in distributed parameter systems and in infinite dimensional systems described by partial differential equations are much more complicated than state estimation methods for lumped parameter systems [98, 232, 278, 281, 289]. Of particular interest is state estimation of diffusion-type and wave-type nonlinear phenomena, appearing in several engineering applications [47, 63, 94, 95, 106].

The following nonlinear diffusion equation is considered

$$\frac{\partial \phi}{\partial t} = K\frac{\partial^2 \phi}{\partial x^2} + f(\phi) \tag{6.15}$$

Using the approximation for the partial derivative

$$\frac{\partial^2 \phi}{\partial x^2} \simeq \frac{\phi_{i+1} - 2\phi_i + \phi_{i-1}}{\Delta x^2} \tag{6.16}$$

and considering spatial measurements of variable ϕ along axis x at points $x_0 + i\Delta x,\ i = 1, 2, \cdots, N$ one has

$$\frac{\partial \phi_i}{\partial t} = \frac{K}{\Delta x^2}\phi_{i+1} - \frac{2K}{\Delta x^2}\phi_i + \frac{K}{\Delta x^2}\phi_{i-1} + f(\phi_i) \tag{6.17}$$

By taking the associated samples of ϕ given by $\phi_0, \phi_1, \cdots, \phi_N, \phi_{N+1}$ one has

$$\begin{aligned}
\frac{\partial \phi_1}{\partial t} &= \frac{K}{\Delta x^2}\phi_2 - \frac{2K}{\Delta x^2}\phi_1 + \frac{K}{\Delta x^2}\phi_0 + f(\phi_1) \\
\frac{\partial \phi_2}{\partial t} &= \frac{K}{\Delta x^2}\phi_3 - \frac{2K}{\Delta x^2}\phi_2 + \frac{K}{\Delta x^2}\phi_1 + f(\phi_2) \\
\frac{\partial \phi_3}{\partial t} &= \frac{K}{\Delta x^2}\phi_4 - \frac{2K}{\Delta x^2}\phi_3 + \frac{K}{\Delta x^2}\phi_2 + f(\phi_3) \\
&\cdots \\
\frac{\partial \phi_{N-1}}{\partial t} &= \frac{K}{\Delta x^2}\phi_N - \frac{2K}{\Delta x^2}\phi_{N-1} + \frac{K}{\Delta x^2}\phi_{N-2} + f(\phi_{N-1}) \\
\frac{\partial \phi_N}{\partial t} &= \frac{K}{\Delta x^2}\phi_{N+1} - \frac{2K}{\Delta x^2}\phi_N + \frac{K}{\Delta x^2}\phi_{N-1} + f(\phi_N)
\end{aligned} \tag{6.18}$$

By defining the following state vector

$$x^T = (\phi_1, \phi_2, \cdots, \phi_N) \tag{6.19}$$

one obtains the following state-space description

$$\begin{aligned}
\dot{x}_1 &= \tfrac{K}{\Delta x^2}x_2 - \tfrac{2K}{\Delta x^2}x_1 + \tfrac{K}{\Delta x^2}\phi_0 + f(x_1)\\
\dot{x}_2 &= \tfrac{K}{\Delta x^2}x_3 - \tfrac{2K}{\Delta x^2}x_2 + \tfrac{K}{\Delta x^2}x_1 + f(x_2)\\
\dot{x}_3 &= \tfrac{K}{\Delta x^2}x_4 - \tfrac{2K}{\Delta x^2}x_3 + \tfrac{K}{\Delta x^2}x_2 + f(x_3)\\
&\cdots\\
\dot{x}_{N-1} &= \tfrac{K}{\Delta x^2}x_N - \tfrac{2K}{\Delta x^2}x_{N-1} + \tfrac{K}{\Delta x^2}x_{N-2} + f(x_{N-1})\\
\dot{x}_N &= \tfrac{K}{\Delta x^2}\phi_{N+1} - \tfrac{2K}{\Delta x^2}x_N + \tfrac{K}{\Delta x^2}x_{N-1} + f(x_N)
\end{aligned} \tag{6.20}$$

Next, the following state variables are defined

$$y_{1,i} = x_i \tag{6.21}$$

and the state-space description of the system becomes as follows

$$\begin{aligned}
\dot{y}_{1,1} &= \tfrac{K}{\Delta x^2}y_{1,2} - \tfrac{2K}{\Delta x^2}y_{1,1} + \tfrac{K}{\Delta x^2}\phi_0 + f(y_{1,1})\\
\dot{y}_{1,2} &= \tfrac{K}{\Delta x^2}y_{1,3} - \tfrac{2K}{\Delta x^2}y_{1,2} + \tfrac{K}{\Delta x^2}y_{1,1} + f(y_{1,2})\\
\dot{y}_{1,3} &= \tfrac{K}{\Delta x^2}y_{1,4} - \tfrac{2K}{\Delta x^2}y_{1,3} + \tfrac{K}{\Delta x^2}y_{1,2} + f(y_{1,3})\\
&\cdots\\
&\cdots\\
\dot{y}_{1,N-1} &= \tfrac{K}{\Delta x^2}y_{1,N} - \tfrac{2K}{\Delta x^2}y_{1,N-1} + \tfrac{K}{\Delta x^2}y_{1,N-2} + f(y_{1,N-1})\\
\dot{y}_{1,N} &= \tfrac{K}{\Delta x^2}\phi_{N+1} - \tfrac{2K}{\Delta x^2}y_{1,N} + \tfrac{K}{\Delta x}y_{1,N-1} + f(y_{1,N})
\end{aligned} \tag{6.22}$$

The dynamical system described in Eq. (6.22) is a differentially flat one with flat output defined as the vector $\tilde{y} = [y_{1,1}, y_{1,2}, \cdots, y_{1,N}]$. Indeed all state variables can be written as functions of the flat output and its derivatives.

Denoting $a = \frac{K}{\Delta x^2}$ and $b = -\frac{2K}{\Delta x^2}$, the initial description of the system is rewritten as follows

$$\begin{pmatrix} \dot{y}_{1,1}\\ \dot{y}_{1,2}\\ \cdots\\ \dot{y}_{1,N-1}\\ \dot{y}_{1,N} \end{pmatrix} = \begin{pmatrix} b\,0\,a\,0\,0\,0\,0 \cdots 0\,0\,0\,0\,0\,0\\ a\,0\,b\,0\,a\,0\,0 \cdots 0\,0\,0\,0\,0\,0\\ 0\,0\,a\,0\,b\,0\,a \cdots 0\,0\,0\,0\,0\,0\\ \cdots\cdots\cdots\cdots\cdots\cdots\cdots\\ 0\,0\,0\,0\,0\,0\,0 \cdots a\,0\,b\,0\,a\,0\\ 0\,0\,0\,0\,0\,0\,0 \cdots 0\,0\,a\,0\,b\,0 \end{pmatrix} \begin{pmatrix} y_{1,1}\\ y_{1,2}\\ \cdots\\ y_{1,N-1}\\ y_{1,N} \end{pmatrix} + \\ + \begin{pmatrix} 1\,0\,0 \cdots 0\,0\\ 0\,1\,0 \cdots 0\,0\\ 0\,0\,1 \cdots 0\,0\\ \cdots\cdots\cdots\cdots\\ 0\,0\,0 \cdots 1\,0\\ 0\,0\,0 \cdots 0\,1 \end{pmatrix} \begin{pmatrix} v_1\\ v_2\\ v_3\\ \cdots\\ v_{N-1}\\ v_N \end{pmatrix} \tag{6.23}$$

The associated control inputs are defined as

$$\begin{aligned} v_1 &= \frac{K}{\Delta x^2}\phi_0 + f(y_{1,1}) \\ v_2 &= f(y_{1,2}) \\ v_3 &= f(y_{1,3}) \\ &\cdots \\ v_{N-1} &= f(y_{1,N-1}) \\ v_N &= \frac{K}{\Delta x^2}\phi_{N+1} + f(y_{1,N}) \end{aligned} \tag{6.24}$$

By selecting measurements from a subset of points x_j $j\in[1, 2, \cdots, m]$, the associated observation (measurement) equation becomes

$$\begin{pmatrix} z_1 \\ z_2 \\ \cdots \\ z_m \end{pmatrix} = \begin{pmatrix} 1\,0\,0 \cdots 0\,0 \\ \cdots \cdots \cdots \cdots \\ 0\,0\,0 \cdots 0\,1 \end{pmatrix} \begin{pmatrix} y_{1,1} \\ y_{1,2} \\ \cdots \\ y_{1,N} \end{pmatrix} \tag{6.25}$$

Thus, in matrix form one has the following state-space description of the system

$$\begin{aligned} \dot{\tilde{y}} &= A\tilde{y} + Bv \\ \tilde{z} &= C\tilde{y} \end{aligned} \tag{6.26}$$

For the linear description of the system in the form of Eq. (6.26) one can perform estimation using the standard Kalman Filter recursion. The discrete-time Kalman filter can be decomposed into two parts: (i) time update (prediction stage), and (ii) measurement update (correction stage) [10, 198],f [205]. By denoting as A_d, B_d and C_d the discrete-time equivalents of matrices A, B and C one has that

Measurement update:

$$\begin{aligned} K(k) &= P^-(k){C_d}^T[C_d{\cdot}P^-(k){C_d}^T + R]^{-1} \\ \hat{y}(k) &= \hat{y}^-(k) + K(k)[z(k) - C_d\hat{y}^-(k)] \\ P(k) &= P^-(k) - K(k)C_dP^-(k) \end{aligned} \tag{6.27}$$

Time update:

$$\begin{aligned} P^-(k+1) &= A_d(k)P(k){A_d}^T(k) + Q(k) \\ \hat{y}^-(k+1) &= A_d(k)\hat{y}(k) + B_d(k)u(k) \end{aligned} \tag{6.28}$$

6.4 State Estimation for the Black–Scholes PDE

6.4.1 Modeling in Canonical Form of the Nonlinear Black–Scholes Equation

Next, the following nonlinear Black–Scholes PDE is considered:

$$\frac{\partial V(t,S)}{\partial t} + r_t S\frac{\partial V(t,S)}{\partial S} + \frac{1}{2}\sigma_t^2 S^2\frac{\partial^2 V(t,S)}{\partial S^2} - r_t V(t,S) = 0 \tag{6.29}$$

It is considered that the volatility σ_t has a nonlinear dependence on V_{ss}, that is

$$\sigma_t = \sigma(V_{ss}) \tag{6.30}$$

where $V_{ss} = \frac{\partial^2 V(t,s)}{\partial S^2}$ and σ is a nonlinear function. Again, a grid of N points is considered, that is $\{s_1, s_2, \cdots, s_{N-1}, s_N\}$ which are placed at equal distances on the S axis. At the points of spatial discretization it holds

$$\frac{\partial V(t,s_i)}{\partial t} = r_t V(t,s_i) - r_t s_i \frac{\partial V(t,s_i)}{\partial S} - \frac{1}{2}\sigma_t^2(V_{ss}) s_i^2 \frac{\partial^2 V(t,s_i)}{\partial S^2} \tag{6.31}$$

where $i = 1, 2, \cdots, N$. The following state vector is defined

$$\begin{array}{c} \tilde{V} = [V(t,s_1), V(t,s_2), V(t,s_3), \cdots, V(t,s_{N-1}), V(t,s_N)] \text{ or} \\ \tilde{V} = [V_1, V_2, V_3, \cdots, V_{N-1}, V_N] \end{array} \tag{6.32}$$

thus one has

$$\begin{array}{c} \frac{\partial V_1}{\partial t} = r_t V_1 - r_t s_1 \frac{V_2 - V_1}{\Delta S} - \frac{1}{2}\sigma_t^2(V_{ss}) s_1^2 \frac{V_2 - 2V_1 + V_0}{\Delta S^2} \\ \frac{\partial V_2}{\partial t} = r_t V_2 - r_t s_2 \frac{V_3 - V_2}{\Delta S} - \frac{1}{2}\sigma_t^2(V_{ss}) s_2^2 \frac{V_3 - 2V_2 + V_1}{\Delta S^2} \\ \frac{\partial V_3}{\partial t} = r_t V_3 - r_t s_3 \frac{V_4 - V_3}{\Delta S} - \frac{1}{2}\sigma_t^2(V_{ss}) s_3^2 \frac{V_4 - 2V_3 + V_2}{\Delta S^2} \\ \cdots \\ \cdots \\ \frac{\partial V_{N-1}}{\partial t} = r_t V_{N-1} - r_t s_{N-1} \frac{V_N - V_{N-1}}{\Delta S} - \frac{1}{2}\sigma_t^2(V_{ss}) s_{N-1}^2 \frac{V_N - 2V_{N-1} + V_{N-2}}{\Delta S^2} \\ \frac{\partial V_N}{\partial t} = r_t V_N - r_t s_N \frac{V_{N+1} - V_N}{\Delta S} - \frac{1}{2}\sigma_t^2(V_{ss}) s_N^2 \frac{V_{N+1} - 2V_N + V_{N-1}}{\Delta S^2} \end{array} \tag{6.33}$$

Next, the following control inputs are defined

$$\begin{array}{c} u_1 = -\frac{1}{2}\sigma_t^2(V_{ss}) s_1^2 \frac{V_2 - 2V_1 + V_0}{\Delta S^2} \\ u_2 = -\frac{1}{2}\sigma_t^2(V_{ss}) s_2^2 \frac{V_3 - 2V_2 + V_1}{\Delta S^2} \\ u_3 = -\frac{1}{2}\sigma_t^2(V_{ss}) s_3^2 \frac{V_4 - 2V_3 + V_2}{\Delta S^2} \\ \cdots \\ u_{N-1} = -\frac{1}{2}\sigma_t^2(V_{ss}) s_{N-1}^2 \frac{V_N - 2V_{N-1} + V_{N-2}}{\Delta S^2} \\ u_N = -\frac{1}{2}\sigma_t^2(V_{ss}) s_N^2 \frac{V_{N+1} - 2V_N + V_{N-1}}{\Delta S^2} \end{array} \tag{6.34}$$

Therefore, the dynamics of the Black–Scholes PDE is written as

$$\dot{\tilde{V}} = \tilde{A}\tilde{V} + \tilde{B}\tilde{U} \tag{6.35}$$

where $\tilde{U} = [u_1, u_2, \cdots, u_{N-1}, u_N]^T$ and

$$\tilde{A}=\begin{pmatrix} \tilde{a}_{11} & \tilde{a}_{12} & \tilde{a}_{13} & \tilde{a}_{14} & \cdots & \tilde{a}_{1,N-2} & \tilde{a}_{1,N-1} & \tilde{a}_{1,N} \\ \tilde{a}_{21} & \tilde{a}_{22} & \tilde{a}_{23} & \tilde{a}_{24} & \cdots & \tilde{a}_{2,N-2} & \tilde{a}_{2,N-1} & \tilde{a}_{2,N} \\ \tilde{a}_{31} & \tilde{a}_{32} & \tilde{a}_{33} & \tilde{a}_{34} & \cdots & \tilde{a}_{3,N-2} & \tilde{a}_{3,N-1} & \tilde{a}_{3,N} \\ \cdots & \cdots & \cdots & \cdots & \cdots & \cdots & \cdots & \cdots \\ \tilde{a}_{N-1,1} & \tilde{a}_{N-1,2} & \tilde{a}_{N-1,3} & \tilde{a}_{N-1,4} & \cdots & \tilde{a}_{N-1,N-2} & \tilde{a}_{N-1,N-1} & \tilde{a}_{N-1,N} \\ \tilde{a}_{N,1} & \tilde{a}_{N,2} & \tilde{a}_{N,3} & \tilde{a}_{N,4} & \cdots & \tilde{a}_{N,N-2} & \tilde{a}_{N,N-1} & \tilde{a}_{N,N} \end{pmatrix} \tag{6.36}$$

where
$\tilde{a}_{11}=r_t+r_t\frac{s_1}{\Delta s}$, $\tilde{a}_{12}=-r_t\frac{s_1}{\Delta s}$, $\tilde{a}_{13}=0$, $\tilde{a}_{14}=0$, $\cdots$, $\tilde{a}_{1,N-2}=0$, $\tilde{a}_{1,N-1}=0$, $\tilde{a}_{1,N}=0$.

$\tilde{a}_{21}=0$, $\tilde{a}_{22}=r_t+r_t\frac{s_1}{\Delta s}$, $\tilde{a}_{23}=-r_t\frac{s_1}{\Delta s}$, $\tilde{a}_{24}=0$, $\cdots$, $\tilde{a}_{2,N-2}=0$, $\tilde{a}_{2,N-1}=0$, $\tilde{a}_{2,N}=0$.

$\tilde{a}_{31}=0$, $\tilde{a}_{32}=0$, $\tilde{a}_{33}=r_t+r_t\frac{s_1}{\Delta s}$, $\tilde{a}_{34}=-r_t\frac{s_1}{\Delta s}$, $\cdots$, $\tilde{a}_{3,N-2}=0$, $\tilde{a}_{3,N-1}=0$, $\tilde{a}_{3,N}=0$.

$\cdots$

$\tilde{a}_{N-1,1}=0$, $\tilde{a}_{N-1,2}=0$, $\tilde{a}_{N-1,3}=0$, $\tilde{a}_{N-1,4}=0$, $\cdots$, $\tilde{a}_{N-1,N-2}=0$, $\tilde{a}_{N-1,N-1}=r_t+r_t\frac{s_{N-1}}{\Delta s}$, $\tilde{a}_{N-1,N}=-r_t\frac{s_{N-1}}{\Delta s}$.

$\tilde{a}_{N,1}=0$, $\tilde{a}_{N,2}=0$, $\tilde{a}_{N,3}=0$, $\tilde{a}_{N,4}=0$, $\cdots$, $\tilde{a}_{N,N-2}=0$, $\tilde{a}_{N,N-1}=0$, $\tilde{a}_{N,N}=r_t+r_t\frac{s_{N-1}}{\Delta s}$.

Matrix $\tilde{B}$ is defined as

$$\tilde{B}=\begin{pmatrix} 1 & 0 & 0 & 0 & \cdots & 0 & 0 & 0 \\ 0 & 1 & 0 & 0 & \cdots & 0 & 0 & 0 \\ 0 & 0 & 1 & 0 & \cdots & 0 & 0 & 0 \\ \cdots & \cdots & \cdots & \cdots & \cdots & \cdots & \cdots & \cdots \\ 0 & 0 & 0 & 0 & \cdots & 0 & 1 & 0 \\ 0 & 0 & 0 & 0 & \cdots & 0 & 0 & 1 \end{pmatrix} \tag{6.37}$$

The measurement (observation) matrix $\tilde{C}$ is taken to be the identity matrix. Therefore, one obtains again a description of the Black–Scholes PDE in the linear canonical form. For the state-space model of the Black–Scholes nonlinear PDE given in Eq. (6.35) one can perform estimation using the Derivative-free nonlinear Kalman Filter.

The system of Eq. (6.33) is actually a differentially flat one, with the elements of the state vector $\tilde{V}=[V_1, V_2, V_3, \cdots, V_{N-1}, V_N]^T$ to stand for a flat output. Actually, the auxiliary control inputs which have been defined in Eq. (6.34) can be expressed as function of the state vector elements, and this means that all variables of the state-space description of the Black–Scholes PDE are finally written as functions of the flat output and its derivatives. Therefore, the Black–Scholes PDE model is shown to be a differentially flat one.

It has been proven that nonlinear differential flat models can be finally transformed into the linear canonical form. As explained in Sect. 4.3, the necessary condition for succeeding this is the nonlinear model to be differentially flat [34, 216]. Systems that admit static feedback linearization are differentially flat and can be transformed to the canonical form. This was also confirmed in the case of the Black–Scholes PDE model and finally the state-space description of Eq. (6.35) stands for a linear canonical form.

6.4.2 State Estimation with the Derivative-Free Nonlinear Kalman Filter

As mentioned above, for the system of Eq. (6.35), state estimation is possible by applying the standard Kalman Filter. The system is first turned into discrete-time form using common discretization methods and then the recursion of the linear Kalman Filter is applied.

To implement Kalman Filtering for the state-space model of the Black–SCholes PDE one has to substitute the previously defined matrices $\tilde{A}$, $\tilde{B}$ and $\tilde{C}$ by their discrete-time equivalents $\tilde{A}_d$, $\tilde{B}_d$ and $\tilde{C}_d$. Matrices A_d, B_d and C_d can be computed using established discretization methods. Moreover, the covariance matrices $P(k)$ and $P^{-}(k)$ are the ones obtained from the linear Kalman Filter update equations.

For the linear description of the system one can perform estimation using the standard Kalman Filter recursion. The discrete-time Kalman filter can be decomposed into two parts: (i) time update (prediction stage), and (ii) measurement update (correction stage).

Measurement update:

$$\begin{aligned}
K(k) &= P^{-}(k)\tilde{C}_d^T[\tilde{C}_d \cdot P^{-}(k)\tilde{C}_d^T + R]^{-1} \\
\hat{V}(k) &= \hat{V}^{-}(k) + K(k)[\tilde{C}_d V(k) - \tilde{C}_d \hat{V}^{-}(k)] \\
P(k) &= P^{-}(k) - K(k)\tilde{C}_d P^{-}(k)
\end{aligned} \tag{6.38}$$

Time update:

$$\begin{aligned}
P^{-}(k+1) &= \tilde{A}_d(k)P(k)\tilde{A}_D^T(k) + Q(k) \\
\hat{V}^{-}(k+1) &= \tilde{A}_d(k)\hat{V}(k) + \tilde{B}_d(k)\tilde{U}(k)
\end{aligned} \tag{6.39}$$

6.4.3 Consistency Checking of the Option Pricing Model

The residuals' sequence (differences between the real option values and the ones estimated by the Kalman Filter) is a discrete error process e_k with dimension $m \times 1$ (here $m = N$ and N is the number of grid points). Actually, it is a zero-mean Gaussian white-noise process with covariance given by E_k. A conclusion can be stated based on a measure of certainty that the Black–Scholes PDE model, with the specific choice of parameter values, is accurate.

To this end, the following *normalized error square* (NES) is defined [212]

$$\varepsilon_k = e_k^T E_k^{-1} e_k \tag{6.40}$$

The normalized error square follows a χ^2 distribution. An appropriate test for the normalized error sum is to numerically show that the following condition is met within a level of confidence (according to the properties of the χ^2 distribution)

$$E\{\varepsilon_k\} = m \tag{6.41}$$

This can be succeeded using statistical hypothesis testing, which is associated with confidence intervals. A 95% confidence interval is frequently applied, which is specified using $100(1-a)$ with $a = 0.05$. Actually, a two-sided probability region is considered cutting-off two end tails of 2.5% each. For M runs the normalized error square that is obtained is given by

$$\bar{\varepsilon}_k = \frac{1}{M}\sum_{i=1}^{M} \varepsilon_k(i) = \frac{1}{M}\sum_{i=1}^{M} e_k^T(i) E_k^{-1}(i) e_k(i) \tag{6.42}$$

where ε_i stands for the i-th run at time t_k. Then $M\bar{\varepsilon}_k$ will follow a χ^2 density with Mm degrees of freedom. This condition can be checked using a χ^2 test. The hypothesis holds true if the following condition is satisfied

$$\bar{\varepsilon}_k \in [\zeta_1, \zeta_2] \tag{6.43}$$

where ζ_1 and ζ_2 are derived from the tail probabilities of the χ^2 density. For example, for $m = 20$ and $M = 100$ one has $\chi^2_{Mm}(0.025) = 1878$ and $\chi^2_{Mm}(0.975) = 2126$. Using that $M = 100$ one obtains $\zeta_1 = \chi^2_{Mm}(0.025)/M = 18.78$ and $\zeta_2 = \chi^2_{Mm}(0.975)/M = 21.26$.

The probability density functions of the χ^2 distribution for different degrees of freedom are depicted in Fig. 6.1.

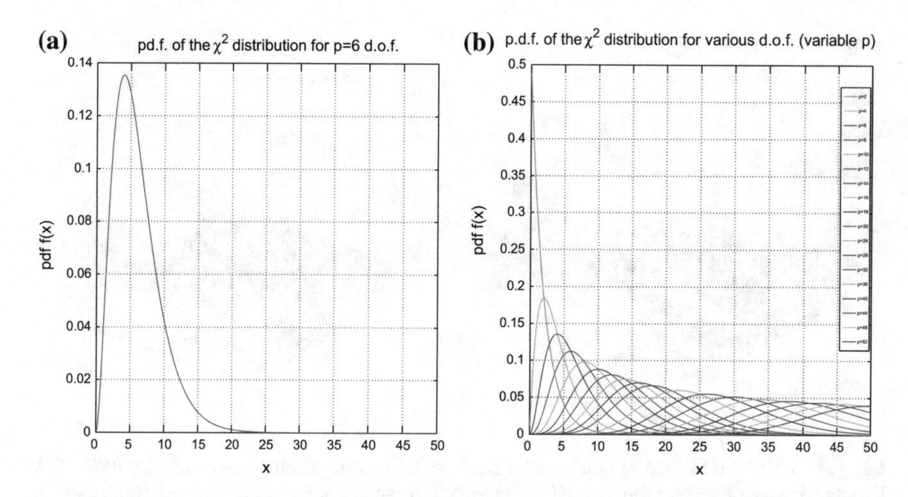

Fig. 6.1 **a** Probability density function of the χ^2 distribution for $p = 6$ degrees of freedom, **b** Probability density function of the distribution for several values of the degrees of freedom (variable p)

6.5 Simulation Tests

6.5.1 Estimation with the Use of an Accurate Black–Scholes Model

Simulation results are provided about the estimation of the value of the option $\hat{V}(t, S)$ as a function of time t and security index S. Indicative values for the model's parameters were: interest were $r_t = 0.02$ and volatility $s_t = 0.05$. As it can be seen in Fig. 6.2 the estimate provided by the Derivative-free nonlinear Kalman Filter is very accurate (although there is no knowledge about initial conditions $\tilde{V}(0, S)$). Moreover, results about the accuracy of estimation in particular points s_i, $i = 1, 2, \cdots, N$ of the spatial grid corresponding to the security index S are given. As it can be observed in Fig. 6.3a and Fig. 6.3b the estimates provided by the Derivative-free nonlinear Kalman Filter converge to the real values $V(t, s_i)$.

6.5.2 Detection of Mispricing in the Black–Scholes Model

As mentioned above, an issue that arises in the use of the Black–Scholes model is to check its accuracy. Frequently, the real dynamics of the option's value change, since there can be changes in parameters such as the interest rate or the volatility. In the latter case the Black–Scholes model loses its efficiency in predicting the variations

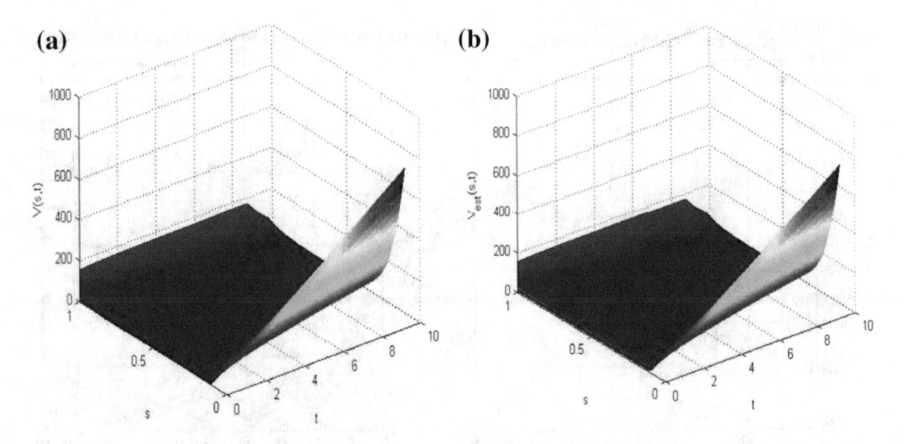

Fig. 6.2 **a** Variation of the option's cost $V(t, S)$ as a function of time t and security index S, **b** Estimated value of the cost function $\hat{V}(s, t)$ provided by the Derivative-free nonlinear Kalman Filter

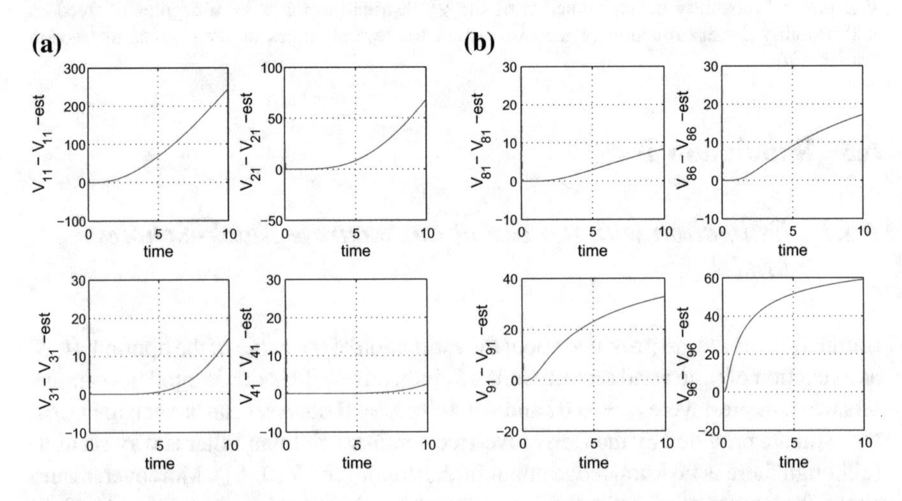

Fig. 6.3 Variation of the option's cost $V(t, S)$ at specific grid points in the case of an accurate Black–Scholes model: real value (*red continuous line*) versus estimated value (*dashed blue line*)

of the financial system. Thus, there is need to find-out if the Black–Scholes model remains reliable or if its efficiency is lost and retuning is required.

In the simulation experiments that follow it is assumed that the volatility σ_t used by the Derivative-free nonlinear Kalman Filter, is not the one of the real financial system. The associated results are presented in Fig. 6.4. It can be noticed, that in case of parametric changes and deviation from the nominal parameter values the residuals' sequence (differences between the real option prices and the ones predicted by the Derivative-free nonlinear Kalman Filter) has clearly a non-zero mean value. A more advanced processing of the residuals sequence can be based on elaborated

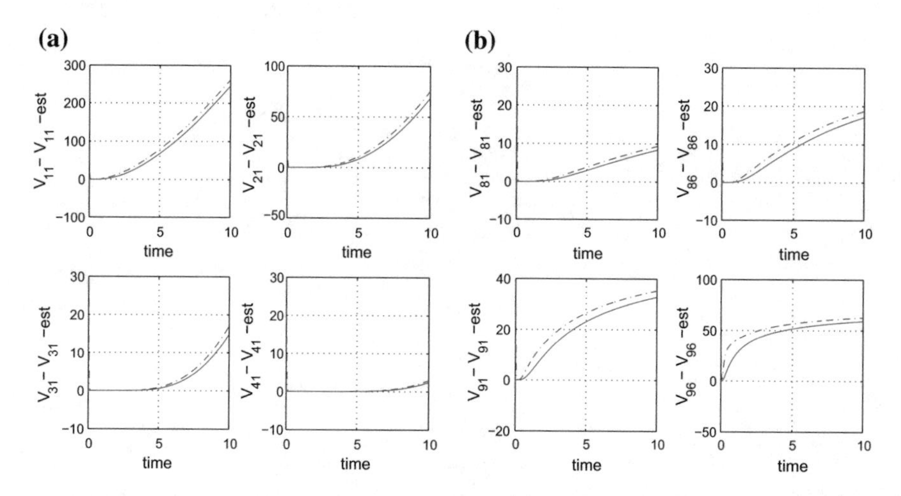

Fig. 6.4 Variation of the option's cost $V(t, S)$ at specific grid points in the case of an inconsistent Black–Scholes model: real value (*red continuous line*) versus estimated value (*dashed blue line*)

χ^2 distribution change detection tests such as the *local statistical approach to fault diagnosis* [10, 205, 212, 290].

The present chapter develops a method for statistical validation of forecasting and risk assessment tools used in finance. The Black–Scholes PDE can model the variation of options price provided that it is precisely parameterized. However, such a model can often get invalidated because market's condition changes and in such a case its parameters (such as interest rate and volatility) no longer reflect the actual dynamics of the monitored financial system. Consequently, there is need for frequent validation of the model and update of the values of its parameters. The proposed Kalman Filtering-based approach and the associated χ^2 distribution-based decision criterion enable to complete this task within specific confidence intervals. Validation methods for models used in financial engineering have not been sufficiently developed. Moreover, validation of models of distributed parameter systems, such as the Black–Scholes PDE appears to be complicated and hardly any results exist about it. By exploiting results from systems and estimation theory the chapter arrives at a systematic and complete solution for such problems [23, 24, 57, 141].

It is also noted that the Kalman Filter is a real-time sequential estimation method and thus it is computationally fast. Unlike to estimation methods processing data in batch mode in Kalman Filtering the estimate of the state vector of the monitored system is obtained at each iteration of the algorithm. As in the case of recursive least squares suitable fast convergence properties also hold for the Kalman Filter. About the accuracy of estimation, the Kalman Filter is known to be an optimal estimator (provided that the measurement noise is white and Gaussian), which means it provides state vector estimates of minimum error variance. According to the above, the Kalman Filter exhibits significant advantages when used in estimation problems met in financial engineering [60, 88].

Chapter 7
Kalman Filtering Approach to the Detection of Option Mispricing in Elaborated PDE Finance Models

7.1 Outline

In the previous chapter, the Kalman Filter has been used for solving the state estimation problem in PDE models of finance, such as the Black–Scholes PDE. In the present chapter the use of this method is extended to elaborated finance PDE models such as those described by partial integrodifferential equations. The method can be used for the detection of option mispricing in electric power markets.

Certain features of electricity prices distinguish electricity markets from other financial markets, and to a lesser degree, from most other commodity markets. A first key point is that spot pricing methods are proven to be inadequate. The spot price of electricity is determined by the intersection by the aggregate demand and supply functions for the electric power and an outage of a major power plant or a sudden surge of demand due to extreme natural conditions can cause the spot price to rise to extremely high levels, with negative consequences in terms of losses for the participants in the electricity market. Actually, the non-storability of electricity, together with real-time matching mechanism of supply and demand in spot pricing lead to disturbances that make spot pricing be inadequate [25, 44, 99, 123, 174]. The need for transition to option pricing models with more secure trading mechanisms has generated a new research field. Another key point is that models for electric power systems can be negatively affected by (i) high and varying volatility (fluctuations of the prices of the electricity market options and derivatives), (ii) seasonality, (iii) abrupt variations known as jumps which are due to disruptions in transmission and generation outages [30, 37, 176].

The Black–Scholes model of option pricing is considered to be a major tool in financial engineering since it explains the changes of the options value as a function of time and of an underlying pricing index (security index). The Black–Scholes model is actually a diffusion partial differential equation and its numerical solution gives a precise view of how the value of options varies [102, 189]. In case of electric power markets these models are complemented with integral terms which describe

G.G. Rigatos, *State-Space Approaches for Modelling and Control in Financial Engineering*, Intelligent Systems Reference Library 125,
DOI 10.1007/978-3-319-52866-3_7

the effects of jumps and changes in the diffusion process and which are associated with the variations in the production rates, with the condition of the transmission and distribution system, with the pay-off capability, etc. Considering the latter case, that is a partial integrodifferential equation for the option's price, a new filtering method, is developed for estimating option prices variations without knowledge of initial conditions. The proposed filtering method is the so-called Derivative-free nonlinear Kalman Filter and is based on a transformation of the initial option price dynamics into a state-space model of the linear canonical form. The transformation is shown to be in accordance to differential flatness theory [34, 77, 134, 169]. The method finally provides a model of the option price dynamics for which state estimation is possible by applying the standard Kalman Filter recursion [216, 222, 225, 228].

The application of the proposed filtering method is computationally efficient and fast. Through simulation experiments it is confirmed that the Derivative-free nonlinear Kalman Filter approximated with accuracy the two-dimensional model of the Black–Scholes partial integrodifferential equation and consequently enables to estimate successfully variations of the energy market option prices (despite no knowledge of initial conditions). Based on these estimates, validation of the Black–Scholes partial integrodifferential model can be performed. Thus one can find out if the parameters appearing in the model, such as interest rate and volatility are precise enough, to capture the real dynamics of the energy market pricing system [58, 74, 137]. If it is observed that the accuracy of the model is lost a retuning of its parameters can be carried out.

7.2 Option Pricing in the Energy Market

7.2.1 Energy Market and Swing Options

Option price modelling with the use of the Black–Scholes PDE has been analyzed in Sect. 6.2. In the present chapter, options price modelling with the use of partial integrodifferential equations (PIDE) will be presented. In the last years the role of the electricity market has significantly changed, now allowing trading between generators, retailers and other financial intermediaries. Wholesale transactions in the physical commodity are typically settled by a supervisory authority which is the grid operator. In most European countries, a power exchange is now based on a voluntary marketplace. A power exchange provides a spot market (mainly day-ahead) for transactions in the electricity grid to be carried out on the next day. Participants submit their purchase or sale orders electronically. The supply and demand are aggregated and compared in the power exchange to decide the market price for each hour of the following day.

Apart from spot pricing in the electricity market, options and derivatives pricing have been established as a mean to manage more efficiently and profitably the financial risk associated with electricity price volatility (which in turn can be caused by variations in the production rates, with the condition of the transmission and distribution system, pay-off capability etc.) [25, 44, 99, 123, 174]. Derivatives contracts traded in power exchanges include the so-called forward contracts, futures, swaps and options. These new pricing schemes need to be supported by efficient computational forecasting and accuracy assessment tools.

As already explained, the deregulation in the energy markets and at the same time the need for avoidance of extreme price fluctuations has led to the establishment of more elaborated pricing methods in the electric power trade. Thus in upgrade to spot pricing methods, more advanced trading schemes based on the use of financial derivatives and options have emerged. These schemes aim at assuring profits for both the consumer and the producer during the transaction of electric power trading. Swing options are one type of the contracts these trading schemes offer with respect to the time and amount of energy delivered. Such agreements offer flexibility with respect to limitations in the amount of electric energy that can be delivered at a specific time period, production costs and pay-off conditions [30, 37, 176].

A swing option is an agreement to buy or to sell electric energy, according to a previously agreed strike price K, at different times $(t_1, t_2, \ldots, t_{N_s})$ during the contract period $[0, T]$. This approach gives flexibility to the quantity of electric energy to be purchased or delivered. A minimum amount of q_{min} and a maximum amount q_{max} of energy is delivered for each swing action time, while for the whole contract period the overall minimum and maximum amount of energy are denoted as Q_{min} and Q_{max} respectively. The term "swing" comes from the flexibility in the amount of electric power that can be delivered at the intermediate time instances, since the purchaser swings between the lower and upper consumption boundaries. The flexibility in delivery is often constrained by introducing penalty payments, and for this reason swing options are also called take-or-pay options. Such penalties become effective if the overall amount of energy purchased in the contract's period $[0, T]$ exceeds the predefined quantity in the contract [25, 44, 99, 123, 174].

Another characteristic of swing options is the refraction (or recovery) time δ_R which separates two different exercise rights and avoids exercising all the rights at the same time. In case that there is no refraction time set in the contract, the problem of pricing swings options is simplified. Two particular cases are: (i) one swing right which means equivalence to the problem of an American option, (ii) full swing which means that the number or rights is equal to the number of exercise dates, then the value of the swing option is equivalent to a strip of European options expiring at the exercise dates $t_1, t_2, \ldots, t_N$.

7.2.2 Energy Options Pricing Models

Modeling tools for pricing in energy markets are based on extended forms of the Black–Scholes partial differential equation. In the Black–Scholes PDE the variation of the underlying security index is according to the stochastic differential equation [174]

$$dS = \mu S dt + \sigma S dX \tag{7.1}$$

To apply the Black–Scholes pricing model

$$\frac{\partial V}{\partial t} + rS\frac{\partial V}{\partial S} + \frac{\sigma^2 S^2}{2}\frac{\partial^2 V}{\partial S^2} = rV \tag{7.2}$$

in the case of swing options met in the energy market, the initial partial differential equation has to be transformed into a Partial Integrodifferential Equation (r is the interest rate, δ is the dividend payment, σ is the volatility). The previous SDE given in Eq. (7.1) in the case of option market swing options becomes

$$dS = (\mu - \delta)S dt + \sigma S dX \tag{7.3}$$

and the corresponding PDE becomes

$$\frac{\partial V}{\partial t} + (r - \delta)S\frac{\partial V}{\partial S} + \frac{\sigma^2 S^2}{2}\frac{\partial^2 V}{\partial S^2} = rV \tag{7.4}$$

The stock price does not have only continuous changes but exhibits also jumps. In such a case the assumption about a constant volatility σ is proven to be inefficient. For modeling spikes in the price's variation the following SDE of the underlying asset has been proposed [174]

$$dS = (\mu - \delta)S dt + \sigma S dX + (q - 1)S dQ \tag{7.5}$$

where dQ indicates a Poisson process, of intensity λ with

$$\begin{array}{c} P[dQ = 0] = 1 - \lambda dt \\ P[dQ = 1] = \lambda dt \end{array} \tag{7.6}$$

and $q \in [0, \infty]$ is a random variable that is log-normally distributed with the p.d.f

$$\Gamma_{\gamma,\mu_j}(q) = \frac{1}{\sqrt{2\pi}\gamma} exp\left(-\frac{1}{2}\frac{((log(q) - \mu_j)}{\gamma}\right)^2 \tag{7.7}$$

satisfying also

$$\int_0^\infty \Gamma_\gamma(q) dq = 1 \; \forall q \in [0, \infty] \tag{7.8}$$

The term $q - 1$ is called *impulse function* which produces a jump from S to Sq while it holds $E(q - 1) = \bar{k}$. If $0 \leq q < 1$ the negative impulse function causes the spot price to decrease. If $q = 1$, the spot price remains the same. If $q > 1$ the impulse function causes the spot price to increase. In such a case the option price dynamics can be described by the partial integrodifferential equation [174]

$$\frac{\partial V}{\partial t} + \frac{\sigma^2 S^2}{2}\frac{\partial^2 V}{\partial S^2} + (r - \delta - \lambda\bar{k})S\frac{\partial V}{\partial S} - rV + \lambda\int_0^{\infty}(V(qS,t) - V(S,t))\Gamma_{\gamma}(q)dq = 0 \quad (7.9)$$

Next, the following change of variable is performed [174]

$$\tilde{V}(x,\tau) = V(e^x, T-\tau) = V(S,t) \quad (7.10)$$

$$\frac{\partial \tilde{V}}{\partial \tau} = -\frac{\partial V}{\partial t}\frac{\partial V}{\partial S} = \frac{1}{S}\frac{\partial \tilde{V}}{\partial x}\frac{\partial^2 V}{\partial S^2} = \frac{1}{S^2}\frac{\partial^2 \tilde{V}}{\partial x^2} \quad (7.11)$$

with $q \in [0,\infty]$, $S{\geq}0{\rightarrow}Sq$ and $ln(q), x \in R$. Following this change of variables the partial integrodifferential equation is written as

$$\frac{\partial \tilde{V}}{\partial \tau} = \frac{\sigma^2}{2}\frac{\partial^2 \tilde{V}}{\partial x^2} + (r - \delta - \lambda\bar{k})\frac{\partial \tilde{V}}{\partial x} - r\tilde{V} + \lambda(\int_{-\infty}^{\infty}\tilde{V}(z+x,\tau)\Gamma_{\gamma}(z)dz - \tilde{V}) \quad (7.12)$$

there the associated initial condition is $\tilde{V}(x,0)$. The partial integrodifferential equation can be also written as [174]:

$$\frac{\partial \tilde{V}}{\partial \tau} = \frac{\sigma^2}{2}\frac{\partial^2 \tilde{V}}{\partial x^2} + (r - \delta - \lambda\bar{k})\frac{\partial \tilde{V}}{\partial x} - (r+\lambda)\tilde{V} + \lambda(\int_{-\infty}^{\infty}\tilde{V}(z+x,\tau)\Gamma_{\gamma}(z)dz) \quad (7.13)$$

where the following coefficients are denoted $c = \frac{\sigma^2}{2}$, $a = r - \delta - \lambda\bar{k}$, $b = r + \lambda$ and about the integral term one has $I\tilde{V} = \lambda(\int_{-\infty}^{\infty}\tilde{V}(z+x,\tau)\Gamma_{\gamma}(z)dz - \tilde{V})$. To proceed to the discretization of the option price dynamic model the following operator is defined

$$L = a\frac{\partial}{\partial x} - b + c\frac{\partial^2}{\partial x^2} \quad (7.14)$$

Thus, the dynamics of the electric power pricing model can be written in the form

$$\frac{\partial \tilde{V}}{\partial \tau} = L\tilde{V} + I{*}\tilde{V} \quad (7.15)$$

7.3 Validation of the Energy Options Pricing Model

7.3.1 *State Estimation with the Derivative-Free Nonlinear Kalman Filter*

Next, state estimation for the dynamic model of the options' price will be studied. Previous approaches to the problem of state estimation for distributed parameter systems, that is for systems described by partial differential equations, can be found in [47, 63, 95, 106, 232, 278]. The problem treated here is more complicated because the dynamic model of the options' price is no longer described by Black–Scholes

PDE, but it also takes into account jump diffusions which finally lead to the form of the partial integrodifferential equation described in Eq. (7.12).

The following computations of partial derivatives are used again:

$$\frac{\partial \tilde{V}(t,x_i)}{\partial x} = \frac{\tilde{V}(t,x_{i+1}) - \tilde{V}(t,x_i)}{\Delta x} \tag{7.16}$$

$$\frac{\partial^2 \tilde{V}(t,x_i)}{\partial x^2} = \frac{\tilde{V}(t,x_{i+1}) - 2\tilde{V}(t,x_i) + \tilde{V}(t,x_{i-1})}{\Delta x^2} \tag{7.17}$$

The partial integrodifferential equation that describes the option price dynamics is

$$\frac{\partial \tilde{V}}{\partial \tau} = \frac{\sigma^2}{2}\frac{\partial^2 \tilde{V}}{\partial x^2} + (r - \delta - \lambda\bar{k})\frac{\partial \tilde{V}}{\partial x} - (r+\lambda)\tilde{V} + \lambda\int_{\infty}^{+\infty}\tilde{V}(z+x,\tau)\Gamma_{\gamma}(z)dz \tag{7.18}$$

Substituting Eqs. (7.16) and (7.17) into Eq. (7.18) one has

$$\begin{aligned} \frac{\partial \tilde{V}}{\partial \tau}(\tau,x_i) = \frac{\sigma_2}{2}[\frac{\tilde{V}(\tau,x_{i+1}) - 2\tilde{V}(\tau,x_i) + \tilde{V}(\tau,x_{i+1})}{\Delta x^2}]+ \\ +(r-\delta-\lambda\bar{k})[\frac{\tilde{V}(\tau,x_{i+1}) - \tilde{V}(\tau,x_i)}{\Delta x}] - (r+\lambda\tilde{V}(\tau,x_i)) \\ +\lambda\int_{-\infty}^{+\infty}\tilde{V}(z+x_i,\tau)\Gamma_{\gamma}(z)dz \end{aligned} \tag{7.19}$$

and denoting the integral

$$\tilde{I}{*}\tilde{V}(\tau,x_i) = \lambda\int_{-\infty}^{+\infty}\tilde{V}(z+x_i,\tau)\Gamma_{\gamma}(z)dz \tag{7.20}$$

the previous relation is rewritten as

$$\begin{aligned} \frac{\partial \tilde{V}}{\partial \tau}(\tau,x_i) = \frac{\sigma_2}{2}[\frac{\tilde{V}(\tau,x_{i+1}) - 2\tilde{V}(\tau,x_i) + \tilde{V}(\tau,x_{i+1})}{\Delta x^2}]+ \\ +(r-\delta-\lambda\bar{k})[\frac{\tilde{V}(\tau,x_{i+1}) - \tilde{V}(\tau,x_i)}{\Delta x}] - (r+\lambda\tilde{V}(\tau,x_i)) \\ +\lambda\tilde{I}{*}\tilde{V}(\tau,x_i) \end{aligned} \tag{7.21}$$

Considering a grid partitioning $[x_0, x_1, x_2, x_3, \ldots, x_{N-1}, x_N, x_{N+1}]$ along the x-axis in equidistant points with boundary conditions $\tilde{V}(\tau, x_0)$ and $\tilde{V}(\tau, x_{N+1})$ one has

$$\frac{\partial\tilde{V}}{\partial\tau}(\tau,x_1)=\left[\frac{\sigma^2}{2\Delta x^2}+(r-\delta-\lambda\tilde{k})\right]\tilde{V}(\tau,x_2)+\left[-\frac{\sigma^2}{\Delta x^2}-(r-\delta-\lambda\tilde{k})\right]\tilde{V}(\tau,x_1)+$$
$$+\frac{\sigma^2}{2\Delta x^2}\tilde{V}(\tau,x_0)+\lambda\tilde{I}*\tilde{V}(\tau,x_1)$$

$$\frac{\partial\tilde{V}}{\partial\tau}(\tau,x_2)=\left[\frac{\sigma^2}{2\Delta x^2}+(r-\delta-\lambda\tilde{k})\right]\tilde{V}(\tau,x_3)+\left[-\frac{\sigma^2}{\Delta x^2}-(r-\delta-\lambda\tilde{k})\right]\tilde{V}(\tau,x_2)+$$
$$+\frac{\sigma^2}{2\Delta x^2}\tilde{V}(\tau,x_1)+\lambda\tilde{I}*\tilde{V}(\tau,x_2)$$

$$\frac{\partial\tilde{V}}{\partial\tau}(\tau,x_3)=\left[\frac{\sigma^2}{2\Delta x^2}+(r-\delta-\lambda\tilde{k})\right]\tilde{V}(\tau,x_4)+\left[-\frac{\sigma^2}{\Delta x^2}-(r-\delta-\lambda\tilde{k})\right]\tilde{V}(\tau,x_3)+$$
$$+\frac{\sigma^2}{2\Delta x^2}\tilde{V}(\tau,x_2)+\lambda\tilde{I}*\tilde{V}(\tau,x_3)$$
$$\cdots$$
$$\frac{\partial\tilde{V}}{\partial\tau}(\tau,x_{N-1})=\left[\frac{\sigma^2}{2\Delta x^2}+(r-\delta-\lambda\tilde{k})\right]\tilde{V}(\tau,x_N)+\left[-\frac{\sigma^2}{\Delta x^2}-(r-\delta-\lambda\tilde{k})\right]\tilde{V}(\tau,x_{N-1})+$$
$$+\frac{\sigma^2}{2\Delta x^2}\tilde{V}(\tau,x_{N-2})+\lambda\tilde{I}*\tilde{V}(\tau,x_{N-1})$$

$$\frac{\partial\tilde{V}}{\partial\tau}(\tau,x_N)=\left[\frac{\sigma^2}{2\Delta x^2}+(r-\delta-\lambda\tilde{k})\right]\tilde{V}(\tau,x_{N+1})+\left[-\frac{\sigma^2}{\Delta x^2}-(r-\delta-\lambda\tilde{k})\right]\tilde{V}(\tau,x_N)+$$
$$+\frac{\sigma^2}{2\Delta x^2}\tilde{V}(\tau,x_{N-1})+\lambda\tilde{I}*\tilde{V}(\tau,x_N) \tag{7.22}$$

Considering the state vector definition

$$\tilde{V}=\left[\tilde{V}(\tau,x_1),\,\tilde{V}(\tau,x_2),\,\tilde{V}(\tau,x_3),\ldots,\tilde{V}(\tau,x_{N-1}),\,\tilde{V}(t,x_N)\right] \tag{7.23}$$

and the control input definition

$$\tilde{V}=\left[\frac{\sigma^2}{2\Delta x^2}\tilde{V}(\tau,x_0)+\lambda\tilde{I}*\tilde{V}(\tau,x_1),\,\lambda\tilde{I}*\tilde{V}(\tau,x_2),\,\lambda\tilde{I}*\tilde{V}(\tau,x_3),\ldots,\right.$$
$$\left.\cdots,\lambda\tilde{I}*\tilde{V}(\tau,x_{N-1}),\,(\frac{\sigma^2}{2\Delta x^2}+(r-\delta-\lambda\tilde{k}))\tilde{V}(\tau,x_{N+1})+\lambda\tilde{I}*\tilde{V}(t,x_N)\right] \tag{7.24}$$

the following state-space description is obtained for the option pricing model

$$\dot{\tilde{V}}=\tilde{A}\tilde{V}+\tilde{B}\tilde{U} \tag{7.25}$$

where

$$\tilde{A}=\begin{pmatrix} \tilde{a}_{11} & \tilde{a}_{12} & \tilde{a}_{13} & \tilde{a}_{14} & \cdots & \tilde{a}_{1,N-2} & \tilde{a}_{1,N-1} & \tilde{a}_{1,N} \\ \tilde{a}_{21} & \tilde{a}_{22} & \tilde{a}_{23} & \tilde{a}_{24} & \cdots & \tilde{a}_{2,N-2} & \tilde{a}_{2,N-1} & \tilde{a}_{2,N} \\ \tilde{a}_{31} & \tilde{a}_{32} & \tilde{a}_{33} & \tilde{a}_{34} & \cdots & \tilde{a}_{3,N-2} & \tilde{a}_{3,N-1} & \tilde{a}_{3,N} \\ \cdots & \cdots & \cdots & \cdots & \cdots & \cdots & \cdots & \cdots & \cdots \\ \tilde{a}_{N-1,1} & \tilde{a}_{N-1,2} & \tilde{a}_{N-1,3} & \tilde{a}_{N-1,4} & \cdots & \tilde{a}_{N-1,N-2} & \tilde{a}_{N-1,N-1} & \tilde{a}_{N-1,N} \\ \tilde{a}_{N,1} & \tilde{a}_{N,2} & \tilde{a}_{N,3} & \tilde{a}_{N,4} & \cdots & \tilde{a}_{N,N-2} & \tilde{a}_{N,N-1} & \tilde{a}_{N,N} \end{pmatrix} \tag{7.26}$$

with
$\tilde{a}_{1,1}=-\frac{\sigma^2}{\Delta x^2}-(r-\delta-\lambda\bar{k}), \tilde{a}_{1,2}=\frac{\sigma^2}{\Delta x^2}+(r-\delta-\lambda\bar{k}), \tilde{a}_{1,3}=0, \tilde{a}_{1,4}, \dots \tilde{a}_{1,N-2}=0,$
$\tilde{a}_{1,N-1}=0, \tilde{a}_{1,N}=0$
$\tilde{a}_{2,1}=\frac{\sigma^2}{2\Delta x^2}, \tilde{a}_{2,2}=-\frac{\sigma^2}{\Delta x^2}-(r-\delta-\lambda\bar{k}), \tilde{a}_{2,3}=\frac{\sigma^2}{\Delta x^2}+(r-\delta-\lambda\bar{k}), \tilde{a}_{2,4}=0, \dots$
$\tilde{a}_{2,N-2}=0, \tilde{a}_{2,N-1}=0, \tilde{a}_{2,N}=0$
$\tilde{a}_{3,1}=0, \tilde{a}_{3,2}=\frac{\sigma^2}{2\Delta x^2}, \tilde{a}_{3,3}=(-\frac{\sigma^2}{\Delta x^2}-(r-\delta-\lambda\bar{k})), \tilde{a}_{3,4}=0, \dots \tilde{a}_{3,N-2}=0,$
$\tilde{a}_{3,N-1}=0, \tilde{a}_{3,N}=0$
$\dots$
$\tilde{a}_{N-1,1}=0, \tilde{a}_{N-1,2}=0, \tilde{a}_{N-1,3}=0, \tilde{a}_{N-1,4}=0, \dots \tilde{a}_{N-1,N-2}=\frac{\sigma^2}{2\Delta x^2}, \tilde{a}_{N-1,N-1}=$
$-\frac{\sigma^2}{\Delta x^2}-(r-\delta-\lambda\bar{k}), \tilde{a}_{N-1,N}=\frac{\sigma^2}{2\Delta x^2}+(r-\delta-\lambda\bar{k})$
$\tilde{a}_{N,1}=0, \tilde{a}_{N,2}=0, \tilde{a}_{N,3}=0, \tilde{a}_{N,4}=0, \dots \tilde{a}_{N,N-2}=0, \tilde{a}_{N,N-1}=\frac{\sigma^2}{2\Delta x^2},$
$\tilde{a}_{N,N}=-\frac{\sigma^2}{\Delta x^2}-(r-\delta-\lambda\bar{k})$

$$\tilde{B}=\begin{pmatrix} 1 & 0 & 0 & 0 & \cdots & 0 & 0 \\ 0 & 1 & 0 & 0 & \cdots & 0 & 0 \\ 0 & 0 & 1 & 0 & \cdots & 0 & 0 \\ \cdots & \cdots & \cdots & \cdots & \cdots & \cdots & \cdots \\ \cdots & \cdots & \cdots & \cdots & \cdots & \cdots & \cdots \\ 0 & 0 & 0 & 0 & \cdots & 1 & 0 \\ 0 & 0 & 0 & 0 & \cdots & 0 & 1 \end{pmatrix} \tag{7.27}$$

The measurement/observation equation is

$$\tilde{V}_m=\tilde{C}\tilde{V} \tag{7.28}$$

where matrix $\tilde{C}$ is selected such that the observability of the system is preserved. e.g. $\tilde{C}=I_{N\times N}$ which is the identity matrix. By defining the flat output for the system of Eq. (7.25) to be $\tilde{Y}=[V_1, V_2, V_3, \dots, V_n]^T$ it is obvious that the option price dynamic model is a differentially flat one since all its state variables and the control inputs can be written as functions of the flat output and its derivatives [34, 77, 134, 169]. For the previous state-space description of the system one can perform state estimation using the Kalman Filter recursion [10, 198, 205]. Previously, matrices $\tilde{A}$, $\tilde{B}$, and $\tilde{C}$ are substituted by their discrete-time equivalents $\tilde{A}_d$, $\tilde{B}_d$ and $\tilde{C}_d$ respectively.

Although estimation for a nonlinear system is carried out, the method does not require the computation of partial derivatives and Jacobian matrices. This stands for the *Derivative-free nonlinear Kalman Filter*:

Measurement update:

$$\begin{aligned} K(k)&=P^-(k)\tilde{C}_d^T[[\tilde{C}_dP^-(k)\tilde{C}_d^T+R]^{-1} \\ \hat{V}(k)&=\hat{V}^-(k)+K(k)[\tilde{C}_d\tilde{V}(k)-\tilde{C}_d\hat{V}^-(k)] \\ P(k)&=P^-(k)-K(k)\tilde{C}_dP^-(k) \end{aligned} \tag{7.29}$$

Time update:

$$\begin{aligned} P^-(k+1) &= \tilde{A}_d(k)P(k)\tilde{A}_d^T(k) + Q(k) \\ \hat{V}^-(k+1) &= \tilde{A}_d(k)\hat{V}(k) + \tilde{B}_d\tilde{U}(k) \end{aligned} \tag{7.30}$$

Consistency checking for the pricing model can be performed by generating residuals through the comparison between the real prices values and the estimates provided by the Kalman Filter, and subsequently by applying χ^2 change detection tests. The concept of parametric change detection in the Kalman Filter, with the use of the χ^2 distribution and followed by the processing of the filter's residuals was explained in Chap. 6.

7.3.2 Consistency Checking of the Option Pricing Model

The residuals' sequence (differences between the real option values and the ones estimated by the Kalman Filter) is a discrete error process e_k with dimension $m \times 1$ (here $m = N$). Actually, it is a zero-mean Gaussian white-noise process with covariance given by E_k. A conclusion can be stated based on a measure of certainty that the Black–Scholes PDE model of Eq. (9.13), with the specific choice of parameter values, is accurate.

To this end, the following *normalized error square* (NES) is defined [212]

$$\varepsilon_k = e_k^T E_k^{-1} e_k \tag{7.31}$$

The normalized error square follows a χ^2 distribution. An appropriate test for the normalized error sum is to numerically show that the following condition is met within a level of confidence (according to the properties of the χ^2 distribution)

$$E\{\varepsilon_k\} = m \tag{7.32}$$

This can be succeeded using statistical hypothesis testing, which are associated with confidence intervals. A 95% confidence interval is frequently applied, which is specified using $100(1-a)$ with $a = 0.05$. Actually, a two-sided probability region is considered cutting-off two end tails of 2.5% each. For M runs the normalized error square that is obtained is given by

$$\bar{\varepsilon}_k = \frac{1}{M}\sum_{i=1}^{M}\varepsilon_k(i) = \frac{1}{M}\sum_{i=1}^{M}e_k^T(i)E_k^{-1}(i)e_k(i) \tag{7.33}$$

where ε_i stands for the i-th run at time t_k. Then $M\bar{\varepsilon}_k$ will follow a χ^2 density with Mm degrees of freedom. This condition can be checked using a χ^2 test. The hypothesis holds true if the following condition is satisfied

$$\bar{\varepsilon}_k \in [\zeta_1, \zeta_2] \tag{7.34}$$

where ζ_1 and ζ_2 are derived from the tail probabilities of the χ^2 density. For example, for $m = 20$ and $M = 100$ one has $\chi^2_{Mm}(0.025) = 1878$ and $\chi^2_{Mm}(0.975) = 2126$. Using that $M = 100$ one obtains $\zeta_1 = \chi^2_{Mm}(0.025)/M = 18.78$ and $\zeta_2 = \chi^2_{Mm}(0.975)/M = 21.26$.

7.4 Simulation Tests

7.4.1 *Estimation with the Use of an Accurate Energy Pricing Model*

Simulation results are provided about the estimation of the value of the option $\hat{V}(t, S)$ of the electricity market as a function of time t and security index S. Indicative values for the model's parameters were: interest were $r_t = 0.005$ and volatility $s_t = 0.006$. As it can be seen in Fig. 7.1 the estimate provided by the Derivative-free nonlinear Kalman Filter is very accurate (although there is no knowledge about initial conditions $\tilde{V}(0, S)$). Moreover, results about the accuracy of estimation in particular points s_i, $i = 1, 2, \ldots, N$ of the spatial grid corresponding to the security index S are given. As it can be observed in Fig. 7.2a and b the estimates provided by the Derivative-free nonlinear Kalman Filter converge to the real values $V(t, s_i)$.

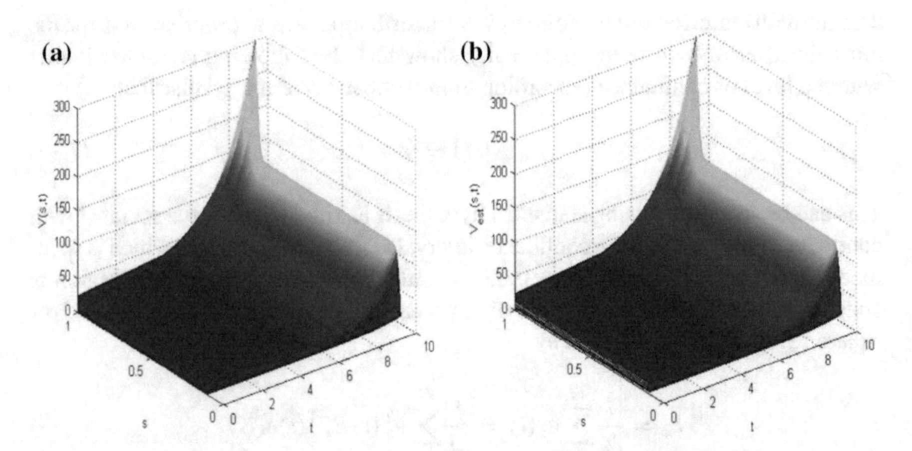

Fig. 7.1 **a** Variation of the option's cost $V(t, S)$ as a function of time t and spot price S, **b** Estimated value of the cost function $\hat{V}(t, S)$ provided by the Derivative-free nonlinear Kalman filter

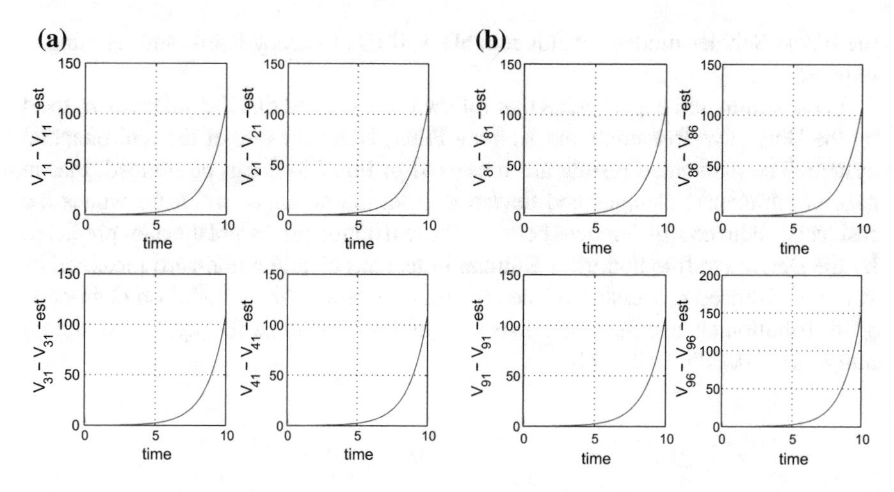

Fig. 7.2 Variation of the option's cost $V(t, S)$ at specific grid points in the case of an accurate Black–Scholes model: real value (*red continuous line*) versus estimated value (*dashed blue line*)

7.4.2 Detection of Mispricing in the Energy Pricing Model

As mentioned above, an issue that arises in the use of the Black–Scholes integrodifferential model is to check its accuracy. Frequently, the real dynamics of the option's value change, since there can be changes in parameters such as the interest rate or the volatility. In the latter case the Black–Scholes model loses its efficiency in predicting the variations of the financial system. Thus, there is need to find-out if

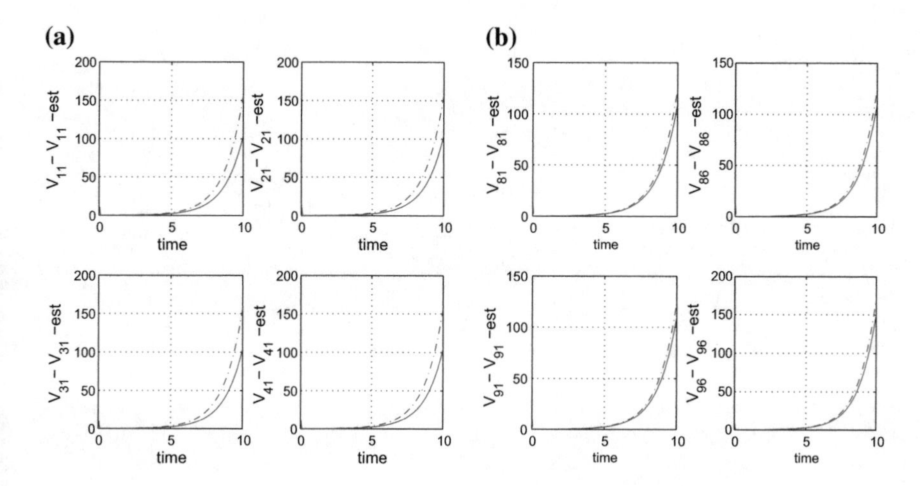

Fig. 7.3 Variation of the option's cost $V(t, S)$ at specific grid points in the case of an inconsistent Black–Scholes model: real value (*red continuous line*) versus estimated value (*dashed blue line*)

the Black–Scholes model remains reliable or if its efficiency is lost and retuning is required.

In the simulation experiments that follow it is assumed that the volatility σ_t used by the Derivative-free nonlinear Kalman Filter, is not the one of the real financial system. The associated results are presented in Fig. 7.3. It can be noticed, that in case of parametric changes and deviation from the nominal parameter values the residuals' sequence (differences between the real option prices and the ones predicted by the Derivative-free nonlinear Kalman Filter) has clearly a non-zero mean value. A more advanced processing of the residuals sequence can be based on elaborated χ^2 distribution change detection tests such as the *local statistical approach to fault diagnosis* [10, 205, 212, 290].

Chapter 8
Corporations' Default Probability Forecasting Using the Derivative-Free Nonlinear Kalman Filter

8.1 Outline

The purpose of this chapter is to demonstrate how state-space models of finance systems can be used in risk management and particularly in firms' default probability forecasting. In the recent years the problems of option pricing, dynamic hedging and risk management in firms, including the energy sector, have been widely studied [79–81]. Investments planning in electric power generation and in the electricity grid is highly affected by the credit risk assessment of power corporations and may be canceled or severely modified in case that electric power corporations undergo financial distress [1, 136, 144, 235, 271]. Of particular importance is the problem of bankruptcy prediction and of the early detection of indexes that reveal the financial stresses of such a company, as well as the company's distance to default [6, 22, 125, 194, 229, 280, 287]. Software and computational packages for monitoring and assessing the default risk make to a great extent use of Merton's model for a company's proximity to bankruptcy [22, 56, 253, 260, 279].

The present chapter proposes a nonlinear filtering method for forecasting default probabilities for financial firms with particular interest in risk management for electric power corporations. First, results from credit risk theory are used [6, 22, 125, 194, 229, 280, 287]. The probability or default of equivalently the distance to default is computed as a function of the company's assets value, of the assets value volatility and of the company's accumulated debts [22, 56, 253, 260, 279]. Since the first two parameters are not directly measurable they are estimated through the processing of measurements of the company's market value, that is of the company's options price. Therefore, if one is in position to forecast the company's market value he can finally forecast the company's probability of default or the company's distance to default.

Forecasting of the company's market (options) value is a non-trivial problem. The dynamics of options' value is known to be described by the Black–Scholes partial differential equation [102, 111, 189, 217, 218, 222]. This theory has been used for credit risk assessment [29, 160]. It has been shown however, that starting

G.G. Rigatos, *State-Space Approaches for Modelling and Control in Financial Engineering*, Intelligent Systems Reference Library 125,
DOI 10.1007/978-3-319-52866-3_8

from the Black–Scholes PDE and by applying a new nonlinear filtering method, the so-called Derivative-free nonlinear Kalman Filter, forecasting of the options' value can be succeeded [216, 222, 225, 228]. The proposed filtering method is based on a transformation of the initial option price dynamics into a state-space model of the linear canonical form. The transformation has been proven to be in accordance to differential flatness theory, and finally provides a model of the option price dynamics for which state estimation is possible by applying the standard Kalman Filter recursion [34, 77, 134, 156, 191, 220, 231, 247, 265]. Moreover, by redesigning the Kalman Filter as a m-step ahead predictor it becomes possible to obtain estimates of the future options' price. This estimated future market value of the company can be finally used to compute bankruptcy risk and the probability of default.

8.2 Company's Credit Risk Models

8.2.1 The Merton-KMV Credit-Risk Model

This model was developed in 2001 and its name comes from the initials of its developers (Kealhofer, McQuown and Vasicek). The default probability of a firm is assessed with respect to the following parameters [28, 46, 64, 110, 112, 237, 301, 307]: (i) The *value of assets*, that is the market's value of the firm's assets. (ii) The *asset risk*, that is uncertainty about the asset's value. This indicates the firm's business and industry risk, (iii) The *Leverage* which denotes the extent of the firm's contractual liabilities. The value of liabilities is computed relatively to the market value of assets.

The default risk of the firm increases when the value of the assets drops and approaches the value of the liabilities. The firm defaults when the value of the assets becomes smaller than the value of the liabilities. The model computes also the Default Point (DFT) which is the assets' value at which the firm defaults.

The default probability of a firm can be determined through the following steps [22, 56, 253, 260, 279]:

1. Estimate the assets value and the asset volatility. In this step, the asset value and the asset volatility of the firm is estimated from the market value and volatility of equity (stocks and options) and from the value of liabilities (accumulated debts).
2. Calculate the distance to default. The distance to default is calculated from the asset value and asset volatility (according to step 1) and from the value of liabilities.
3. Calculate the default probabilities. The default probability is determined directly from the distance to default (DD) and from the default rate for given levels of distance to default.

An approach to calculate the distance to default, based on Merton's model is as follows. It is assumed that the capital structure of the firm includes both equity and debt. Merton's model is based on the Black–Scholes option pricing model. The value of the firm's underlying asset is denoted by V_A and is assumed to evolve in time according to a stochastic process. V_A is considered not to be directly measurable.

$$dV_A = \mu V_A dt + \sigma_A V_A dW \tag{8.1}$$

where μ is a drift coefficient and σ_A is the volatility variable, while W is a Wiener process. As previously explained, in the framework of the Black–Scholes–Merton option pricing theory it has been proven that this stochastic differential equation is equivalent to a parabolic differential equation for function $V_E = F((V_A, t))$, where V_E is the market value of the company (option) and F is a function of both the values of the assets V_A and of the time t. This partial differential equation is given by

$$\frac{\partial V_E}{\partial t} = -\frac{1}{2}\sigma_t^2 V_A^2 \frac{\partial^2 V_E}{\partial V_A^2} - r_t V_A \frac{\partial V_E}{\partial V_A} + r_t V_E \tag{8.2}$$

where r_t is the instantaneous riskless rate of interest, while the partial differential equation is also subjected to appropriate boundary conditions. The market value (options value V_E) of the equity of the firm at time instant T (maturing period) is the difference between the value of its assets and the value of its debt X (the latter decaying in an exponential rate). Thus one has

$$V_E = V_A \cdot N(d_1) - Xe^{-rT} N(d_2) \tag{8.3}$$

where $N()$ denotes the normal distribution, r is the risk-free rate and variables d_1, d_2 are defined as

$$d_1 = \frac{ln(\frac{V_A}{X}) + (r + \frac{\sigma_A^2}{2})T}{\sigma_A\sqrt{T}} \tag{8.4}$$

$$d_2 = d_1 - \sigma_A\sqrt{T} \tag{8.5}$$

The KMV-Merton model for computing the market value of a company is based on two important equations. The first one comes from the Black–Scholes theory and is the equation for finding the company's equity value given in Eq. (8.3). The second is that the volatility of the firm's market value is related to the volatility of its equity. Therefore it holds

$$\sigma_E = (V_A/V_E)\frac{\partial V_E}{\partial V_A}\sigma_A \tag{8.6}$$

In the Black–Scholes–Merton model and using Eq. (8.3), it holds that $\frac{\partial V_E}{\partial V_A} = N(d_1)$, that the previous equation which connects the volatility of the market's value to the volatility of the equity, is rewritten as

$$\sigma_E = (V_A/V_E)N(d_1)\sigma_A \tag{8.7}$$

where d_1 has been defined in Eq. (8.4).

8.2.2 Computation of a Company's Distance to Default

The computation of the distance to default of a firm is based on knowledge of the asset's value of the company V_A and on knowledge of the assets volatility σ_A. However, these variables are not directly measurable and have to be estimated through the processing of sequential measurements of the company's market value (options price) V_E. Thus one has the set of equations

$$\begin{matrix} V_A N(d_1) - e^{-rT} X N(d_2) - V_E = 0 \\ \sigma_E V_E - V_A N(d_1)\sigma_A = 0 \end{matrix} \tag{8.8}$$

Once the company's asset value V_A and the assets value volatility have been computed from numerical data the distance to default (DD) can be also found. The probability of default is given by the relation

$$P = Prob\{V_A(T) < X\} = N(-d_2) = N\left(\frac{ln(\frac{V_A}{X}) + (r - \frac{\sigma_A^2}{2})T}{\sigma_A\sqrt{T}}\right) \tag{8.9}$$

while the distance to default is computed by the following relation

$$DD = \frac{ln(\frac{V_A}{X}) + (r - \frac{\sigma_A^2}{2})T}{\sigma_A\sqrt{T}} \tag{8.10}$$

The implied expected default frequency (EDF) is given by

$$EDF = N(-DD) = N\left(-\frac{ln(\frac{V_A}{X}) + (r - \frac{\sigma_A^2}{2})T}{\sigma_A\sqrt{T}}\right) \tag{8.11}$$

8.3 Estimation of the Market Value of the Company Using the Black–Scholes PDE

8.3.1 State-Space Description of the Black–Scholes Equation

The results, previously presented in Chap. 6, about state-space modelling of the Black–Scholes PDE are overviewed again. State estimation for systems described by partial differential equations (as is the case of the Black–Scholes PDE) has been studied in [95, 232]. Next, the following nonlinear Black–Scholes PDE is considered:

$$\frac{\partial V(t,S)}{\partial t} + r_t S \frac{\partial V(t,S)}{\partial S} + \frac{1}{2}\sigma_t^2 S^2 \frac{\partial^2 V(t,S)}{\partial S^2} - r_t V(t,S) = 0 \tag{8.12}$$

It is considered that the volatility σ_t has a nonlinear dependence on V_{ss}, that is $\sigma_t = \sigma(V_{ss})$, where $V_{ss} = \frac{\partial^2 V(t,s)}{\partial S^2}$ and σ is a nonlinear function. A grid of N points is considered, that is $\{s_1, s_2, \ldots, s_{N-1}, s_N\}$ which are placed at equal distances on the S axis. At the points of spatial discretization it holds

$$\frac{\partial V(t,s_i)}{\partial t} = r_t V(t,s_i) - r_t s_i \frac{\partial V(t,s_i)}{\partial S} - \frac{1}{2}\sigma_t^2(V_{ss})s_i^2 \frac{\partial^2 V(t,s_i)}{\partial S^2} \tag{8.13}$$

where $i = 1, 2, \ldots, N$. The following state vector is defined

$$\begin{array}{c} \tilde{V} = [V(t,s_1), V(t,s_2), V(t,s_3), \ldots, V(t,s_{N-1}), V(t,s_N)] \text{ or} \\ \tilde{V} = [V_1, V_2, V_3, \ldots, V_{N-1}, V_N] \end{array} \tag{8.14}$$

Moreover, by defining the following variables (control inputs)

$$\begin{aligned} u_1 &= -\frac{1}{2}\sigma_t^2(V_{ss})s_1^2 \frac{V_2 - 2V_1 + V_0}{\Delta S^2} \\ u_2 &= -\frac{1}{2}\sigma_t^2(V_{ss})s_2^2 \frac{V_3 - 2V_2 + V_1}{\Delta S^2} \\ u_3 &= -\frac{1}{2}\sigma_t^2(V_{ss})s_3^2 \frac{V_4 - 2V_3 + V_2}{\Delta S^2} \\ &\cdots \\ u_{N-1} &= -\frac{1}{2}\sigma_t^2(V_{ss})s_{N-1}^2 \frac{V_N - 2V_{N-1} + V_{N-2}}{\Delta S^2} \\ u_N &= -\frac{1}{2}\sigma_t^2(V_{ss})s_N^2 \frac{V_{N+1} - 2V_N + V_{N-1}}{\Delta S^2} \end{aligned} \tag{8.15}$$

one has

$$\begin{aligned} \frac{\partial V_1}{\partial t} &= r_t V_1 - r_t s_1 \frac{V_2 - V_1}{\Delta S} + u_1 \\ \frac{\partial V_2}{\partial t} &= r_t V_2 - r_t s_2 \frac{V_3 - V_2}{\Delta S} + u_2 \\ \frac{\partial V_3}{\partial t} &= r_t V_3 - r_t s_3 \frac{V_4 - V_3}{\Delta S} + u_3 \\ &\cdots \\ \frac{\partial V_{N-1}}{\partial t} &= r_t V_{N-1} - r_t s_{N-1} \frac{V_N - V_{N-1}}{\Delta S} + u_{N-1} \\ \frac{\partial V_N}{\partial t} &= r_t V_N - r_t s_N \frac{V_{N+1} - V_N}{\Delta S} + u_N \end{aligned} \tag{8.16}$$

Thus, the dynamics of the Black–Scholes PDE is written as

$$\dot{\tilde{V}} = \tilde{A}\tilde{V} + \tilde{B}\tilde{U} \tag{8.17}$$

where $\tilde{U} = [u_1, u_2, \ldots, u_{N-1}, u_N]^T$ and

$$\tilde{A} = \begin{pmatrix} \tilde{a}_{11} & \tilde{a}_{12} & \tilde{a}_{13} & \tilde{a}_{14} \ldots \tilde{a}_{1,N} \\ \tilde{a}_{21} & \tilde{a}_{22} & \tilde{a}_{23} & \tilde{a}_{24} \ldots \tilde{a}_{2,N} \\ \tilde{a}_{31} & \tilde{a}_{32} & \tilde{a}_{33} & \tilde{a}_{34} \ldots \tilde{a}_{3,N} \\ \cdots & \cdots & \cdots & \cdots \\ \tilde{a}_{N-1,1} & \tilde{a}_{N-1,2} & \tilde{a}_{N-1,3} & \tilde{a}_{N-1,4} \ldots \tilde{a}_{N-1,N} \\ \tilde{a}_{N,1} & \tilde{a}_{N,2} & \tilde{a}_{N,3} & \tilde{a}_{N,4} \ldots \tilde{a}_{N,N} \end{pmatrix} \tag{8.18}$$

where $\tilde{a}_{11} = r_t + r_t \frac{s_1}{\Delta s}$, $\tilde{a}_{12} = -r_t \frac{s_2}{\Delta s}$, $\tilde{a}_{13} = 0$, $\tilde{a}_{14} = 0, \ldots$, $\tilde{a}_{1,N-2} = 0$, $\tilde{a}_{1,N-1} = 0$, $\tilde{a}_{1,N} = 0$.

$\tilde{a}_{21}=0,\ \tilde{a}_{22}=r_t+r_t\frac{s_2}{\Delta s},\ \tilde{a}_{23}=-r_t\frac{s_3}{\Delta s},\ \tilde{a}_{24}=0,\ \ldots,\ \tilde{a}_{2,N-2}=0,\ \tilde{a}_{2,N-1}=0,$
$\tilde{a}_{2,N}=0.$
$\tilde{a}_{31}=0,\ \tilde{a}_{32}=0,\ \tilde{a}_{33}=r_t+r_t\frac{s_3}{\Delta s},\ \tilde{a}_{34}=-r_t\frac{s_4}{\Delta s},\ \ldots,\ \tilde{a}_{3,N-2}=0,\ \tilde{a}_{3,N-1}=0,$
$\tilde{a}_{3,N}=0.$
...
$\tilde{a}_{N-1,1}=0,\ \tilde{a}_{N-1,2}=0,\ \tilde{a}_{N-1,3}=0,\ \tilde{a}_{N-1,4}=0,\ \ldots,\ \tilde{a}_{N-1,N-2}=0,\ \tilde{a}_{N-1,N-1}=$
$r_t+r_t\frac{s_{N-1}}{\Delta s},\ \tilde{a}_{N-1,N}=-r_t\frac{s_N}{\Delta s}.$
$\tilde{a}_{N,1}=0,\ \tilde{a}_{N,2}=0,\ \tilde{a}_{N,3}=0,\ \tilde{a}_{N,4}=0,\ \ldots,\ \tilde{a}_{N,N-2}=0,\ \tilde{a}_{N,N-1}=0,\ \tilde{a}_{N,N}=$
$r_t+r_t\frac{s_N}{\Delta s}.$

Matrix $\tilde{B}$ is defined as

$$\tilde{B}=\begin{pmatrix} 1 & 0 & 0 & 0 & \cdots & 0 & 0 & 0 \\ 0 & 1 & 0 & 0 & \cdots & 0 & 0 & 0 \\ 0 & 0 & 1 & 0 & \cdots & 0 & 0 & 0 \\ \cdots & \cdots & \cdots & \cdots & \cdots & \cdots & \cdots & \cdots \\ 0 & 0 & 0 & 0 & \cdots & 0 & 1 & 0 \\ 0 & 0 & 0 & 0 & \cdots & 0 & 0 & 1 \end{pmatrix} \tag{8.19}$$

The measurement (observation) matrix $\tilde{C}$ is taken to be the identity matrix (or less measurement points can be considered provided that the observability condition holds). For the state-space model of the Black–Scholes nonlinear PDE given in Eq. (8.17) one can perform estimation using the Derivative-free nonlinear Kalman Filter.

8.4 Forecasting Default with the Derivative-Free Nonlinear Kalman Filter

8.4.1 State Estimation with the Derivative-Free Nonlinear Kalman Filter

The Derivative-free nonlinear Kalman Filter has been shown to consist of the application of the Kalman Filter to the linearized equivalent state-space model of the Black–Scholes PDE given in Eq. (8.17). The system of Eq. (8.17) is first turned into discrete-time form using common discretization methods and then the recursion of the linear Kalman Filter is applied. The recursion of the Derivative-free nonlinear Kalman Filter requires to substitute matrices $\tilde{A}$, $\tilde{B}$ and $\tilde{C}$ which have been defined Eqs. (8.17)–(8.19) with their discrete-time equivalents $\tilde{A}_d$, $\tilde{B}_d$ and $\tilde{C}_d$. Matrices $\tilde{A}_d$, $\tilde{B}_d$ and $\tilde{C}_d$ can be computed using established discretization methods. Moreover, the covariance matrices $P(k)$ and $P^-(k)$ are the ones obtained from the linear Kalman Filter update equations. The Derivative-free nonlinear Kalman Filter can be decomposed into two parts: (i) time update (prediction stage), and (ii) measurement update (correction stage) [10, 198, 212, 215].

Measurement update:

$$\begin{aligned} K(k) &= P^-(k)\tilde{C}_d^T[\tilde{C}_d \cdot P^-(k)\tilde{C}_d^T + R]^{-1} \\ \hat{V}(k) &= \hat{V}^-(k) + K(k)[\tilde{C}_d\tilde{V}(k) - \tilde{C}_d\hat{V}^-(k)] \\ P(k) &= P^-(k) - K(k)\tilde{C}_d P^-(k) \end{aligned} \quad (8.20)$$

Time update:

$$\begin{aligned} P^-(k+1) &= \tilde{A}_d(k)P(k)`\tilde{A}_d^T(k) + Q(k) \\ \hat{V}^-(k+1) &= \tilde{A}_d(k)\hat{V}(k) + \tilde{B}_d(k)\tilde{U}(k) \end{aligned} \quad (8.21)$$

As explained above, the Kalman filter recursion can be decomposed into two parts: (i) time update (prediction stage), and (ii) measurement update (correction stage). The first part employs an estimate of the state vector $\tilde{V}(k)$ made before the output measurement $\tilde{C}_d\tilde{V}(k)$ is available (a priori estimate). The second part estimates $\tilde{V}(k)$ after $\tilde{C}_d\tilde{V}(k)$ has become available (a posteriori estimate).

8.4.2 The Derivative-Free Nonlinear Kalman Filter as Extrapolator

The Kalman Filter can be designed as m-step ahead predictor. Consider the initial state-space model of a dynamical system in the discrete-time form

$$x(k+1) = Ax(k) + Bu(k) \quad (8.22)$$

It can be easily proven that it holds

$$x(k+m) = A^m x(k) + \sum_{j=1}^{m} A^{j-1}Bu(k-j+m) \quad (8.23)$$

Then by denoting the new matrices $A_m = A^m$ and $B_m = I_{n\times n}$, as well as the new control input $v(k) = \sum_{j=1}^{m} A^{j-1}Bu(k-j+m)$ one has the new state-space description of the system that is projected m-steps ahead in time

$$x(k+m) = A_m x(k) + B_m v(k) \quad (8.24)$$

For the latter model it is possible to perform a m-step ahead estimation of the system's state vector, using Kalman Filtering.

8.4.3 Forecasting of the Market Value Using the Derivative-Free Nonlinear Kalman Filter

Once the dynamic model of the company's market value (option), that was initially described by the Black–Scholes partial differential equation, has been transformed into a canonical state-space form it is possible to forecast future values of the option prices using Kalman Filtering. The dynamics of the Black–Scholes PDE has been described by Eq. (8.17)

$$\dot{\tilde{V}} = \tilde{A}\tilde{V} + \tilde{B}\tilde{U} \tag{8.25}$$

where matrices $\tilde{A}$ and $\tilde{B}$ have been defined in Eq. (8.18) and in Eq. (8.19). After performing discretization the state-space description of the system becomes

$$\tilde{V}(k+1) = \tilde{A}_d\tilde{V}(k) + \tilde{B}_d\tilde{U}(k) \tag{8.26}$$

where $\tilde{A}_d$ and $\tilde{B}_d$ are the discrete-time equivalents of matrices $\tilde{A}_d$ and $\tilde{B}_d$ respectively. Using the above notation one can design the Kalman Filter as a m-step ahead predictor, which takes the form

$$V(k+m) = \tilde{A}_m x(k) + \tilde{B}_m\tilde{U}^*(k) \tag{8.27}$$

where $\tilde{A}_m = \tilde{A}_d^m$ and $\tilde{B}_m = I_{n\times n}$ and $\tilde{U}^*(k) = \sum_{j=1}^m \tilde{A}_d^{j-1}\tilde{B}_d\tilde{U}(k-j+m)$. After having defined the m-step ahead state-space model of the options' dynamics it is possible to perform forecasting by applying the Kalman Filter recursion.

8.4.4 Assessment of the Accuracy of Forecasting with the Use of Statistical Criteria

As previously analyzed, the residuals' sequence, that is the differences between the real output of the monitored financial system and the one estimated by the Kalman Filter (Fig. 8.1) is a discrete error process e_k with dimension $m \times 1$ (here $m = N$ is the number of grid points and is equal to the dimension of the state vector on the Black–Scholes state-space model). Actually, it is a zero-mean Gaussian white-noise process with covariance given by E_k.

A conclusion can be stated based on a measure of certainty that the Kalman Filter provides accurate estimates for the state-variables of the Black–Scholes PDE model. To this end, the following *normalized error square* (NES) is defined [222]

$$\varepsilon_k = e_k^T E_k^{-1} e_k \tag{8.28}$$

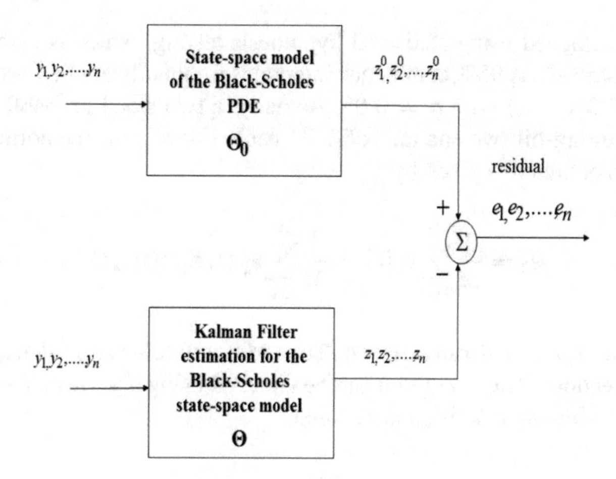

Fig. 8.1 Residuals'generation for the Black–Scholes PDE state-space model, with the use of the Kalman Filtering

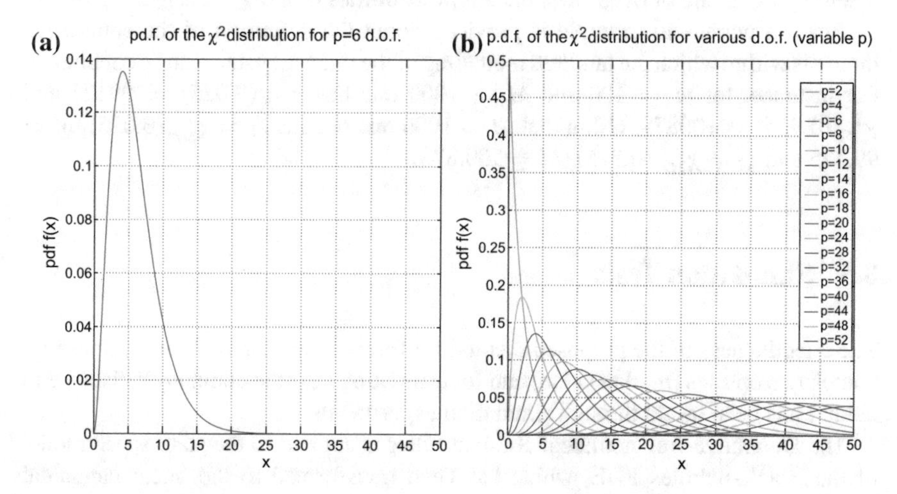

Fig. 8.2 **a** Probability density function of the χ^2 distribution for $p = 6$ degrees of freedom, **b** Probability density function of the distribution for several values of the degrees of freedom (variable p)

As already shown in Chap. 6, the normalized error square follows a χ^2 distribution (Fig. 8.2). An appropriate test for the normalized error sum is to numerically show that the following condition is met within a level of confidence (according to the properties of the χ^2 distribution)

$$E\{\varepsilon_k\} = m \tag{8.29}$$

This can be achieved using statistical hypothesis testing, which is associated with confidence intervals. A 95% confidence interval is frequently applied, which is specified using $100(1-a)$ with $a=0.05$. Actually, a two-sided probability region is considered cutting-off two end tails of 2.5% each. For M runs the normalized error square that is obtained is given by

$$\bar{\varepsilon}_k = \frac{1}{M}\sum_{i=1}^{M}\varepsilon_k(i) = \frac{1}{M}\sum_{i=1}^{M}e_k^T(i)E_k^{-1}(i)e_k(i) \tag{8.30}$$

where ε_i stands for the i-th run at time t_k. Then $M\bar{\varepsilon}_k$ will follow a χ^2 density with Mm degrees of freedom. This condition can be checked using a χ^2 test. The hypothesis holds, if the following condition is satisfied

$$\bar{\varepsilon}_k \in [\zeta_1, \zeta_2] \tag{8.31}$$

where ζ_1 and ζ_2 are derived from the tail probabilities of the χ^2 density.

Again a numerical example is provided about the definition of the confidence intervals within which the faultless modelling of the finance system can be concluded. For instance, for $m=100$ and $M=1000$ one has $\chi^2_{Mm}(0.025)=99125$ and $\chi^2_{Mm}(0.975)=100878$. Using that $M=1000$ one obtains $\zeta_1=\chi^2_{Mm}(0.025)/M=99.125$ and $\zeta_2=\chi^2_{Mm}(0.975)/M=100.878$.

8.5 Simulation Tests

The effectiveness of the proposed method for forecasting first the future values of a company's options (market value) and for estimating next the company's distance to default has been tested through simulation experiments.

The Derivative-free nonlinear Kalman Filter is applied to the state-space model of the Black–Scholes PDE, which has been transformed to the linear canonical (Brunovsky) form. The recursion of the filter consists of two stages, the measurement update (correction stage) and the time update (prediction stage). In the measurement update stage, a new measurement is received about the monitored financial system. The difference of the measurement from the real value of the options value generates the innovation or residual of the filter. This difference is multiplied with the Kalman Filter's gain and is used to correct the estimate about the model's state vector. In the time update (prediction stage) a new value for the state vector is computed one step ahead in time, using the state-space model of the Black–Scholes PDE and the present estimate of the state-vector. This two-stage procedure is computed within one sampling period. The procedure is repeated in the subsequent sampling periods and until the estimated state vector of the system converges to its real value. Convergence and stability of the filter is assured because the state-space model of the Black–Scholes PDE is an observable one.

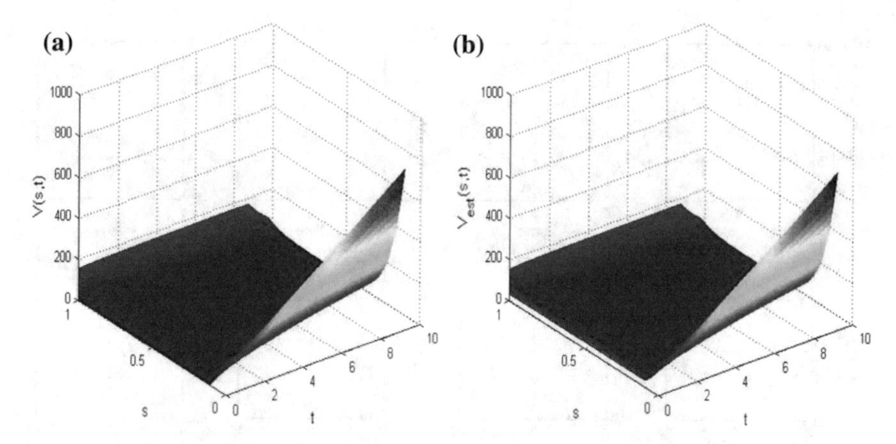

Fig. 8.3 Forecasting of $m = 5$ steps ahead: **a** variation of the option's cost $V(t, S)$ as a function of time t and security index S, **b** predicted value of the cost function $\hat{V}(s, t)$ provided by the Derivative-free nonlinear Kalman Filter

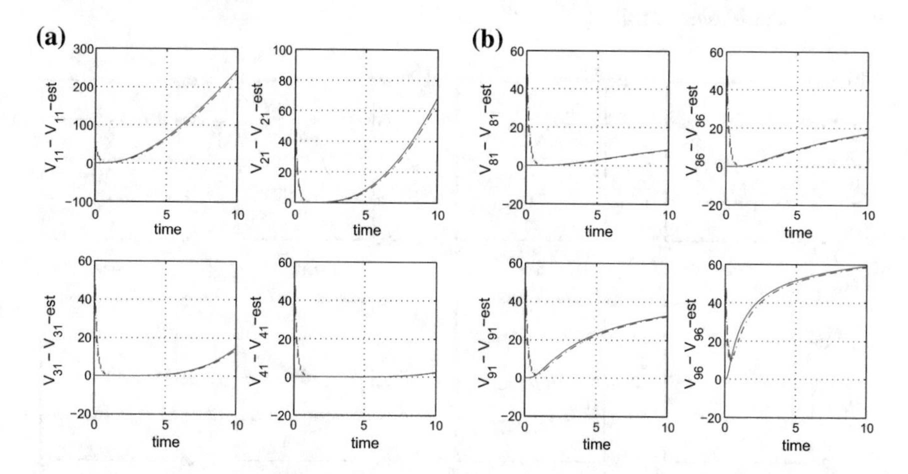

Fig. 8.4 Forecasting of $m = 5$ steps ahead: variation of the option's cost $V(t, S)$ at specific grid points in the case of an accurate Black–Scholes model: real value (*red continuous line*) versus predicted value (*dashed blue line*)

Simulation results are provided about the forecasting of the option value (market value) $\hat{V}(t, S)$ as a function of time t and security index S. First, an $m = 5$ step-ahead predictor of the variations of the company's options has been implemented. As it can be seen in Fig. 8.3 the forecasting provided by the Derivative-free nonlinear Kalman Filter is very accurate (although there is no knowledge about initial conditions $\tilde{V}(0, S)$). Moreover, results about the accuracy of the forecasting in particular points s_i, $i = 1, 2, \ldots, N$ of the spatial grid corresponding to the security index S are given. As it can be observed in Fig. 8.4a, b the forecasting provided by the Derivative-free nonlinear Kalman Filter converges to the real values $V(t, s_i)$.

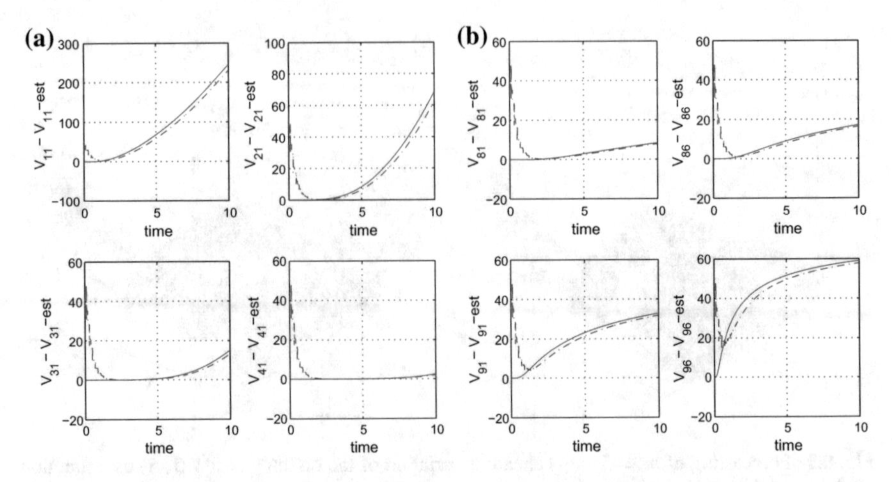

Fig. 8.5 Forecasting of $m = 10$ steps ahead: variation of the option's cost $V(t, S)$ at specific grid points in the case of an accurate Black–Scholes model: real value (*red continuous line*) versus predicted value (*dashed blue line*)

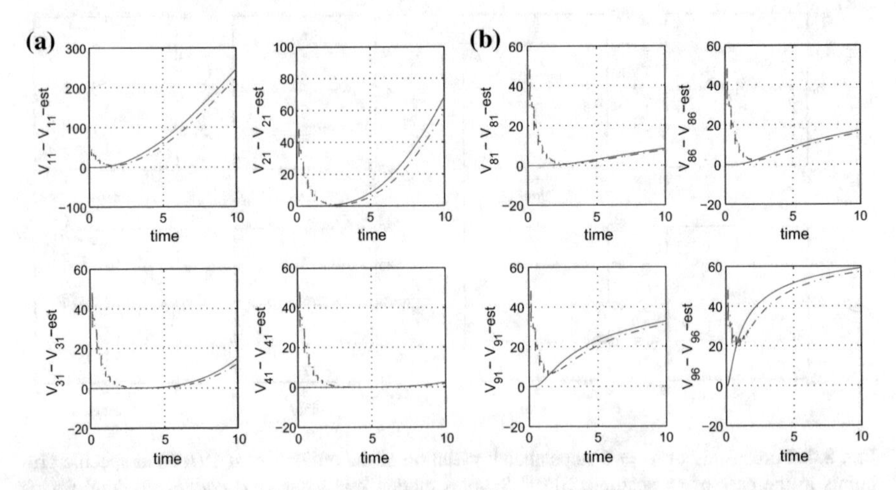

Fig. 8.6 Forecasting of $m = 15$ steps ahead: variation of the option's cost $V(t, S)$ at specific grid points in the case of an accurate Black–Scholes model: real value (*red continuous line*) versus predicted value (*dashed blue line*)

Next, the simulation tests have been repeated in the case of a $m = 10$ and a $m = 15$ step-ahead predictor. The obtained results confirming the accuracy of forecasting are depicted in Figs. 8.5 and 8.6 respectively. In both cases, it can be noticed that although the look-ahead time interval has been increased the accuracy of forecasting remained satisfactory. This also implies that the accuracy of estimation of the company's probability of default is again satisfactory.

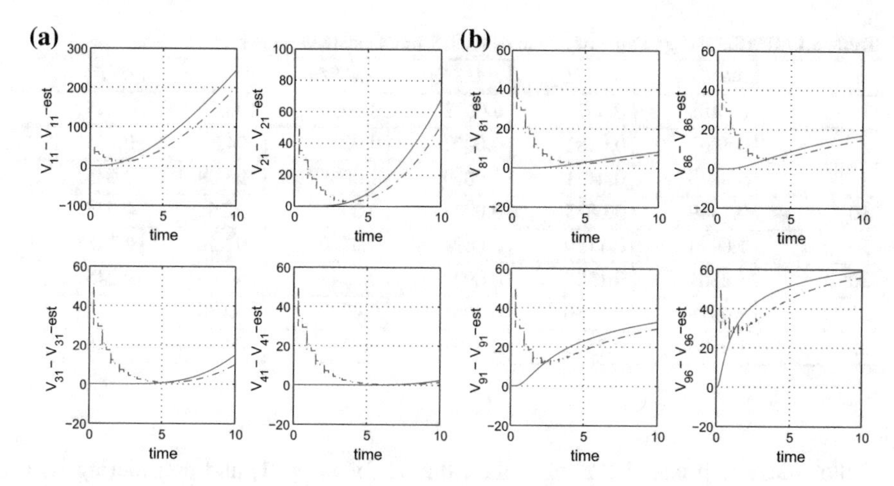

Fig. 8.7 Forecasting of $m = 30$ steps ahead: variation of the option's cost $V(t, S)$ at specific grid points in the case of an accurate Black–Scholes model: real value (*red continuous line*) versus predicted value (*dashed blue line*)

Finally, the simulation tests have been repeated in the case of a $m = 30$ step-ahead predictor. The obtained results confirming the accuracy of forecasting are depicted in Fig. 8.7. Although the prediction interval in this case is quite large the obtained forecasting remains within satisfactory accuracy levels. Therefore, the proposed method can still provide a reliable estimation of the company's default probability.

After computing the estimate of the future value of the company's market price (option) $\hat{V}(t, S)$ (or equivalently V_E) one can solve the system of Eq. (8.8) with respect to the asset's value V_A and to the asset's volatility σ_E. To this end nonlinear optimization methods can be used, e.g. the Gauss–Newton or the Levenberg–Marquardt gradient search algorithms. Finally, by knowing the estimates of V_A and σ_A and using Eq. (8.10) one can find the company's distance to default and the associated probability of default.

The accuracy of forecasting provided by the Derivative-free nonlinear Kalman Filter has been evaluated with the use of statistical measures, as shown in Table 8.1. Actually, the root mean square error of estimation has been computed at four different grid points, namely $s_1 = 30$, $s_2 = 40$, $s_3 = 50$ and $s_4 = 60$. Additionally, a χ^2 statistical test has been implemented for variable forecasting horizon m. This test consisted of the square of the estimation error vector,weighted by the inverse of the covariance matrix of the Kalman Filter.According to the properties of the χ^2 distribution, and a 95% confidence interval, boundaries were computed within which the estimation of the Kalman Filter could be considered as highly accurate.

Table 8.1 RMSE and χ^2 value of residuals for BS-model state variables

m	$RMSE_1$	$RMSE_2$	$RMSE_3$	$RMSE_4$	χ^2	$tr(P)$
1	0.0485	0.0077	0.0052	0.0051	100.19	0.2451
2	0.1367	0.0181	0.0057	0.0058	94.33	0.2451
5	0.3840	0.0451	0.0059	0.0065	93.89	0.2451
10	0.7449	0.0835	0.0074	0.0095	92.61	0.2451
15	1.0474	0.1139	0.0094	0.0119	92.58	0.2451
20	1.2867	0.1306	0.0962	0.0344	91.89	0.2451
30	1.4978	0.3686	0.4081	0.4037	90.70	0.2451
35	1.4833	0.7502	0.8198	0.8147	89.85	0.2451

For the 1-step ahead Kalman Filter, that is for $m = 1$, and considering (i) a state-space description of the χ^2 distribution of dimension $N = 100$, and (ii) confidence intervals equal to 95%, the residuals' χ^2 test value should vary in the interval [99, 12, 100, 78]. As it can be noticed from the above table, for $m = 1$ one can confirm that the value of the χ^2 test lies in the previously noted confidence interval. When a larger prediction horizon is used the value of the residuals χ^2 test exceed the previously defined boundaries of the 95% confidence intervals.

Chapter 9
Validation of Financial Options Models Using Neural Networks with Invariance to Fourier Transform

9.1 Outline

Previously, validation of forecasting and estimation tools used in finance systems (as for instance the Kalman Filter) has been analyzed. Next, the validation procedures will be extended to models of finance systems in the form of neural networks. As already explained, in the recent years there has been a wide deployment of option pricing as a mean for trading several commodities, such as electricity and gas, coal and petroleum, as well as mining products [184, 189]. In particular, in electricity markets, certain features distinguish them from most other commodity markets. A first key point is that spot pricing methods are proven to be inadequate. The spot price of electricity is determined by the intersection by the aggregate demand and supply functions for the electric power and an outage of a major power unit or a sudden surge of demand due to extreme natural conditions can cause the spot price to rise to extremely high levels, with negative consequences in terms of losses for the participants in the electricity market. Actually, the non-storability of electricity, together with real-time matching mechanism of supply and demand in spot pricing lead to disturbances that make spot pricing be inadequate, [25, 44, 99, 123, 174]. The need for transition to option pricing models with more secure trading mechanisms has generated a new research field. Another key point is that models for electric power systems can be negatively affected by (i) high and varying volatility (fluctuations of the prices of the electricity market options and derivatives), (ii) seasonality, (iii) abrupt variations known as jumps which are due to disruptions in transmission and generation outages [30, 37, 176].

It is known that numerical solution of the Black–Scholes PDE enables to compute with precision the values of financial options, within a finite time horizon. Among various computational approaches, Fourier transform methods have been widely used for obtaining a solution for the option pricing problem in a closed form. To this end one can note methods based on the Fourier transform and on the Fast Fourier Transform [32, 129, 186, 297, 298], methods based on the computation of the coefficients

G.G. Rigatos, *State-Space Approaches for Modelling and Control in Financial Engineering*, Intelligent Systems Reference Library 125,
DOI 10.1007/978-3-319-52866-3_9

of Fourier-cosine expansions [299, 300] and finally methods based on the computation of the coefficients of Fourier–Hermite expansions [250]. In parallel, several approaches have emerged about modelling of the dynamics of option prices with the use of neural networks [3, 100, 305].

In this chapter, a neural network with 2D Gauss–Hermite polynomial activation functions is used for modeling the dynamics of financial options. The output of the considered neural network provides also a series expansion that takes the form of a weighted sum of Gauss–Hermite basis functions. Knowing that the Gauss–Hermite basis functions satisfy the orthogonality property and remain unchanged under the Fourier transform, subjected only to a change of scale, one has that the considered neural network provides the spectral analysis of the options' dynamics model [193, 219]. Actually, the squares of the weights of the output layer of the neural network denote the spectral components for the monitored options' dynamics. by observing changes in the amplitude of the aforementioned spectral components one can have also an indication about deviations from nominal values, for parameters that affect the options' dynamics, such as interest rate, dividend payment and volatility. Moreover, since specific parametric changes are associated with amplitude changes of specific spectral components of the options' model, isolation of the distorted model parameters can be also performed [10, 205, 212, 290].

9.2 Option Pricing in the Energy Market

Neural networks will be used for modelling the dynamics of a finance system described by the Black–Scholes partial integrodifferential equation. To this end, the concept of Black–Scholes partial integrodifferential equation is overviewed, with focus on its use for modelling of options'pricing of electric power corporations.

As explained in the previous chapter, modeling tools for pricing in energy markets are based on extended forms of the Black–Scholes partial differential equation [58, 73, 74, 137]. In the Black–Scholes PDE the variation of the underlying security index is according to the stochastic differential equation [174]

$$dS = \mu S dt + \sigma S dX \tag{9.1}$$

To apply the Black–Scholes pricing model

$$\frac{\partial V}{\partial t} + rS\frac{\partial V}{\partial S} + \frac{\sigma^2 S^2}{2}\frac{\partial^2 V}{\partial S^2} = rV \tag{9.2}$$

in the case swing options met in the energy market, the initial partial differential equation has to be transformed into a Partial Integrodifferential Equation (r is the interest rate, δ is the dividend payment, σ is the volatility). The previous SDE given in Eq. (9.1) in the case of option market swing options becomes

$$dS = (\mu - \delta)S dt + \sigma S dX \tag{9.3}$$

and the corresponding PDE becomes

$$\frac{\partial V}{\partial t} + (r - \delta)S\frac{\partial V}{\partial S} + \frac{\sigma^2 S^2}{2}\frac{\partial^2 V}{\partial S^2} = rV \tag{9.4}$$

The stock price does not have only continuous changes but exhibits also jumps. In such a case the assumption about a constant volatility σ is proven to be inefficient. For modeling spikes in the price's variation the following SDE of the underlying asset has been proposed [174]

$$dS = (\mu - \delta)Sdt + \sigma SdX + (q - 1)SdQ \tag{9.5}$$

where dQ indicates a Poisson process, of intensity λ with

$$\begin{array}{c} P[dQ = 0] = 1 - \lambda dt \\ P[dQ = 1] = \lambda dt \end{array} \tag{9.6}$$

and $q \in [0, \infty]$ is a random variable that is log-normally distributed with the p.d.f

$$\Gamma_{\gamma,\mu_j}(q) = \frac{1}{\sqrt{2\pi}\gamma} exp\left(-\frac{1}{2}\frac{((log(q) - \mu_j)}{\gamma}\right)^2 \tag{9.7}$$

satisfying also

$$\int_0^\infty \Gamma_\gamma(q)dq = 1 \; \forall q \in [0, \infty] \tag{9.8}$$

The term $q - 1$ is called *impulse function* which produces a jump from S to Sq while it holds $E(q - 1) = \bar{k}$. If $0 \leq q < 1$ the negative impulse function causes the spot price to decrease. If $q = 1$, the spot price remains the same. If $q > 1$ the impulse function causes the spot price to increase. In such a case the option price dynamics can be described by the partial integrodifferential equation [174]

$$\frac{\partial V}{\partial t} + \frac{\sigma^2 S^2}{2}\frac{\partial^2 V}{\partial S^2} + (r - \delta - \lambda\bar{k})S\frac{\partial V}{\partial S} - rV + \lambda\int_0^\infty (V(qS, t) - V(S, t))\Gamma_\gamma(q)dq = 0 \tag{9.9}$$

Next, the following change of variable is performed [174]

$$\tilde{V}(x, \tau) = V(e^x, T - \tau) = V(S, t) \tag{9.10}$$

$$\frac{\partial \tilde{V}}{\partial \tau} = -\frac{\partial V}{\partial t}\frac{\partial V}{\partial S} = \frac{1}{S}\frac{\partial \tilde{V}}{\partial x}\frac{\partial^2 V}{\partial S^2} = \frac{1}{S^2}\frac{\partial^2 \tilde{V}}{\partial x^2} \tag{9.11}$$

with $q \in [0, \infty]$, $S \geq 0 \rightarrow Sq$ and $ln(q)$, $x \in R$. Following this change of variables the partial integrodifferential equation (PIDE) is written as

$$\frac{\partial \tilde{V}}{\partial \tau} = \frac{\sigma^2}{2}\frac{\partial^2 \tilde{V}}{\partial x^2} + (r - \delta - \lambda\bar{k})\frac{\partial \tilde{V}}{\partial x} - r\tilde{V} + \lambda\left(\int_{-\infty}^{\infty}\tilde{V}(z+x,\tau)\Gamma_\gamma(z)dz - \tilde{V}\right) \tag{9.12}$$

there the associated initial condition is $\tilde{V}(x, 0)$. The partial integrodifferential equation can be also written as [174]:

$$\frac{\partial \tilde{V}}{\partial \tau} = \frac{\sigma^2}{2}\frac{\partial^2 \tilde{V}}{\partial x^2} + (r - \delta - \lambda\bar{k})\frac{\partial \tilde{V}}{\partial x} - (r+\lambda)\tilde{V} + \lambda\left(\int_{-\infty}^{\infty}\tilde{V}(z+x,\tau)\Gamma_\gamma(z)dz\right) \tag{9.13}$$

where the following coefficients are denoted $c = \frac{\sigma^2}{2}$, $a = r - \delta - \lambda\bar{k}$, $b = r + \lambda$ and about the integral term one has $I\tilde{V} = \lambda\left(\int_{-\infty}^{\infty}\tilde{V}(z+x,\tau)\Gamma_\gamma(z)dz - \tilde{V}\right)$. To proceed to the discretization of the option price dynamic model the following operator is defined

$$L = a\frac{\partial}{\partial x} - b + c\frac{\partial^2}{\partial x^2} \tag{9.14}$$

Thus, the dynamics of the electric power pricing model can be written in the form

$$\frac{\partial \tilde{V}}{\partial \tau} = L\tilde{V} + I{*}\tilde{V} \tag{9.15}$$

9.3 Neural Networks Using Hermite Activation Functions

9.3.1 Generalized Fourier Series

A neural network will be used for modeling the options' price dynamics. Modeling with neural networks is widely reckoned to be equivalent to Fourier series expansion. It is known that every function $\psi(x)$ in the interval $(0, L)$ can be approximated by the Fourier series expansion [219]

$$\psi(x) = \sum_{k=1}^{\infty} A_k sin\left(\frac{k\pi x}{L}\right) \quad k = 0, 1, 2, \cdots \tag{9.16}$$

The above relation can be considered as a solution of the harmonic oscillator differential equation $\frac{\partial^2\psi(x)}{\partial x^2} + k^2\psi(x) = 0$. Index $k = 0, 1, 2, \cdots$ denotes the normal modes of oscillation. To define coefficients A_k the orthogonality of the sinusoidal basis functions is used, i.e.

$$\int_0^L sin\left(\frac{k\pi x}{L}\right) sin\left(\frac{m\pi x}{L}\right) = 0 \ if \ k \neq m \tag{9.17}$$

Therefore it holds

$$\int_0^L \psi(x)sin\left(\frac{k\pi x}{L}\right) = \int_0^L A_k sin^2\left(\frac{k\pi x}{L}\right)dx \tag{9.18}$$

where $sin^2(a) = \frac{1}{2} - \frac{1}{2}cos(2a)$. Since it holds $\int_0^L A_k\left[\frac{1}{2} - \frac{1}{2}cos\left(\frac{2k\pi x}{L}\right)\right]dx = A_k\frac{L}{2}$ one gets

$$A_k = \frac{2}{L}\int_0^L \psi(x)sin\left(\frac{k\pi x}{L}\right) \tag{9.19}$$

It follows that, the solution $\psi(x)$ of any diffusion differential equation can be written in the form of a Fourier series expansion which is given by Eq. (9.16), for $0 < x < L$. Similarly, the solution $\psi(x)$ can be written in the form of a Fourier series containing cosine basis functions, i.e.

$$\psi(x) = \sum_{k=0}^{\infty} A_k cos\left(\frac{k\pi x}{L}\right) = \frac{1}{2}A_0 + \sum_{k=1}^{\infty} A_k cos\left(\frac{k\pi x}{L}\right) \tag{9.20}$$

where the coefficients A_k are calculated using again the orthogonality principle

$$A_k = \frac{2}{L}\int_0^L \phi(x)cos\left(\frac{k\pi x}{L}\right)dx \tag{9.21}$$

Finally, the Fourier series of function $\psi(x)$ can be written in both sine and cosine basis functions (Fig. 9.1), i.e.

$$\psi(x) = \sum_{k=0}^{\infty}\left[A_k cos\left(\frac{k\pi x}{L}\right) + B_k sin\left(\frac{k\pi x}{L}\right)\right] = \frac{1}{2}A_0 + \sum_{k=1}^{\infty}\left[A_k cos\left(\frac{k\pi x}{L}\right) + B_k sin\left(\frac{k\pi x}{L}\right)\right] \tag{9.22}$$

where the coefficients A_k and B_k are calculated again using the orthogonality property, i.e.

$$\begin{array}{l} A_k = \frac{1}{L}\int_{-L}^{L}\phi(x)cos\left(\frac{k\pi x}{L}\right)dx \ k = 0, 1, 2, \cdots \\ B_k = \frac{1}{L}\int_{-L}^{L}\phi(x)sin\left(\frac{k\pi x}{L}\right)dx \ k = 0, 1, 2, \cdots \end{array} \tag{9.23}$$

Assume for example function $\psi(x) = 1, x \in [0, L]$. The expansion in sinusoidal Fourier series gives

$$\psi(x) = \sum_{k=0}^{\infty} A_k sin\left(\frac{k\pi x}{L}\right) \tag{9.24}$$

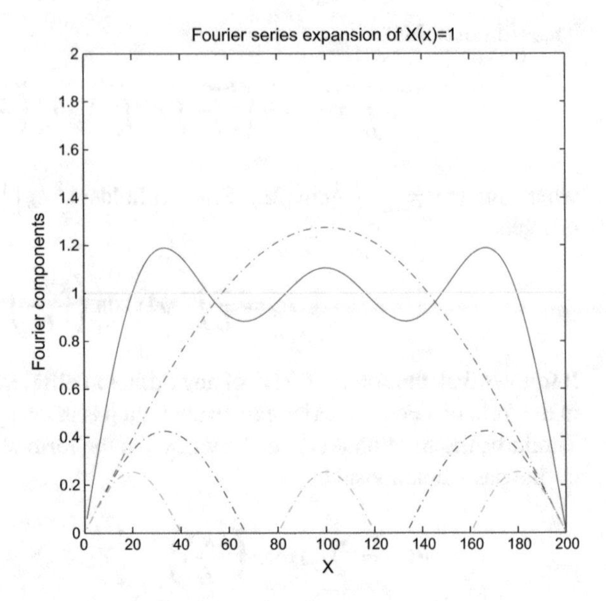

Fig. 9.1 Fourier series expansion of X(x) = 1. *Continuous line*: sum of the first three terms of the Fourier series, *Dashed lines*: harmonics

where $A_k = \frac{2}{L}\int_0^L \psi(x) sin\left(\frac{k\pi x}{L}\right) dx = \frac{2}{k\pi}[1-(-1)^k]$. Therefore $A_k = \frac{4}{k\pi}$ if k is odd and $A_k = 0$ if k is even.

Thus $1 = \frac{4}{\pi} sin\left(\frac{\pi x}{L}\right) + \frac{4}{3\pi} sin\left(\frac{3\pi x}{L}\right) + \frac{4}{5\pi} sin\left(\frac{5\pi x}{L}\right) + \cdots$, i.e. $1 = \sum_{k=0}^{\infty} \frac{4}{k\pi} sin\left(\frac{k\pi x}{L}\right)$.

The complex form of Fourier series is based on Euler's formula, i.e. $sin\theta = \frac{e^{i\theta} - e^{-i\theta}}{2i}$ and $cos\theta = \frac{e^{i\theta}+e^{-i\theta}}{2}$.

It holds that

$$\psi(x) = \sum_{k=-\infty}^{+\infty} c_k e^{\frac{ik\pi x}{L}} \tag{9.25}$$

$$c_k = \frac{1}{2L}\int_{-L}^{L} \psi(x) e^{\frac{-ik\pi x}{L}} \tag{9.26}$$

The property of orthogonality is of primary importance for generalized Fourier series.

9.3.2 The Gauss–Hermite Series Expansion

Next, as orthogonal basis functions of a feed-forward neural network Gauss–Hermite activation functions are considered. These are the spatial components $X_k(x)$ of the solution of Schrödinger's differential equation and describe a stochastic oscillation [197]:

$$X_k(x) = H_k(x) e^{\frac{-x^2}{2}}, \; k = 0, 1, 2, \cdots \tag{9.27}$$

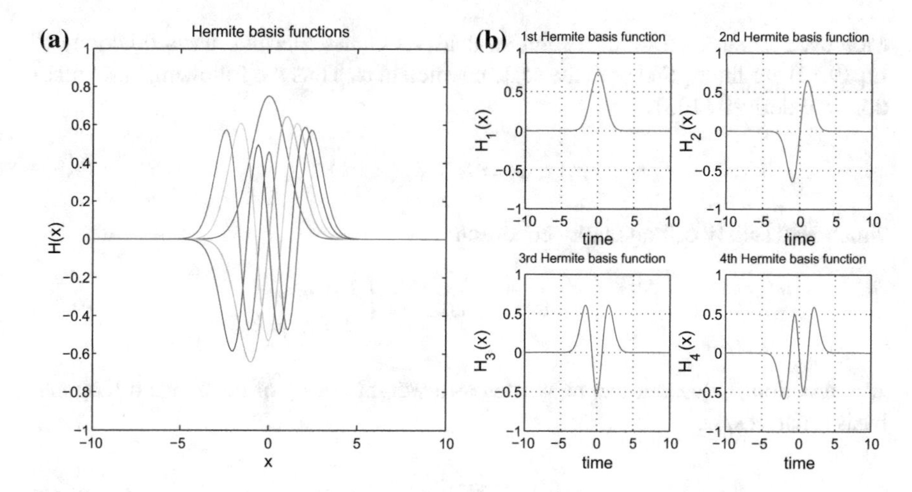

Fig. 9.2 **a** First five one-dimensional Hermite basis functions **b** Analytical representation of the 1D Hermite basis function

where $H_k(x)$ are the Hermite orthogonal functions (Fig. 9.2). The Hermite functions $H_k(x)$ are the eigenstates of the quantum harmonic oscillator. The general relation for the Hermite polynomials is

$$H_k(x) = (-1)^k e^{x^2} \frac{d^{(k)}}{dx^{(k)}} e^{-x^2} \tag{9.28}$$

According to Eq. (9.28) the first five Hermite polynomials are:

$H_0(x) = 1$, $H_1(x) = 2x$, $H_2(x) = 4x^2 - 2$, $H_3(x) = 8x^3 - 12x$, $H_4(x) = 16x^4 - 48x^2 + 12$

It is known that Hermite polynomials are orthogonal, i.e. it holds

$$\int_{-\infty}^{+\infty} e^{-x^2} H_m(x) H_k(x) dx = \begin{cases} 2^k k! \sqrt{\pi} \ if \ m = k \\ 0 \ if \ m \neq k \end{cases} \tag{9.29}$$

Using now, Eq. (9.29), the following basis functions can be defined [193]:

$$\psi_k(x) = [2^k \pi^{\frac{1}{2}} k!]^{-\frac{1}{2}} H_k(x) e^{-\frac{x^2}{2}} \tag{9.30}$$

where $H_k(x)$ is the associated Hermite polynomial. From Eq. (9.29), the orthogonality of basis functions of Eq. (9.30) can be concluded, which means

$$\int_{-\infty}^{+\infty} \psi_m(x) \psi_k(x) dx = \begin{cases} 1 \ if \ m = k \\ 0 \ if \ m \neq k \end{cases} \tag{9.31}$$

Moreover, to succeed multi-resolution analysis Gauss–Hermite basis functions of Eq. (9.30) are multiplied with the scale coefficient α. Thus the following basis functions are derived [193]

$$\beta_k(x,\alpha) = \alpha^{-\frac{1}{2}}\psi_k(\alpha^{-1}x) \tag{9.32}$$

which also satisfy orthogonality condition

$$\int_{-\infty}^{+\infty}\beta_m(x,\alpha)\beta_k(x,\alpha)dx = \begin{cases} 1 \text{ if } m = k \\ 0 \text{ if } m \neq k \end{cases} \tag{9.33}$$

Any function $f(x)$, $x \in R$ can be written as a weighted sum of the above orthogonal basis functions, i.e.

$$f(x) = \sum_{k=0}^{\infty} c_k\beta_k(x,\alpha) \tag{9.34}$$

where coefficients c_k are calculated using the orthogonality condition

$$c_k = \int_{-\infty}^{+\infty} f(x)\beta_k(x,\alpha)dx \tag{9.35}$$

Assuming now that instead of infinite terms in the expansion of Eq. (9.34), M terms are maintained, then an approximation of $f(x)$ is succeeded. The expansion of $f(x)$ using Eq. (9.34) is a Gauss–Hermite series. Equation (9.34) is a form of Fourier expansion for $f(x)$. Equation (9.34) can be considered as the Fourier transform of $f(x)$ subject only to a scale change. Indeed, the Fourier transform of $f(x)$ is given by

$$F(s) = \frac{1}{2\pi}\int_{-\infty}^{+\infty} f(x)e^{-jsx}dx \Rightarrow f(x) = \frac{1}{2\pi}\int_{-\infty}^{+\infty} F(s)e^{jsx}ds \tag{9.36}$$

The Fourier transform of the basis function $\psi_k(x)$ of Eq. (9.30) satisfies [193]

$$\Psi_k(s) = j^k\psi_k(s) \tag{9.37}$$

while for the basis functions $\beta_k(x,\alpha)$ using scale coefficient α it holds that

$$B_k(s,\alpha) = j^k\beta_k(s,\alpha^{-1}) \tag{9.38}$$

Therefore, it holds

$$f(x) = \sum_{k=0}^{\infty} c_k\beta_k(x,\alpha) \overset{F}{\rightarrow} F(s) = \sum_{k=0}^{\infty} c_k j^n\beta_k(s,\alpha^{-1}) \tag{9.39}$$

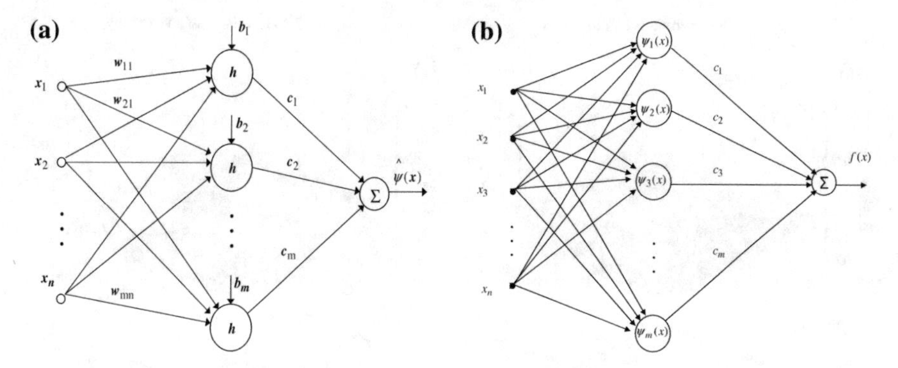

Fig. 9.3 **a** Feed-forward neural network **b** Neural network with Gauss–Hermite basis functions

which means that the Fourier transform of Eq. (9.34) is the same as the initial function, subject only to a change of scale. The structure of a a feed-forward neural network with Hermite basis functions is depicted in Fig. 9.3b.

9.3.3 *Neural Networks Using 2D Hermite Activation Functions*

Two-dimensional Hermite polynomial-based neural networks can be constructed by taking products of the one dimensional basis functions $B_k(x,\alpha)$ [193]. Thus, setting $x=[x_1,x_2]^T$ one can define the following basis functions [193]

$$B_k(x,\alpha)=\frac{1}{\alpha}B_{k_1}(x_1,\alpha)B_{k_2}(x_2,\alpha) \tag{9.40}$$

These two dimensional basis functions are again orthonormal, i.e. it holds

$$\int d^2xB_n(x,\alpha)B_m(x,\alpha)=\delta_{n_1m_1}\delta_{n_2m_2} \tag{9.41}$$

The basis functions $B_k(x)$ are the eigenstates of the two dimensional harmonic oscillator and form a complete basis for integrable functions of two variables. A two dimensional function $f(x)$ can thus be written in the series expansion:

$$f(x)=\sum_{k_1,k_2}^{\infty}c_kB_k(x,\alpha) \tag{9.42}$$

The choice of an appropriate scale coefficient α and maximum order k_{max} is of practical interest. The coefficients c_k are given by

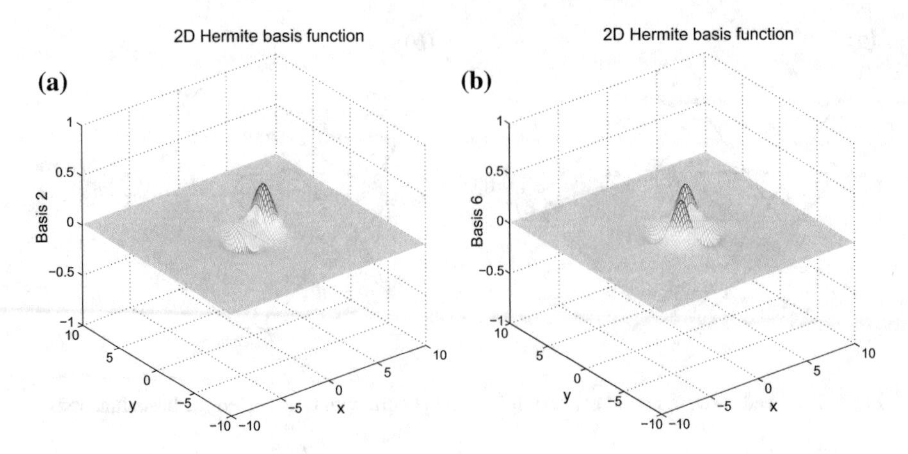

Fig. 9.4 2D Hermite polynomial activation functions: **a** basis function $B_2(x, \alpha)$ **b** basis function $B_6(x, \alpha)$

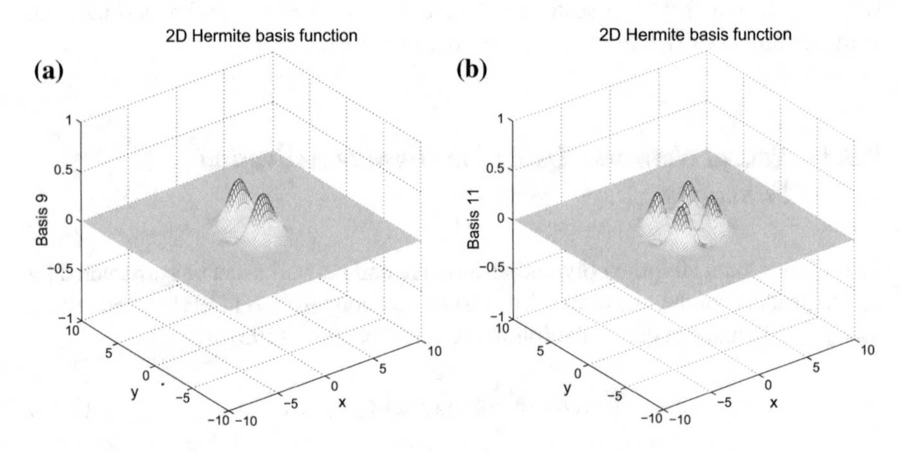

Fig. 9.5 2D Hermite polynomial activation functions: **a** basis function $B_9(x, \alpha)$ **b** basis function $B_{11}(x, \alpha)$

$$c_k = \int dx^2 f(x) B_k(x, \alpha) \tag{9.43}$$

Indicative basis functions $B_2(x, \alpha)$, $B_6(x, \alpha)$, $B_9(x, \alpha)$, $B_{11}(x, \alpha)$ and $B_{13}(x, \alpha)$, $B_{15}(x, \alpha)$ of a 2D feed-forward neural network with Hermite basis functions are depicted in Figs. 9.4, 9.5 and 9.6. Following, the same method N-dimensional Hermite polynomial-based neural networks ($N > 2$) can be constructed. The associated high-dimensional Gauss–Hermite activation functions preserve the properties of orthogonality and invariance to Fourier transform.

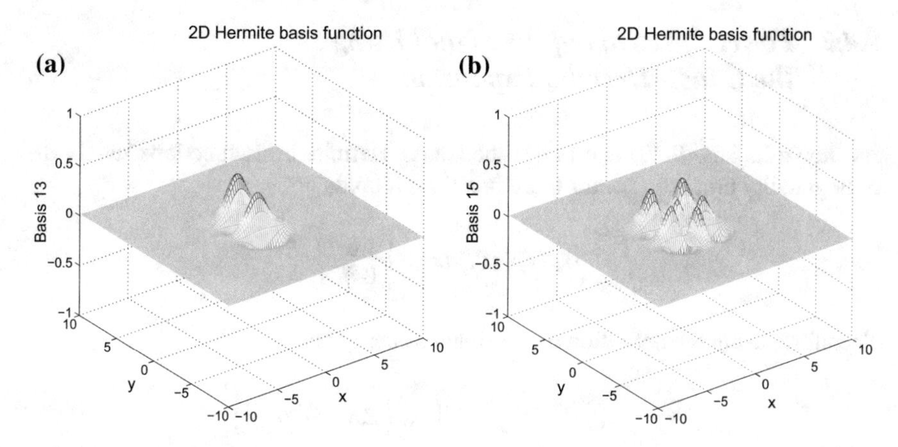

Fig. 9.6 2D Hermite polynomial activation functions: **a** basis function $B_{13}(x,\alpha)$ **b** basis function $B_{15}(x,\alpha)$

9.4 Signals Power Spectrum and the Fourier Transform

9.4.1 *Parseval's Theorem*

To find the spectral density of a signal $\psi(t)$ with the use of its Fourier transform $\Psi(j\omega)$, the following definition is used:

$$\begin{aligned} E_\psi = \int_{-\infty}^{+\infty}(\psi(t))^2 dt = \frac{1}{2\pi}\int_{-\infty}^{+\infty}\psi(t)\left(\int_{-\infty}^{+\infty}\Psi(j\omega)e^{j\omega}d\omega\right)dt \text{ i.e.} \\ E = \frac{1}{2\pi}\int_{-\infty}^{+\infty}\Psi(j\omega)\Psi(-j\omega)d\omega \end{aligned} \quad (9.44)$$

Taking that $\psi(t)$ is a real signal it holds that $\Psi(-j\omega) = \Psi^*(j\omega)$ which is the signal's complex conjugate. Using this in Eq. (9.44) one obtains

$$\begin{aligned} E_\psi = \frac{1}{2\pi}\int_{-\infty}^{+\infty}\Psi(j\omega)\Psi^*(j\omega)d\omega \text{ or} \\ E_\psi = \frac{1}{2\pi}\int_{-\infty}^{+\infty}|\Psi(j\omega)|^2 d\omega \end{aligned} \quad (9.45)$$

This means that the energy of the signal is equal to $\frac{1}{2\pi}$ times the integral over frequency of the square of the magnitude of the signal's Fourier transform. This is *Parseval's theorem*. The integrated term $|\Psi(j\omega)|^2$ is the energy density per unit of frequency and has units of magnitude squared per Hertz.

9.4.2 Power Spectrum of the Signal Using the Gauss–Hermite Expansion

As shown in Eqs. (9.29) and (9.41) the Gauss–Hermite basis functions satisfy the orthogonality property, i.e. for these functions it holds

$$\int_{-\infty}^{+\infty} \psi_m(x)\psi_k(x)dx = \begin{cases} 1 \ if \ m = k \\ 0 \ if \ m \neq k \end{cases}$$

Therefore, using the definition of the signal's energy one has

$$E = \int_{-\infty}^{+\infty} (\psi(t))^2 dt = \int_{-\infty}^{+\infty} \left[\sum_{k=1}^{N} c_k \psi_k(t)\right]^2 \tag{9.46}$$

and exploiting the orthogonality property one obtains

$$E = \sum_{k=1}^{N} c_k^2 \tag{9.47}$$

Therefore the square of the coefficients c_k provides an indication of the distribution of the signal's energy to the associated basis functions. One could arrive at the same results using the Fourier transformed description of the signal and Parseval's theorem. It has been shown that the Gauss–Hermite basis functions remain invariant under the Fourier transform subject only to a change of scale. Denoting by $\Psi(j\omega)$ the Fourier transformed signal of $\psi(t)$ and by $\Psi_k(j\omega)$ the Fourier transform of the k-th Gauss–Hermite basis function one obtains

$$\Psi(j\omega) = \sum_{k=1}^{N} c_k \Psi_k(j\omega) \tag{9.48}$$

and the energy of the signal is computed as

$$E_\psi = \frac{1}{2\pi} \int_{-\infty}^{+\infty} |\Psi(j\omega)|^2 d\omega \tag{9.49}$$

Substituting Eq. (9.48) into Eq. (9.49) one obtains

$$E_\psi = \frac{1}{2\pi} \int_{-\infty}^{+\infty} \left|\sum_{k=1}^{N} c_k \Psi_k(j\omega)\right|^2 d\omega \tag{9.50}$$

and using the invariance of the Gauss–Hermite basis functions under Fourier transform one gets

$$E_\psi = \frac{1}{2\pi} \int_{-\infty}^{+\infty} \left|\sum_{k=1}^{N} c_k \alpha^{-\frac{1}{2}} \psi_k(\alpha^{-1} j\omega)\right|^2 d\omega \tag{9.51}$$

while performing the change of variable $\omega_1 = \alpha^{-1}\omega$ it holds that

$$E_\psi = \frac{1}{2\pi} \int_{-\infty}^{+\infty} \left|\sum_{k=1}^{N} c_k \alpha^{\frac{1}{2}} \psi_k(j\omega_1)\right|^2 d\omega_1 \tag{9.52}$$

Next, by exploiting the orthogonality property of the Gauss–Hermite basis functions one gets that the signal's energy is proportional to the sum of the squares of the coefficients c_k which are associated with the Gauss–Hermite basis functions, i.e. a relation of the form

$$E_{\psi} = \sum_{k=1}^{N} c_k^2 \tag{9.53}$$

9.5 Simulation Tests

Gauss–Hermite neural networks were used to learn the dynamics of an options' pricing model, associated with electricity pricing. The training data were produced by the numerical solution of the Black–Scholes PIDE under two different parameter sets. The number of nodes in the hidden layer of the neural network was $n = 25$. The discretization of t axis was in $N_1 = 1000$ equidistant points, while the discretization of the S axis was in $N_2 = 100$ equidistant points. This means that at each epoch of the learning algorithm of the neural network the number of the processed data points was $N = 10^5$.

The model's parameters that could be subjected to change were the interest rate r, the dividend payment δ, and the volatility σ. In the specific simulation tests the parametric variation was associated with δ which changed from the nominal value $\delta^* = 0.10$ to the new value $\delta^* = 0.08$. The variations in the S, t, V coordinates system for the option's price model, for the case before and after change are given in Fig. 9.7a, b, respectively.

Next, processing of the weights vectors was carried out for the two neural networks which were trained with the data coming from the two different option pricing models (before and after the parametric change in δ). As explained, the squares of the weights coefficients provided an indication about the distribution of the spectral content of the option pricing model into spectral components. As it can be observed from Fig. 9.8a, b the parametric change in δ induced also a clear variation of the spectral content of the option pricing model. Thus by processing the aforementioned plots one can get an indication about a parametric change, that will lead to mispricing if the initial options' pricing model is still used.

The evaluation tests were repeated, considering this time that the parametric change was associated with the volatility (which moved from its nominal value $\sigma^* = 0.00612$ into the new value $\sigma = 0.00632$). Again two neural networks were trained with data coming from the option pricing models, before and after the parametric change in σ. As it can be observed from Fig. 9.9a, b the parametric change in r caused once more a clear variation of the spectral content of the option pricing model.

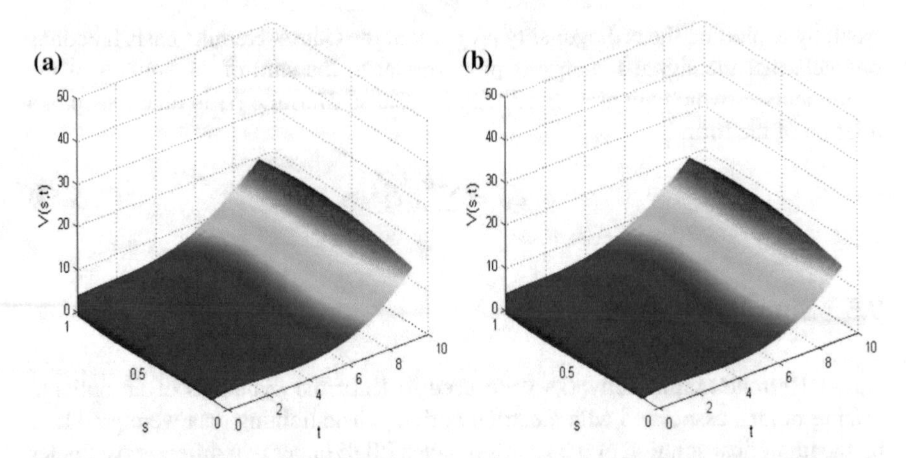

Fig. 9.7 **a** Neural modelling of the exact Black–Scholes PIDE, **b** Neural modelling of the Black–Scholes PIDE after parametric change (there are only slight differences between the *two curves*)

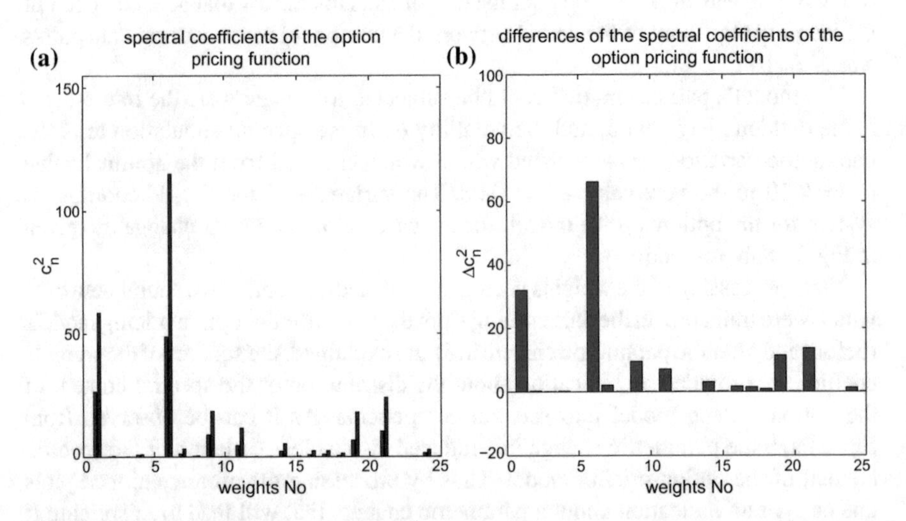

Fig. 9.8 Change in dividend payment δ: **a** Spectral content of the option pricing function before parametric change (*blue bars*) and after parametric change (*red bars*), **b** Changes in the spectral coefficients of the option pricing function, induced by parametric change

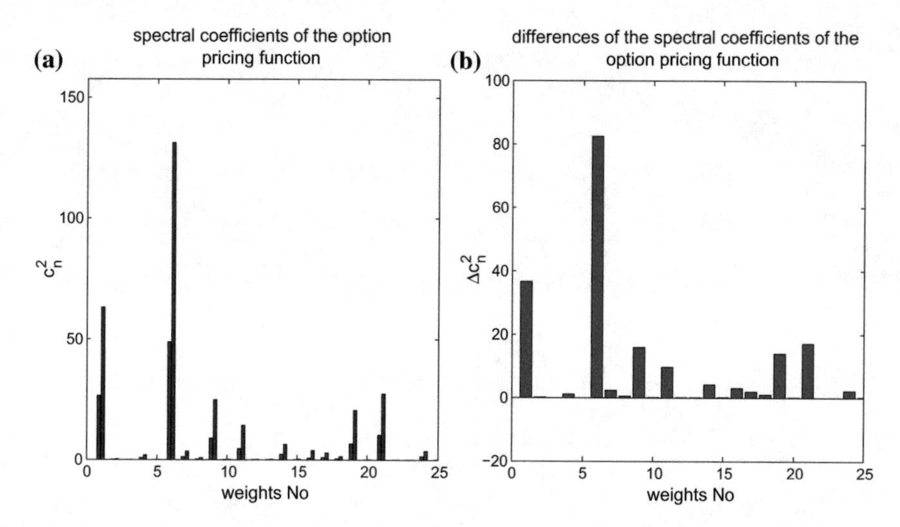

Fig. 9.9 Change in volatility σ: **a** Spectral content of the option pricing function before parametric change (*blue bars*) and after parametric change (*red bars*), **b** Changes in the spectral coefficients of the option pricing function, induced by parametric change

Finally, it is noted that in case that a parametric change is associated with the distortion of particular spectral components one can use the proposed change detection method for isolating the changed parameter of the model. For, instance in the spectral content diagrams of the volatility given in Fig. 9.9a, b, it can be observed that after parametric change specific spectral components are affected (e.g. appearance of spectral component No 4).

Chapter 10
Statistical Validation of Financial Forecasting Tools with Generalized Likelihood Ratio Approaches

10.1 Outline

The present chapter analyzes the problem of validation of non-parametric estimators, such as neuro-fuzzy approximators, used in modelling and forecasting of finance systems with nonlinear dynamics. Neuro-fuzzy modelling is one of the techniques used for the approximation of nonlinear and uncertain systems [256, 268]. Up to now little attention has been paid to the validation of fuzzy rule bases, i.e. to the verification of fuzzy rules consistency with respect to the physical system being modelled. To decide whether the outcome of a fuzzy model is satisfactory or not, the root mean square error (RMSE) or the normalized root mean square error (NRMSE) is commonly used. If the RMSE exceeds a threshold, then retraining is usually recommended [105]. It frequently occurs that a fuzzy model is invalidated, however the parameters that cause the discrepancy between the fuzzy model and the modelled system are rarely identified and a global retraining of the fuzzy rule base is attempted.

This chapter addresses the problem of fuzzy model validation in a systematic way by exploiting a statistical method for Fault Detection and Isolation (FDI). The method is the so called Local Statistical Approach to change detection [10, 21]. The local statistical approach has been successfully applied to several FDI problems. Based on a small parametric disturbance assumption, it aims at transforming complex detection problems concerning a parameterized stochastic process into the problem of monitoring the mean of a Gaussian vector. The local statistical approach consists of two stages: (i) the global χ^2 test which indicates the existence of a change in some parameters of the fuzzy model, (ii) the diagnostics tests (sensitivity or min-max) which isolate the parameter affected by the change [11, 12, 290]. If the changed parameter is isolated, then the partial retraining of the fuzzy system is possible. The local statistical approach is suitable for parametric change detection in financial systems.

Fuzzy model validation with the Local Statistical Approach has two significant advantages: (i) it provides a credible criterion (χ^2 test) to detect if a fuzzy rule baseis

G.G. Rigatos, *State-Space Approaches for Modelling and Control in Financial Engineering*, Intelligent Systems Reference Library 125,
DOI 10.1007/978-3-319-52866-3_10

acceptable or not no matter what the distribution of the training data is. This criterion is more efficient than the RMSE or the NRMSE since it employs the modeling error derivative and records the tendency for change, thus becoming suitable for early change detection; (ii) it recognizes the parameters of the fuzzy model that are responsible for the deviation from the parameters of the reference model. Thus local tuning of the fuzzy model instead of global retraining can be pursued.

Preliminary results on fuzzy model validation with the use of the local statistical approach, which also show the potential of this FDI method for finance systems, can be found in [201]. These results show the suitability of the method for chaotic time series forecasting and early diagnosis of parametric changes in the associated model. In addition to the above, it enables to study fuzzy modeling from the information theory point of view [17]. In this chapter, the Fisher Information Matrix of the fuzzy model is examined and conclusions on the detectability of parameter changes are derived. It is shown that fuzzy models with a strong partition of the universe of discourse may have a singular Fisher matrix. On the other hand fuzzy models with non-strong fuzzy partition have a non-singular Fisher matrix thus in this latter case the changes in the parameters can be distinguished through the statistical test.

10.2 Neuro-Fuzzy Modelling

10.2.1 Problem Statement

Financial time series can be approximated with the use of neuro-fuzzy models. A collection of N data in a $(n+1)$-dimensional space is considered. Then a generic fuzzy model is presented as a set of fuzzy rules in the following form:

$$\begin{array}{l} R^l : IF\ x_1\ is\ A_1^l\ AND\ x_2\ is\ A_2^l\ AND \cdots AND\ x_n\ is\ A_n^l \\ THEN\ \bar{y}^l = z^l(x) \end{array} \tag{10.1}$$

where $x = [x_1, x_2, \ldots, x_n] \in U$ and $\bar{y}^l \in V$ are linguistic variables, A_i^l are fuzzy sets of the universe of discourse $U^i \in R$ and $z^l(x)$ is a function of the input variables. Typically, z can take one of the following three forms: (i) fuzzy set, (ii) singleton and (iii) linear function. The overall data-driven fuzzy modelling problem can be formulated as follows: Given the N input-output patterns $(x; y)$, the model output $\hat{y}$ and a specified model error $\varepsilon > 0$, obtain the minimal number of fuzzy rules and optimal parameters for the fuzzy model such that the error function $E = ||y - \hat{y}||$ satisfies the inequality $E < \varepsilon$ [50]. The modeling procedure begins with the model initialization stage that includes data processing, prior knowledge utilization, and initial rule-base generation. The optimal number of fuzzy rules is determined and the appropriate type of fuzzy rules is selected. Model optimization follows which involves parameter learning and rule-base simplification. Finally, the acquired fuzzy model is validated under certain performance indexes. If the model performance is not satisfactory, further modification including structure and parameter optimization is required.

10.2.2 Determination of the Number and Type of Fuzzy Rules

10.2.2.1 Determination of the Number of Fuzzy Rules

When constructing a fuzzy model for a finance time series, one has to decide on the optimal number and the type of the fuzzy rules that will be used in this model. Two common approaches for the selection of the number of the fuzzy rules are the following [45]: (i) The input space partition, (ii) The input dimension (grid) partition (see Fig. 10.1).

Input space partitioning can be the result of a clustering procedure, as shown in Fig. 10.1a. Two popular clustering algorithms are the fuzzy c-means (FCM) algorithm and the Gustafson-Kessel (GK) algorithm. Furthermore, clustering can be the outcome of an optimization procedure (selection of the centers and spreads of the fuzzy sets from numerical data through a nonlinear least-squares approach). The problem with all clustering methods is that the projections of the obtained partitions on the axes of the input variables x_i may overlap, which implies loss of the interpretability of the fuzzy rules. In input space partition the membership of the i-th rule can be for example a Gaussian membership function, i.e.

$$\mu_{r^l}(x) = e^{-\sum_{i=1}^{n}(\frac{(x_i - c_i^l)}{v_i^l})^2} \tag{10.2}$$

where n is the number of the fuzzy membership functions of the i-th rule, while c_i^l denote the i-th center (spread) in the antecedent part of the l-th rule.

Input dimension partitioning can be the result of the so-called “grid” approach. If the partition of the patterns space is carried out following the “grid” approach the interpretability of the fuzzy rules is maintained. However an increased number of

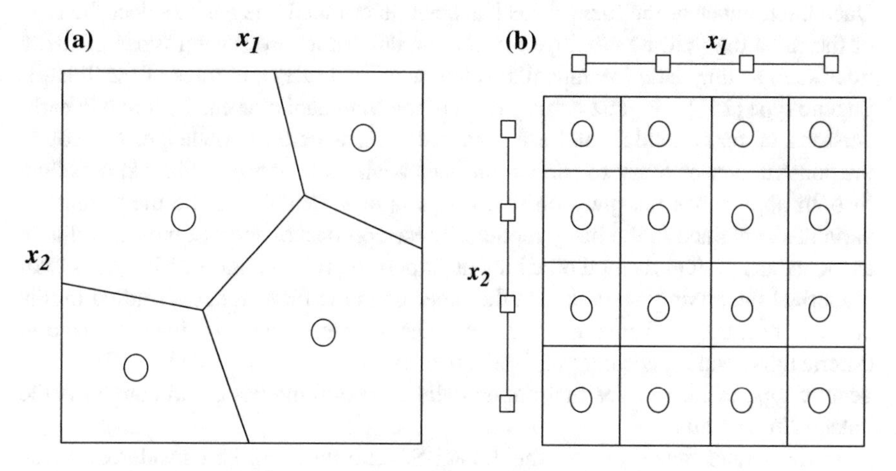

Fig. 10.1 **a** Input space partition **b** Input dimension (grid) partition

fuzzy rules may be obtained and some of the rules maybe practically inactive because they will be associated with empty areas of the patterns space. Therefore, refinement of the fuzzy rule base has to be carried out at a second stage. In input dimension partition, dimension x_i is split into consecutive segments, each one described by a membership function of the form: $\mu_{R^l}(x_i) = exp(-(x_i - c_i^l)/v_i^l)^2$, where c_i^l denotes the center of the i-th fuzzy set in the l-th fuzzy rule, and v_i^l is the associated variance.

When the input variables x_i are split into equal membership functions the input space is covered by the grid shown in Fig. 10.1b. This is the uniform grid partition. It should be noted that the input space partition is equivalent to that of an input dimension (grid) partition. The reason is that Eq. (10.2) can be rewritten as $\mu_{R^l}(x) = e^{-(x-c)^T \Lambda^{-2}(x-c)}$, where $x = [x_1, x_2, \ldots, x_n]^T$, $c^l = [c_1^l, c_2^l, \ldots, c_n^l]^T$ and $\Lambda = diag(v_1^l, v_2^l, \ldots, v_n^l)$. For different Λ^l's, multi-dimensional Gaussian membership functions with different spreads are derived, thus defining a non-uniform grid in the input space. To get a partition similar to the one depicted in Fig. 10.1a, the spread matrix Λ has to be non-diagonal.

In *input dimension (grid) partition*, the data space is split into $N_d = \Pi_{i=1}^n p_i$, where p_i is the number of partitions of each one of the n input dimensions. This is equal to the number of centers (spreads) that have to be tuned. If *input space partition* is applied, the number of centers to be tuned is $N_s = n \cdot p_s$, where n is again the number of dimensions and p_s is the number of the partitions of the input space. Usually $N_s < N_d$. The input space partition can succeed the same approximation accuracy with input dimension partition but with less adaptable parameters. The drawback of input space partition is that it may result in redundant rule bases which are difficult to be linguistically interpreted [90].

10.2.2.2 Determination of the Type of Fuzzy Rules

Once the number of the fuzzy rules has been determined one has to select the type of the rules that will be employed by the model. There are several types of fuzzy rules such as linguistic (Mamdani's) rules, relational rules and rules of the Takagi-Sugeno type [256]. Linguistic fuzzy rules of the Mamdani type can be found in early versions of fuzzy models and are obtained from experts' knowledge. Moreover, the construction of fuzzy relational matrices is also a matter of human knowledge. In both approaches the question that arises is how reliable can be the numerical variables contained in the fuzzy models. These approaches seem at most as reliable as the expert-system method of asking an expert to give condition-action rules with numerical uncertainty weights. On the other hand the fuzzy models studied in this chapter are extracted from numerical data. The rules are obtained using optimization criteria thus assuring accuracy and objectiveness of the obtained rule base. The more generic type of fuzzy rules that numerically extracted models could contain is the Takagi-Sugeno one.

In the sequel fuzzy rules of the Takagi-Sugeno type will be considered. These have the form:

$$R_l \;:\; IF\ x_1\ is\ A_1^l AND\ x_2\ is\ A_2^l\ AND\ \cdots AND\ x_n\ is\ A_n^l \\ THEN\ \bar{y}^l = \sum_{i=1}^{n} w_i^l x_i + b^l\ l = 1, 2, \ldots, L \tag{10.3}$$

where R^l is the l-th rule, $x = [x_1, x_2, \ldots, x_n]^T$ is the input (antecedent) variable, $\bar{y}^l$ is the output (consequent) variable, and w_i^l, b^l are the parameters of the local linear models. The above model is a Takagi-Sugeno model of order 1. Setting $w_i^l = 0$ results in the zero order Takagi-Sugeno model [90]. The output of the Takagi-Sugeno model is given by the consequences:

$$\hat{y} = \frac{\sum_{l=1}^{L} \bar{y}^l \prod_{i=1}^{n} \mu_{A_i^l}(x_i)}{\sum_{l=1}^{L} \prod_{i=1}^{n} \mu_{A_i^l}(x_i)} \tag{10.4}$$

where $\mu_{A_i^l}(x_i) \;:\; R \rightarrow [0, 1]$ is the membership function of the fuzzy set A_i^l in the antecedent part of the rule R^l. In the case of a zero order TS system the output of the l-th local model is $\bar{y}^l = b^l$, while in the case of a first order TS system the output of the l-th local model is given by $\bar{y}^l = \sum_{l=1}^{L} w_i^l x_i + b^l$.

If the numerically extracted fuzzy rule-base does not approximate efficiently the monitored financial (physical) system then a refinement of the partitioning of the patterns space may be required [113, 172].

10.2.3 Stages of Fuzzy Modelling

The individual steps of data-driven fuzzy modelling for nonlinear function approximation are discussed in [45, 114, 115, 239, 240]. These stages are demonstrated in Fig. 10.2.

1. *Initialization of the Fuzzy Model*: As explained in Sect. 10.2.2 the creation of the initial fuzzy model comprises two elements: (i) partition of the data space and formation of data clusters [240]. When clustering is complete a collection of L clusters $C = (c_1, c_2, \ldots, c_L)$ is produced. Each cluster is associated with a fuzzy rule, (ii) Selection of the type of the fuzzy rules. As already mentioned, the fuzzy model may consist of linguistic (Mamdani) rules, relational rules or fuzzy rules of the Takagi-Sugeno type.

2. *Parameter Optimization*: To improve the model performance and achieve higher modeling accuracy, the parameters of the initial fuzzy model should be optimized against a certain performance index, such as the RMSE. The centers and spreads of the membership functions and the weights of the local models are extracted through the recursive solution of a nonlinear least squares problem. This can be carried out with the use, for instance, of Extended Kalman Filtering (EKF), the Gauss-Newton or the Levenberg-Marquardt method [27, 178, 245]. If only the linear weights of the fuzzy system are of interest then the problem reduces to linear least squares and can be efficiently solved by e.g. applying the LMS algorithm.

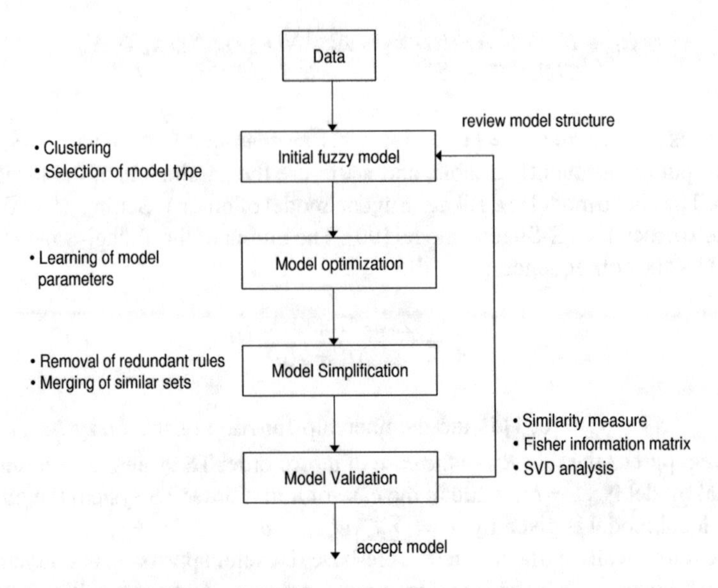

Fig. 10.2 General scheme of data-driven fuzzy modelling

3. *Model Simplification*: The initial rule-base obtained from data is often redundant and unnecessarily complex since it is generated from unconstrained optimization [68]. The absence of constraints on the nonlinear parameters (centers and spreads of the fuzzy sets) may result in a redundant (non interpretable) fuzzy rule base. The complexity is reduced when similarity analysis is used to identify fuzzy sets that represent conflicting or redundant concepts. By merging the redundant fuzzy sets, a more comprehensible fuzzy rule base can be derived. Rule-base simplification can be performed as follows [241]:
(i). Removal of redundant fuzzy rules: If a fuzzy membership function is always near zero over its entire definition set, i.e. $\mu_{A_j}(x_j) \simeq 0$, $\forall x_j \in U_j$ then the rule corresponding to this membership function can be removed because its output is always near zero. This can happen for instance if grid partitioning is followed, and certain segments of the input space are hardly occupied by any input patterns. In that case the degree of activation of the associated fuzzy rules will be negligible.
(ii). Merging similar fuzzy sets: The similarity between the fuzzy sets A_j and A_k is calculated by

$$S(A_j, A_k) = \frac{\sum_{i=1}^{n} \mu_{A_j}(x_i) \wedge \mu_{A_k}(x_i)}{\sum_{i=1}^{n} \mu_{A_j}(x_i) \vee \mu_{A_k}(x_i)} \tag{10.5}$$

for $j, k = 1, \ldots, c$, $j \neq k$. If $S(A_j, A_k) > \lambda$, i.e. fuzzy sets A_j and A_k are highly overlapping, then the two fuzzy sets A_j and A_k can be merged into one new fuzzy set A_j, where $\lambda \in (0, 1)$ is the threshold for merging fuzzy sets that are similar to one another (see Fig. 10.3).

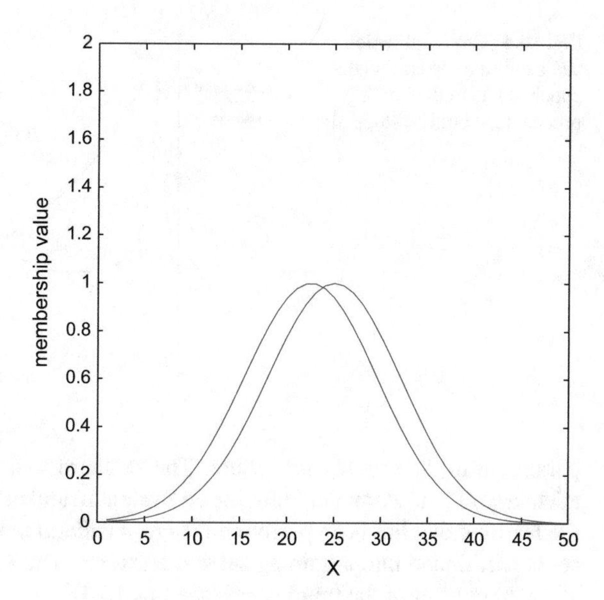

Fig. 10.3 Similar fuzzy sets

4. *Fuzzy model validation*: Using a validation technique the final model is either accepted as suitable for describing the real system or it is rejected and a new training procedure is evoked. Model validation helps to decide which parameters of the fuzzy model need tuning thus avoiding global retraining.

It is noted however that the aforementioned fuzzy modeling stages 1 to 4 are not always necessary. Some of them can be omitted in specific cases. For instance, the use of clustering techniques is not mandatory. Moreover, fuzzy clustering is usually performed to derive an adequate partition of the input space rather than the number of partitions (in several algorithms proposed in the literature the number of clusters, i.e. submodels must be predefined). In this context, if clustering is undertaken, then the nonlinear least squares problem makes sense only if the model designer wishes to refine the membership functions obtained by means of the projection of the resulting clusters (in addition to the optimization of the linear parameters/weights). Furthermore, similarity analysis is not the only way to avoid redundancy and reduce model complexity. For instance linguistic constraints can be included into the optimization problem [178] and Singular Value Decomposition (SVD) can be used in order to prune unnecessary rules [190, 286].

10.2.4 Fuzzy Model Validation for the Avoidance of Overtraining

Neural or fuzzy model validation (cross-validation) enables to determine the number of the fuzzy rules, the size of the training set as well as the size of learning-rate

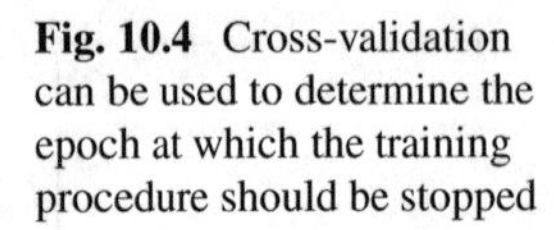

Fig. 10.4 Cross-validation can be used to determine the epoch at which the training procedure should be stopped

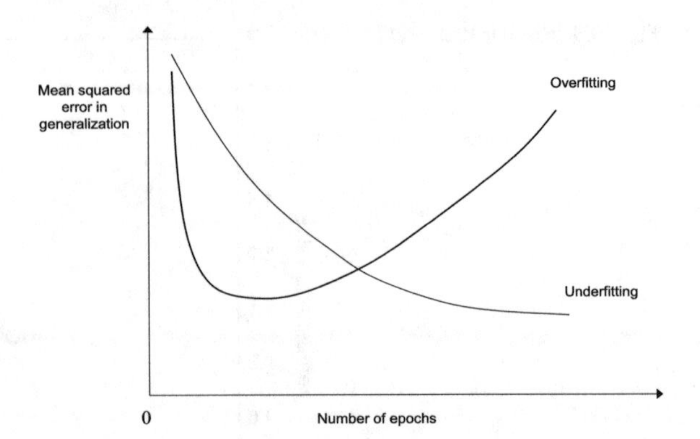

parameter in the training procedure. The standard tool in statistics to evaluate the adequacy of a fuzzy model (or of the equivalent neural network) and the efficiency of the learning that has been performed is *cross validation* [2, 132]. The available data set is partitioned into a training set and a test set. The validation subset is typically 10 to 20 percent of the training set (see Fig. 10.4).

The validation of the fuzzy model can be used to define the complexity of the fuzzy rule base or the neural network in terms of number of fuzzy rules, number of hidden neurons, etc. Statistically the problem may be interpreted as that of choosing the size of the parameter set used to model the data set. Overtraining of a neural network shows up as poorer performance on the validation set. Another way in which cross-validation may be used is to decide when the training of network on the training set should be actually stopped. One can succeed good generalization even if the fuzzy rule base (neural network) is designed to have too many parameters, provided that the training of the network on the training set is stopped at a number of epochs corresponding to the minimum point of the error-performance curve on cross-validation.

Cross-validation may be also used to adjust the size of the learning-rate parameter of a fuzzy rule base (neural network). In particular training of the fuzzy rule base is first performed on the training set, and the cross validation set is used to validate the training after each epoch. If the performance of the fuzzy rule base on the cross validation set fails to improve by a certain amount, the size of the learning-rate parameter is reduced. After each succeeding epoch the learning-rate parameter is further reduced until once again there is no further improvement in performance on the cross-validation set. When this point is reached training of the network is halted.

The usual criterion applied to model validation is the RMSE on the validation set however this criterion is weaker than the proposed χ^2 test used by the local statistical approach for the following reasons: (i) there is no systematic criterion to choose a RMSE value (threshold) that will give an early indication of overtraining and according to which the training procedure can be stopped, (ii) the RMSE is not a sufficient statistic, which means that deviation of certain parameters of the model from their nominal values may not be reflected to a clear change of the RMSE.

10.3 Fuzzy Model Validation with the Local Statistical Approach

10.3.1 The Exact Model

Inconsistencies between the fuzzy model and the monitored finance time series may appear if for some reason the parameters of the financial system change. If the deviation exceeds a threshold then retraining of the fuzzy model can be determined. Furthermore if the parameters subject to change are identified, partial retraining (re-adaptation of a subset of parameters) can be attempted.

Model-based approaches to Fault Detection and Isolation assume that the failures and degradations correspond to changes in some parameters of the so-called "exact model". In simple words the *exact* model is the model that describes the fault-free system up to random noise, or the model within the model set that approximates best the system. Instead of searching for deviation between the output of the physical system and the output of the fuzzy model one can compare the output of the *exact model* to the output of the fuzzy model (see Fig. 10.5). It should be reminded that the assumption that the true system is "exactly" described by a model is also followed in model-based control.

Many financial systems can be written in terms of differential equations and thus provide us with a physical model [40, 41, 263, 304]. However, the physical model is not always available or it can be too complex, or knowledge about it maybe incomplete. In that case, the exact model of the system can be represented in the form of a black-box model (i.e. a neural network, a wavelets network or a fuzzy rule

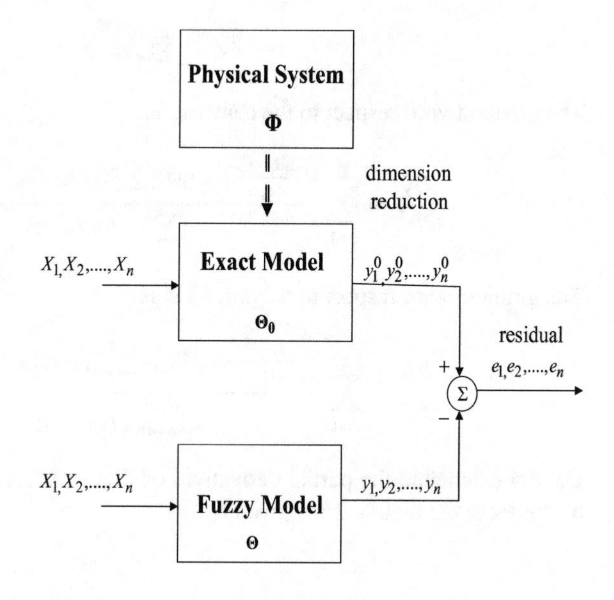

Fig. 10.5 Residual between the fuzzy and the exact model (the exact model is the fuzzy model that is extracted from input/output data of the physical system when the latter is in normal condition

base, trained with the use of input and output measurements of the physical system). Nevertheless, it should be noted that there is no one to one correspondence between the parameters of the physical model and the parameters of a black-box model. Thus, if the FDI method indicates a change in a parameter of the black-box model this cannot be mapped clearly to a failure of a certain component of the physical system. The physical model is usually of higher dimension (parameter vector ϕ) than the black-box model (parameter vector θ) and thus the transformation of the physical model into the black-box model implies dimension reduction [19, 293].

10.3.2 The Change Detection Test

The local statistical approach for change detection in the finance system's model is analyzed first, following closely [10, 12]. First, the residual e_i is defined as the difference between the fuzzy model output $\hat{y}_i$ and the physical system output y_i, i.e. $e_i = \hat{y}_i - y_i$. As explained in Sect. 10.3.1 it is also acceptable to define the residual as the difference between the fuzzy model output and the exact model output, where the exact model replaces the physical system and has the same number of parameters as the fuzzy model (see Fig. 10.5). The partial derivative of the residual square is:

$$H(\theta, y_i) = \frac{\partial e_i^2}{\partial \theta} = e_i \frac{\partial y_i}{\partial \theta} \tag{10.6}$$

The vector H having as elements the above $H(\theta, y_i)$ is called primary residual. In the case of fuzzy models the gradient of the output with respect to the consequent parameters w_i^l is given by

$$\frac{\partial \hat{y}}{\partial w_i^l} = \frac{x_i \mu_{R^l}(x)}{\sum_{l=1}^{L} \mu_{R^l}(x)} \tag{10.7}$$

The gradient with respect to the center c_i^l is

$$\frac{\partial \hat{y}}{\partial c_i^l} = \sum_{l=1}^{L} \frac{\bar{y}^l \frac{2(x_i - c_i^l)}{v_i^l} \mu_{R^l}(x_i)[\sum_{j=1}^{L} \mu_{R^j}(x_i) - \mu_{R^l}(x_i)]}{[\sum_{l=1}^{L} \mu_{R^l}(x_i)]^2} \tag{10.8}$$

The gradient with respect to the spread v_i^l is

$$\frac{\partial \hat{y}}{\partial v_i^l} = \sum_{l=1}^{L} \frac{\bar{y}^l \frac{2(x_i - c_i^l)^2}{v_i^{l^3}} \mu_{R^l}(x_i)[\sum_{j=1}^{L} \mu_{R^j}(x_i) - \mu_{R^l}(x_i)]}{[\sum_{l=1}^{L} \mu_{R^l}(x_i)]^2} \tag{10.9}$$

Having calculated the partial derivatives of Eqs. (10.7), (10.8) and (10.9), the rows of the Jacobian matrix J are given by

$$J(\theta_0, y_k) = \left.\frac{\partial \hat{y}_k(\theta)}{\partial \theta}\right|_{\theta=\theta_0} \tag{10.10}$$

The problem of change detection with the χ^2 test consists of monitoring a change in the mean of the Gaussian variable which for the one-dimensional parameter vector θ is formulated as

$$X = \frac{1}{\sqrt{N}} \cdot \sum_{i=1}^{N} e_k \frac{\partial y_k}{\partial \theta} \sim N(\mu, \sigma^2) \tag{10.11}$$

where $\hat{y}_k$ is the output of the fuzzy model generated by the input pattern x_k, e_k is the associated residual and θ is the vector of the model's parameters. For a multivariable parameter vector θ should hold $X \sim (M\eta, \Sigma)$. In order to decide if the fuzzy model is in good agreement with a given set of data, let θ_* be the value of the parameters vector minimizing the RMSE. The notation is introduced only for the convenience of problem formulation, and its actual value does not need to be known. Then the model validation problem amounts to make a decision between the two hypotheses:

$$\begin{aligned} H_0 &: \theta_* = \theta_0 \\ H_1 &: \theta_* = \theta_0 + \tfrac{1}{\sqrt{N}} \delta\theta \end{aligned} \tag{10.12}$$

where $\delta\theta \neq 0$. It is known from a central limit theorem that for a large data sample, the normalized residual given by Eq. (10.11) asymptotically follows a Gaussian distribution when $N \to \infty$ [10, 21]. More specifically, the hypothesis that has to be tested is:

$$\begin{aligned} H_0 &: X \sim N(0, \Sigma) \\ H_1 &: X \sim N(M\eta, \Sigma) \end{aligned}$$

The sensitivity matrix M of $\frac{1}{\sqrt{N}} X$ is defined as X is defined as $E\{\frac{\partial}{\partial \theta} H(\theta, y_k)\}$ and is approximated by [11]:

$$M(\theta_0) \simeq \frac{\partial}{\partial \theta} \frac{1}{N} \sum_{k=1}^{N} H(\theta_0, y_k) \simeq \frac{1}{N} J^T J \tag{10.13}$$

The covariance matrix S is defined as $E\{H(\theta, y_k) H^T(\theta, y_{k+m})\}$, $m = 0, \pm 1, \ldots$ and is approximated by [12]:

$$\begin{aligned} S = &\simeq \textstyle\sum_{k=1}^{N} [H(\theta_0, y_k) H^T(\theta_0, y_k)] + \\ &+ \textstyle\sum_{k=1}^{I} \frac{1}{N-k} \sum_{k=1}^{N-k} [H(\theta_0, y_k) H^T(\theta_0, y_{k+m}) + H(\theta_0, y_{k+m}) H^T(\theta_0, y_k)] \end{aligned} \tag{10.14}$$

where an acceptable value for I is 3. The decision tool is the likelihood ratio $s(X) = ln \frac{p_{\theta_1(x)}}{p_{\theta_0(x)}}$, where $p_{\theta_1}(X) = e^{[X-\mu(X)]^T S^{-1}} [X - \mu(X)]$ and $p_{\theta_0}(X) = e^{X^T S^{-1} X}$. The center of the Gaussian distribution of the changed system is denoted as $\mu(X) = M\eta$ where η is the parameters vector. The *Generalized Likelihood Ratio* (GLR) is calculated by maximizing the likelihood ratio with respect to η [12]. This means that the most likely case of parameter change is taken into account. This gives the global χ^2 test t:

$$t = X^T S^{-1} M (M^T S^{-1} M)^{-1} M^T S^{-1} X \tag{10.15}$$

Since X asymptotically follows a Gaussian distribution, the statistics defined in Eq. (10.15) follows a χ^2 distribution with n degrees of freedom. Mapping the change detection problem to this χ^2 distribution enables the choice of the change threshold. Assume that the desired probability of false alarm is α then the change threshold λ should be chosen from the relation $\int_\lambda^\infty \chi_n^2(s)ds = \alpha$, where $\chi_n^2(s)$ is the probability density function (p.d.f.) of a variable that follows the χ^2 distribution with n degrees of freedom.

10.3.3 Isolation of Parametric Changes with the Sensitivity Test

A first approach to change isolation is to focus only on a subset of the parameters while considering that the rest of the parameters remain unchanged [12]. The parameters vector η can be written as $\eta = [\phi, \psi]^T$, where ϕ contains those parameters to be subject to the isolation test, while ψ contains those parameters to be excluded from the isolation test. M_ϕ contains the columns of the sensitivity matrix M which are associated with the parameters subject to the isolation test. Similarly M_ϕ contains the columns of M that are associated with the parameters to be excluded from the sensitivity test.

Assume that among the parameters η, it is only the subset ϕ that is suspected to have undergone a change. Thus η is restricted to $\eta = [\phi, 0]^T$. The associated columns of the sensitivity matrix are given by M_ϕ and the mean of the Gaussian to be monitored is $\mu = M_\phi \phi$, i.e.

$$\mu = MA\phi, \quad A = [0, I]^T \tag{10.16}$$

Matrix A is used to select the parameters that will be subject to the fault isolation test. The rows of A correspond to the total set of parameters while the columns of A correspond only to the parameters selected for the test. Thus the fault diagnosis (χ^2) test of Eq. (11.46) can be restated as:

$$t_\phi = X^T S^{-1} M_\phi (M_\phi^T S^{-1} M_\phi)^{-1} M_\phi^T S^{-1} X \tag{10.17}$$

10.3.4 Isolation of Parametric Changes with the Min-Max Test

In this approach the aim is to find a statistic that will be able to detect a change on the part ϕ of the parameters vector η and which will be robust to a change in the non observed part ψ [12]. Assume the vector partition $\eta = [\phi, \psi]^T$. The following notation is used:

$$M^T \Sigma^{-1} M = \begin{pmatrix} I_{\varphi\varphi} & I_{\varphi\psi} \\ I_{\psi\varphi} & I_{\psi\psi} \end{pmatrix} \tag{10.18}$$

$$\gamma = \begin{pmatrix} \varphi \\ \psi \end{pmatrix}^T \cdot \begin{pmatrix} I_{\varphi\varphi} & I_{\varphi\psi} \\ I_{\psi\varphi} & I_{\psi\psi} \end{pmatrix} \cdot \begin{pmatrix} \varphi \\ \psi \end{pmatrix} \tag{10.19}$$

The min-max test aims at minimizing the non-centrality parameter γ with respect to the parameters that are not suspected for change. The minimum of γ with respect to ψ is given for:

$$\psi^* = \arg\min_{\psi} \gamma = \varphi^T (I_{\varphi\varphi} - I_{\varphi\psi} I_{\psi\psi}^{-1} I_{\psi\varphi})\varphi \tag{10.20}$$

and is found to be

$$\begin{aligned} \gamma^* &= \min_{\psi} \gamma = \varphi^T (I_{\varphi\varphi} - I_{\varphi\psi} I_{\psi\psi}^{-1} I_{\psi\varphi})\varphi = \\ &= \begin{pmatrix} \varphi \\ -I_{\psi\psi}^{-1} I_{\psi\varphi}\varphi \end{pmatrix}^T \begin{pmatrix} I_{\varphi\varphi} & I_{\varphi\psi} \\ I_{\psi\varphi} & I_{\psi\psi} \end{pmatrix} \begin{pmatrix} \varphi \\ -I_{\psi\psi}^{-1} I_{\psi\varphi}\varphi \end{pmatrix} \end{aligned} \tag{10.21}$$

which results in

$$\gamma^* = \varphi^T \{[I, -I_{\varphi\psi} I_{\psi\psi}^{-1}] M^T \Sigma^{-1}\} \, \Sigma^{-1} \{\Sigma^{-1} M [I, -I_{\varphi\psi} I_{\psi\psi}^{-1}]\} \varphi \tag{10.22}$$

The following linear transformation of the observations is considered:

$$X_{\phi}^* = [I, -I_{\varphi\psi} I_{\psi\psi}^{-1}] M^T \Sigma^{-1} X \tag{10.23}$$

The transformed variable X_{ϕ}^* follows a Gaussian distribution $N(\mu_{\phi}^*, I_{\phi}^*)$ with mean:

$$\mu_{\varphi}^* = I_{\varphi}^* \varphi \tag{10.24}$$

and with covariance:

$$I_{\varphi}^* = I_{\varphi\varphi} - I_{\varphi\psi} I_{\psi\psi}^{-1} I_{\psi\varphi} \tag{10.25}$$

The max-min test decides between the hypotheses:

$$H_0^* : \mu^* = 0$$
$$H_1^* : \mu^* = I_\varphi^* \varphi$$

and is described by:

$$\tau_\varphi^* = X_\varphi^{*T} I_\varphi^{*-1} X_\varphi^* \tag{10.26}$$

Fuzzy model validation with the use of the local statistical approach is summarized in the following table:

Stages of the local statistical approach
1. Generate the residuals given by Eq. (10.11)
2. Calculate the Jacobian matrix J given by Eq. (10.10)
3. Calculate the sensitivity matrix M given by Eq. (10.13)
4. Calculate the covariance matrix S given by Eq. (10.14)
5. Apply the χ^2 test for change detection of Eq. (10.15)
6. Apply the change isolation tests of Eqs. (10.17) or (10.26)

It should be noted that if the number of faults i is unknown, then solution to change isolation can be obtained from the property that the sensitivity and the min-max tests are connected to each other through a Pythagorean relation $t_0 = t_\phi + t_\psi^*$. In that case the parameter vector is separated into $\eta^T = [\phi, \psi]^T$, where t_ϕ denotes the sensitivity test with respect to the monitored parameters ϕ (the faulty parameters are contained in the set ϕ, thus one should expect the outcome of the sensitivity test to be large), and t_ϕ^* denotes the complementary min-max test with respect to the non-monitored parameters ψ (the faulty parameters are contained in the complementary of the set ψ, thus one should expect the outcome of the min-max test to be small). Using this concept the change isolation test in case of unknown number of faults, can be decomposed in the following steps [14, 149]:

1. The number of the suspected changed parameters is set to $i = 1$.

2. For each set of parameters of dimension i the sensitivity test is carried out. The set that results in the maximum value of the sensitivity test is maintained.

3. For the set of parameters ϕ of dimension i chosen in step 2, the complementary min-max test is carried out, (i.e. the min-max test is performed for the complementary set of parameters ψ). If this test exceeds a threshold (one should expect the score of the min-max test to be small since change was supposed to exist only among parameters ϕ), then this means that change exists also among the parameters ψ. In the latter case, index i is increased by 1, i.e. $i = i + 1$, and the algorithm returns to step 2. Otherwise, the algorithm terminates and the parameters subject to change are those contained in the last selected set.

10.3.5 Model Validation Reduces the Need for Model Retraining

A question that arises is why monitoring cannot be substituted by repeated identification of the parameter vector (on-board detection). In that case a straightforward solution would be to compare two successive estimates of the parameters vector with the aid of a distance measure. This is a valid FDI approach, but has also several weaknesses:
(i) Continuous parameter identification can be computationally expensive, especially in models of complex and large dimensional systems. Of course, dimensionality reduction is possible but this is another open research topic [109]. Thus, there is need for designing FDI algorithms which do not preform re-identification on each incoming data set.
(ii) In case of model retraining, the key issue is how to measure the difference between the new estimated parameter vector and the reference one, and how to decide that this deviation is significant, especially in the presence of disturbances in the system. This requires an estimate of the accuracy of the identification procedure, which ends again to an asymptotic gaussian type of result such as the one included in the local statistical approach. Consequently, one falls again to the problem that has been optimally solved in the local statistical approach.

10.4 Detectability of Changes in Fuzzy Models

A model is said to be detectable when it is possible to find the parameters subject to change through statistical processing of its output. A necessary condition for parameters detectability is given through the positive definiteness of the Fisher matrix [10]. The Fisher information is defined as follows:

$$I(\theta_0) = E\{\frac{\partial L}{\partial \theta^T}\frac{\partial L}{\partial \theta}\}\Big|_{\theta=\theta_0} \tag{10.27}$$

where $L_\theta(Y)$ is the likelihood function $L_\theta(Y) = L(y_1, y_2, \dots, y_n|\theta)$, which is given by $L_\theta(Y) = ln\{p_\theta(y_1, y_2, \dots, y_n)\}$. In the multidimensional case when θ is a vector the Fisher information matrix can be defined as the covariance matrix of the partial derivative of the likelihood ratio L with respect to θ.

It holds that $I(\theta) = E[(\frac{\partial L}{\partial \theta})^2] = -E(\frac{\partial^2 L}{\partial \theta^2})$, i.e. the Fisher information matrix is equal to the negative of mean of the Hessian matrix of L [233]. The Fisher information matrix plays a key role in the sampling distribution of the parameter estimates. Specifically, maximum likelihood parameter estimates are asymptotically normally distributed such that the estimation of $\hat{\theta}_k$ of parameter θ at time instant k satisfies

$(\hat{\theta}_k - \theta) \sim N(0, I^{-1}(\theta))$ as $k \to \infty$. The Fisher information matrix provides a lower bound, called a Cramer-Rao lower bound, for the standard errors of estimates of the model parameters. The following theorem holds: [205].

Theorem *In fuzzy rule bases of strong partition, and of triangular membership functions, the Fisher matrix is a singular one. Therefore some parameter changes may not be detectable.*

Proof The proof is not addressed only to fuzzy rule bases with one variable in the antecedent part of the rule but can also be extended to fuzzy rules with unlimited number of variables in the IF part of the rule. Therefore the generality of the approach is not affected. To provide a generic complete proof, a fuzzy system of N rules (for any integer $N > 1$) is first considered. The l-th fuzzy rule is given by Eq. (10.3).

The parameters vector of the fuzzy rule base is declared as $\theta = [\theta_1, \ldots, \theta_i, \ldots, \theta_m]$. The general element of the sensitivity matrix M is $[m_{i,j}] = \frac{\partial y}{\partial \theta_i}\frac{\partial y}{\partial \theta_j}$. The parameters vector of the fuzzy model of Eq. (11.33) can in turn be written as $\theta = [w_1, \ldots, w_N, c_1 \ldots, c_{N \times n}]$ where N is the number of the fuzzy rules $R^l, l = 1, 2, \ldots,$ N, w^l is the center of the fuzzy set B^l which appears in the consequent part of the l-th rule, n is the number of the fuzzy sets in the antecedent part of the fuzzy rules $i = 1, 2, \ldots, n$, and c_i^l is the center of the i-th fuzzy set in the antecedent part of the l-th rule. In that case the general form of the sensitivity matrix M becomes

$$M = \frac{1}{N}\sum_{k=1}^{N}$$
$$\begin{pmatrix} [\frac{\partial \hat{y}}{\partial w_1}(x_k)]^2 & \cdots \frac{\partial \hat{y}}{\partial w_1}\frac{\partial \hat{y}}{\partial w_N}(x_k) & \cdots \frac{\partial \hat{y}}{\partial w_1}\frac{\partial \hat{y}}{\partial c_1^N}(x_k) & \cdots \frac{\partial \hat{y}}{\partial w_1}\frac{\partial \hat{y}}{\partial c_n^N}(x_k) \\ \frac{\partial \hat{y}}{\partial w_1}\frac{\partial \hat{y}}{\partial w_2}(x_k) & \cdots \frac{\partial \hat{y}}{\partial w_2}\frac{\partial \hat{y}}{\partial w_N}(x_k) & \cdots \frac{\partial \hat{y}}{\partial w_2}\frac{\partial \hat{y}}{\partial c_1^N}(x_k) & \cdots \frac{\partial \hat{y}}{\partial w_2}\frac{\partial \hat{y}}{\partial c_n^N}(x_k) \\ \cdots & \cdots & \cdots & \cdots \\ \frac{\partial \hat{y}}{\partial c_1^N}\frac{\partial \hat{y}(x_k)}{\partial w_1}(x_k) & \cdots \frac{\partial \hat{y}}{\partial c_1^N}\frac{\partial \hat{y}(x_k)}{\partial w_N}(x_k) & \cdots [\frac{\partial \hat{y}}{\partial c_1^N}(x_k)]^2 & \cdots \frac{\partial \hat{y}}{\partial c_1^N}\frac{\partial \hat{y}}{\partial c_n^N}(x_k) \\ \cdots & \cdots & \cdots & \cdots \\ \frac{\partial \hat{y}}{\partial c_n^N}\frac{\partial \hat{y}(x_k)}{\partial w_1}(x_k) & \cdots \frac{\partial \hat{y}}{\partial c_n^N}\frac{\partial \hat{y}(x_k)}{\partial w_N}(x_k) & \cdots \frac{\partial \hat{y}}{\partial c_n^N}\frac{\partial \hat{y}}{\partial c_1^N}(x_k) & \cdots [\frac{\partial \hat{y}}{\partial c_n^N}(x_k)]^2 \end{pmatrix} \quad (10.28)$$

According to Eq. (10.4) the output of the generalized fuzzy model will be

$$\hat{y}(x) = \frac{\sum_{l=1}^{n}\bar{y}^l \prod_{i=1}^{N}\mu_{A_i^l}(x_i)}{\sum_{l=2}^{2}\prod_{i=1}^{N}\mu_{A_l^i}(x_i)}. \quad (10.29)$$

Without loss of generality a minimal fuzzy rule base of two (non-redundant) rules can be considered:

$$\begin{array}{l} R_1 : IF\ x_1\ is\ A_1\ THEN\ y\ is\ B_1 \\ R_2 : IF\ x_2\ is\ A_2\ THEN\ y\ is\ B_2 \end{array} \tag{10.30}$$

The associated fuzzy model will be $\hat{y}(x) = \frac{\sum_{l=1}^{2} \bar{y}^l \mu_{A_l}(x)}{\sum_{l=1}^{2} \mu_{A^l}(x)}$ while the calculation of the elements of the sensitivity matrix gives:

$$M = \frac{1}{N}\sum_{k=1}^{N} \begin{pmatrix} [\frac{\partial \hat{y}}{\partial w_1}(x_k)]^2 & \frac{\partial \hat{y}}{\partial w_1}\frac{\partial \hat{y}}{\partial w_2}(x_k) & \frac{\partial \hat{y}}{\partial w_1}\frac{\partial \hat{y}}{\partial c_1}(x_k) & \frac{\partial \hat{y}}{\partial w_1}\frac{\partial \hat{y}}{\partial c_2}(x_k) \\ \frac{\partial \hat{y}}{\partial w_1}\frac{\partial \hat{y}}{\partial w_2}(x_k) & [\frac{\partial \hat{y}}{\partial w_2}(x_k)]^2 & \frac{\partial \hat{y}}{\partial w_2}\frac{\partial \hat{y}}{\partial c_1}(x_k) & \frac{\partial \hat{y}}{\partial w_2}\frac{\partial \hat{y}}{\partial c_2}(x_k) \\ \frac{\partial \hat{y}}{\partial w_1}\frac{\partial \hat{y}}{\partial c_1}(x_k) & \frac{\partial \hat{y}}{\partial w_2}\frac{\partial \hat{y}}{\partial c_1}(x_k) & [\frac{\partial \hat{y}}{\partial c_1}(x_k)]^2 & \frac{\partial \hat{y}}{\partial c_1}\frac{\partial \hat{y}}{\partial c_2}(x_k) \\ \frac{\partial \hat{y}}{\partial w_1}\frac{\partial \hat{y}}{\partial c_2}(x_k) & \frac{\partial \hat{y}}{\partial w_2}\frac{\partial \hat{y}}{\partial c_2}(x_k) & \frac{\partial \hat{y}}{\partial c_1}\frac{\partial \hat{y}}{\partial c_2}(x_k) & [\frac{\partial \hat{y}}{\partial c_2}(x_k)]^2 \end{pmatrix} \tag{10.31}$$

It holds that the first 2×2 principal sub-matrix is singular. Indeed, the sub-matrix

$$M_{w_2c_1} = \frac{1}{N}\sum_{k=1}^{N} \begin{pmatrix} [\frac{\partial \hat{y}}{\partial w_2}(x_k)]^2 & \frac{\partial \hat{y}}{\partial w_2}\frac{\partial \hat{y}}{\partial c_1}(x_k) \\ \frac{\partial \hat{y}}{\partial c_1}\frac{\partial \hat{y}}{\partial w_2}(x_k) & [\frac{\partial \hat{y}}{\partial c_1}(x_k)]^2 \end{pmatrix} \tag{10.32}$$

has zero determinant. Indeed, substituting

$$\frac{\partial \hat{y}}{\partial w_1} = \frac{\mu_{A_1}}{\mu_{A_1}+\mu_{A_2}} \qquad \frac{\partial \hat{y}}{\partial w_2} = \frac{\mu_{A_2}}{\mu_{A_1}+\mu_{A_2}}$$

$$\frac{\partial \hat{y}}{\partial c_1} = \frac{(w_1-w_2)\mu_{A_2}\frac{\partial \mu_{A_1}}{\partial c_1}}{(\mu_{A_1}+\mu_{A_2})^2} \quad \frac{\partial \hat{y}}{\partial c_2} = \frac{-(w_1-w_2)\mu_{A_1}\frac{\partial \mu_{A_2}}{\partial c_2}}{(\mu_{A_1}+\mu_{A_2})^2}$$

and assuming a strong partition of the fuzzy universe of discourse i.e. $\mu_{A_1}+\mu_{A_2}=1$, thus $\frac{\partial \mu_{A_1}}{\partial c_1} = -\frac{\partial \mu_{A_2}}{\partial c_2} = a$ gives

$$M_{w_2c_1} = \frac{1}{N}\sum_{k=1}^{N} \begin{pmatrix} \mu_{A_2}^2 & (w_1-w_2)\mu_{A_2}{}^2 a \\ (w_1-w_2)\mu_{A_2}^2 a & (w_1-w_2)^2 \mu_{A_2}{}^2 a^2 \end{pmatrix} \tag{10.33}$$

which is a matrix of zero determinant. This means that the sensitivity matrix M is not positive definite.Indeed, the singularity of M is shown by contradiction: assume tha matrix M is positive definite, then according to Sylvester's theorem the associated sub-determinants of this matrix should all be positive. However the sub-determinant $M_{w_2c_1}$ described in Eq. (10.31) is 0 which contradicts the initial argument, thus M cannot be positive definite.

Taking into account the relation $I = \frac{N}{\sigma^2}M$ (the proof can be found in [10]), it can be concluded that the Fisher Information Matrix is not positive definite too. Since the Fisher information matrix is by definition non-negative definite, it is thus singular.

Consequently it must be expected that even in fuzzy rule bases without redundancy the isolation of the changed parameters may be unsuccessful. On the other hand the Fisher information matrix is usually non-singular in fuzzy rule-bases with non-strong partition, which makes the changes in the associated fuzzy models to be detectable by the local statistical approach.

The following corollary can now be stated: In fuzzy models of strong partition the statistical change isolation between linear and nonlinear parameters may be unsuccessful.

Indeed, without loss of generality assume a fuzzy rule base of three rules with output given by:

$$\hat{y} = \frac{w_1\mu_{A_1}(x) + w_2\mu_{A_2}(x) + w_3\mu_{A_3}(x)}{\mu_{A_1}(x) + \mu_{A_2}(x) + \mu_{A_3}(x)} \tag{10.34}$$

which can be equivalently written as

$$\hat{y} = \frac{w_1\varphi(x-c_1) + w_2\varphi(x-c_2) + w_3\varphi(x-c_3)}{\varphi(x-c_1) + \varphi(x-c_2) + \varphi(x-c_3)} \tag{10.35}$$

where $\phi(x-c_i)$ is a membership function of center c_i. Assume a small change on the center c_2 i.e. $c_2 = c_2 + \Delta c$ and take the Taylor expansion of $\phi(x - c2)$ up to the first term. Then the output disturbance will be

$$\hat{y} + \Delta\hat{y} \approx \frac{w_1\varphi(x-c_1) + w_2\varphi(x-c_2) + w_3\varphi(x-c_3) + w_2\varphi(x-c_2)^{'}\Delta c_2}{\varphi(x-c_1) + \varphi(x-c_2) + \varphi(x-c_3) + \varphi(x-c_2)^{'}\Delta c_2} \tag{10.36}$$

In the case of strong fuzzy partition i.e. $\sum_{i=1}^{N}\varphi(x-c_i) = 1$ the previous relation becomes

$$\hat{y} + \Delta\hat{y} \approx \frac{w_1\varphi(x-c_1) + w_2\varphi(x-c_2) + w_3\varphi(x-c_3) + w_2\varphi(x-c_2)^{'}\Delta c_2}{1 + \varphi(x-c_2)^{'}\Delta c_2} \tag{10.37}$$

Since the denominator is close to 1 it can be substituted by its inverse, thus giving

$$\hat{y} + \Delta\hat{y} \approx [y + w_2\varphi(x-c_2)^{'}\Delta c_2][1 + \varphi(x-c_2)^{'}\Delta c_2] \ \ i.e.$$

$$\Delta\hat{y} \approx w_2\varphi(x-c_2)^{'}\Delta c_2$$

In the case of triangular membership functions $\varphi(x-c_2)^{'} = const$ while in the case of Gaussian membership functions with finite support $\varphi(x-c_2)^{'} \simeq const$, i.e.

$$\Delta\hat{y} \approx w_2 \cdot const \cdot \Delta c_2 \tag{10.38}$$

Therefore one can infer two different reasons of the output change Δy: (i) the change on the center Δc_2 and (ii) a change on the weight $w_2{\cdot}const$. This coupling between faults may explain the inefficiency of the diagnostic test when this is applied to both weights and centers of the fuzzy rule base at the same time and the fuzzy rule base consists of fuzzy sets with strong partition.

It is noteworthy that for most black-box models of financial systems the singularity of the Fisher matrix is caused by over-parameterization (redundancy) and the problem can be overcome by reducing the model complexity. For instance, in MLP neural networks the necessary and sufficient condition for a positive definite Fisher information matrix is irreducibility, i.e. the removal of all redundant neurons from the network [84–86]. The previous analysis shows that fuzzy models with strong partition of the fuzzy sets can have a non-positive definite (singular) Fisher matrix even if they are irreducible, i.e. even if all redundant fuzzy sets have been removed from the rule base. However, when strong fuzzy partition does not hold then detectability of the fuzzy model is preserved.

It should be also pointed out that all the information that concerns change detectability of a fuzzy model is contained in the Fisher information matrix. If the columns i and j of the Fisher matrix I are highly correlated, i.e.

$$\rho = \frac{|M_i^T \Sigma^{-1} M_j|}{\sqrt{M_i^T \Sigma^{-1} M_i}\sqrt{M_j^T \Sigma^{-1} M_j}} \simeq 1 \tag{10.39}$$

then one cannot distinguish if the change occurred on the parameter i or on the parameter j.

For fuzzy models with non-triangular membership functions, the Fisher matrix may be close to singular (badly conditioned).

The suitability of triangular fuzzy sets to approximate Gaussian functions is discussed in [69, 159, 185]. It has been shown that the α-cuts of a triangular fuzzy set contain the confidence intervals of any symmetric probability distribution (and thus of the Gaussian distribution) with the same center and support. Thus triangular membership functions yield the optimal distribution-free confidence intervals for symmetric probability distributions with bounded support. Under the assumption of symmetry, the α-cuts of triangular membership functions can be viewed as distribution-free confidence intervals for quantities lying in a specified interval. This result can be approximately extended to unbounded probability distributions, which are characterized by a shape depending on the mean value and the variance (such as the Gaussian distribution).

10.5 Simulation Results

The local statistical approach was applied to the validation of two different fuzzy models: (i) a 1st order Takagi-Sugeno fuzzy rule base that was generated using the input space partition method described in Sect. 10.2.2, (ii) a zero-order Takagi-Sugeno fuzzy rule base that was generated with the input dimension (grid) partition.

10.5.1 Fuzzy Rule Base in Input Space Partitioning

The function $y = 2x_1 + x_2 + 0.2sin(2\pi x_1) + 0.2sin(2\pi x_2)$ was used as a test case. The rule-base generated by the EKF training consisted of the following rules:

$$\begin{aligned}
R^{(1)} &: IF\ x_1\ is\ (c_1^{(1)}, v)\ AND\ x_2\ is\ (c_2^{(1)}, v)\ THEN\ y = a_1^{(1)}x_1 + a_2^{(1)}x_2 \\
R^{(2)} &: IF\ x_1\ is\ (c_1^{(2)}, v)\ AND\ x_2\ is\ (c_2^{(2)}, v)\ THEN\ y = a_1^{(2)}x_1 + a_2^{(2)}x_2 \\
R^{(3)} &: IF\ x_1\ is\ (c_1^{(3)}, v)\ AND\ x_2\ is\ (c_2^{(3)}, v)\ THEN\ y = a_1^{(3)}x_1 + a_2^{(3)}x_2 \\
R^{(4)} &: IF\ x_1\ is\ (c_1^{(4)}, v)\ AND\ x_2\ is\ (c_2^{(4)}, v)\ THEN\ y = a_1^{(4)}x_1 + a_2^{(4)}x_2
\end{aligned}$$

The input vector $[x_1, x_2]^T$ is uniformly distributed in the interval $[0, 6] \times [0, 6]$. The above model implies a local linearization of the initial nonlinear function using 4 sub-models. To initialize the centers $c_i^{(j)}$, $i = 1, 2$ and $j = 1, 2, 3, 4$ the plane $[0, 6] \times [0, 6]$ is split into 4 parts. The following initial values are given: $c_1^{(1)} = 2.0$ $c_1^2 = 2.0$, $c_1^{(2)} = 4.0$, $c_2^2 = 2.0$, $c_1^{(3)} = 2.0$, $c_2^{(3)} = 4.0$, $c_1^{(4)} = 4.0$, $c_2^{(4)} = 4.0$. The spread of the membership functions is denoted by v. The parameters of the consequent part were all initialized to 0. After training the following rules were identified:

$$\begin{aligned}
&R^{(1)}: \text{IF } x_1 \text{ is } (2.86312, v) \text{ AND } x_2 \text{ is } (2.86312, v) \\
&\text{THEN } \hat{y} = 2.13452x_1 + 1.04256x_2 \\
&R^{(2)}: \text{IF } x_1 \text{ is } (4.66312, v) \text{ AND } x_2 \text{ is } (2.66312, v) \\
&\text{THEN } \hat{y} = 2.03898x_1 + 1.20499x_2 \\
&R^{(3)}: \text{IF } x_1 \text{ is } (2.66312, v) \text{ AND } x_2 \text{ is } (4.46312, v) \\
&\text{THEN } \hat{y} = 2.02972x_1 + 0.94512x_2 \\
&R^{(4)}: \text{IF } x_1 \text{ is } (4.76312, v) \text{ AND } x_2 \text{ is } (4.76312, v) \\
&\text{THEN } \hat{y} = 1.89367x_1 + 1.06086x_2
\end{aligned} \tag{10.40}$$

The input space partition is demonstrated in Fig. 10.6.

The spread of the fuzzy membership functions was chosen to be $v = 3.0$. The forgetting factor λ in the EKF training algorithm was initialized to $\lambda = 0.99$ and was progressively raised to 1 to succeed convergence [27].

The fuzzy model consisted of gaussian membership functions. It was initially considered identical to the exact model and consisted of the parameters contained in the previous rule base. The local statistical approach was used to check the deviation between the disturbed exact model and the fuzzy model. White noise of variance $\sigma = 0.05$ was added to the output y of the exact model. The training set contained $N = 3000$ triplets $((x_{1i}, x_{2i}, y,\ i = 1, \ldots, 3000))$. The gradient of the output with respect to the parameters of the fuzzy model is given by Eqs. (10.7), (10.8) and (10.9).

First, the χ^2 test was performed for the case of no change in the parameters of the exact model. The mean value of the χ^2 test, was found to be $t = 16.8524$, while the ideal value would be 16 i.e. equal to the dimension of the parameters vector. The condition number of the Fisher information matrix $M^T \Sigma^{-1} M$, i.e. the

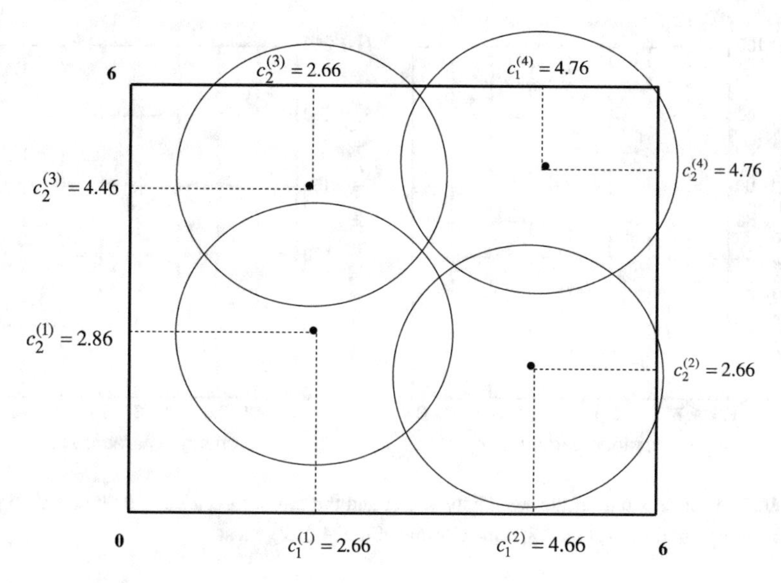

Fig. 10.6 Fuzzy rule base generated with input space partition

ratio of the maximum to the minimum eigenvalue of this matrix was found to be $cond(M^T \Sigma^{-1} M) = 7.14 \times 10^3$.

Then, changes were imposed to linear and nonlinear parameters (weights and centers respectively). The global χ^2 test was able to detect changes of less than 1 % of the nominal parameter values. Two indicative changes were: a change to the weight (linear parameter) $w_2^{(4)}$ and a change to the center (nonlinear parameter) $c_1(^{2)}$.

In the case of the linear parameter the change (fault) was chosen to be $w_2^{(4)}$: 1.05644→1.04644. The value of the χ^2 test was $t = 121.2406$. The condition of the Fisher information matrix was found to be $cond(M^T \Sigma^{-1} M) = 1.83 \times 10^4$. The sensitivity test gave a correct diagnosis about the parameter subject to change. The value of the sensitivity test for the parameter $w_2^{(4)}$ was found to be 115.9481. In the case of one single fault the sensitivity test for the parameter subject to fault returns a value close to the global test.

In the case of the nonlinear parameter the change (fault) was chosen to be $c_1^{(2)}$: 4.66303→4.76303. The value of the χ^2 test was $t = 142.6997$. It can be seen that the global test is less sensitive to a change on a nonlinear parameter.

Detailed results on the performance of the sensitivity and the min-max tests are given in Fig. 10.7a, as well as in Table 10.1. Ten different changes of the nonlinear parameter $c_1^{(1)} = 2.6630$ were generated. The larger the deviation of parameter $c_1^{(1)}$ from its nominal value was, the larger the success rate of the two tests became. Moreover, the mean value of the global χ^2 is depicted in Fig. 10.7b. It can be observed that when $c_1^{(1)}$ is close to its nominal value the χ^2 test takes a value close to the change threshold $\eta = 16$. On the other hand when the deviation of $c_1^{(1)}$ from its nominal value increases the value of the χ^2 grows significantly.

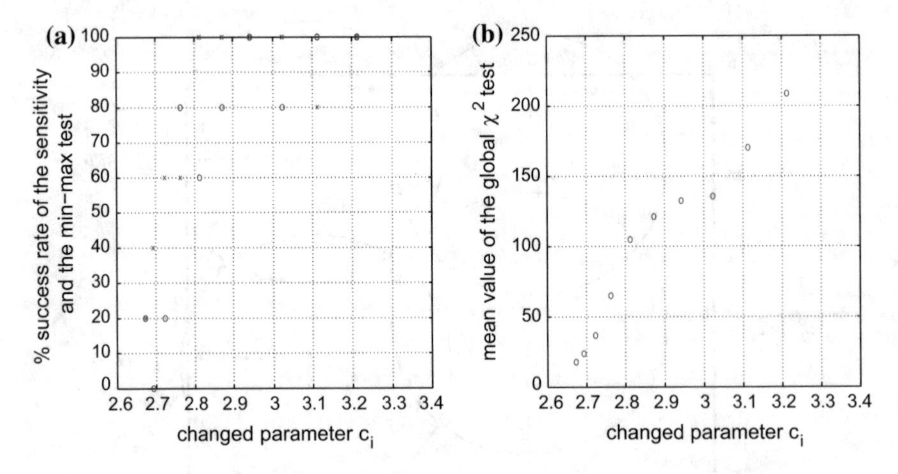

Fig. 10.7 **a** Success rate of the sensitivity × test and the min-max *o* tests in case of a change in parameter $c_1^{(2)}$ of the exact model, **b** mean value of the global χ^2 test

Table 10.1 Success rate of the sensitivity and the min-max test of parameter $c_1^{(1)}$

Parameter value	Parameter change %	Sensitivity test %	Min-max test %
2.668	1.88	20	20
2.673	3.76	40	0
2.678	5.64	60	20
2.683	7.52	60	80
2.688	9.40	100	60
2.693	11.28	100	80
2.690	13.16	100	100
3.030	15.04	100	80
3.080	16.92	100	100
3.130	18.80	100	100

The performance of the sensitivity and the min-max tests was also recorded in the case of a change in parameter $c_2^{(1)} = 2.8630$ of the exact model. As shown in Fig. 10.8a, the magnitude of the parameter change affects the success rate of the change isolation tests. Large parameter changes result also in large values of the global χ^2 test that exceed the change threshold $\eta = 16$ (see Fig. 10.8b).

Furthermore, the performance of the change isolation tests was examined in the case of simultaneous changes in linear and nonlinear parameters of the exact model. Changes took place, at the same time, in center $c_1^{(1)}$ and the weight $\alpha_1^{(1)}$. The results of the sensitivity and min-max test are depicted in Fig. 10.9a. It can be observed that the change isolation tests were not as successful as in the case of changes taking place at only one parameter of the exact model. In Fig. 10.9b the performance of the global χ^2 test is demonstrated. One can see that the global test can always detect the

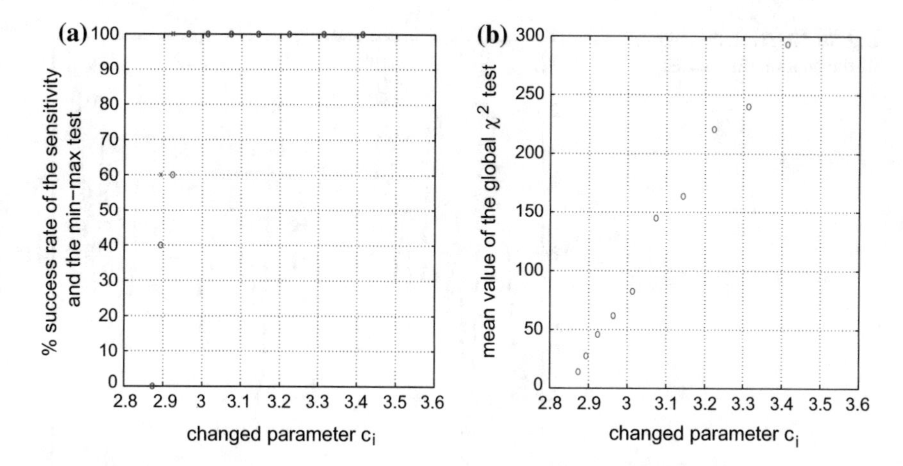

Fig. 10.8 **a** Success rate of the sensitivity (×) and the min-max (*o*) tests in case of a change in parameter $c_2^{(1)}$ of the exact model, **b** mean value of the global χ^2 test

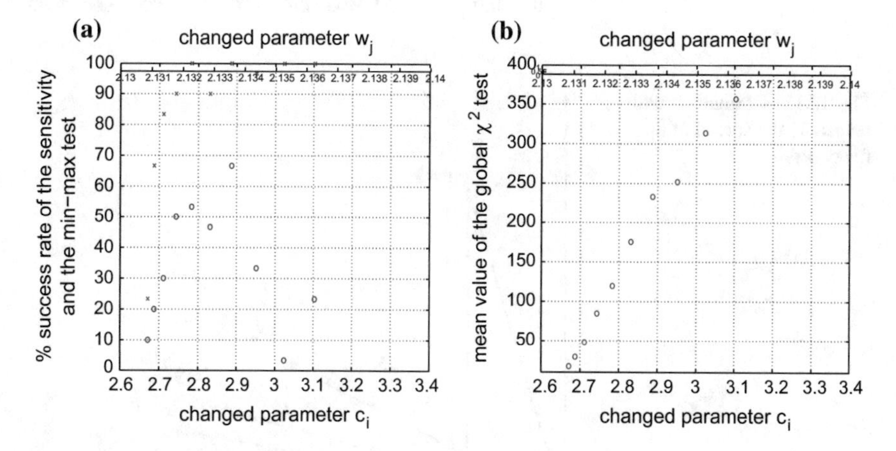

Fig. 10.9 **a** Success rate of the sensitivity (×) and the min-max (*o*) tests in case of simultaneous changes in parameters $c_1^{(1)}$ and $\alpha_1^{(1)}$ of the exact model, **b** mean value of the global χ^2 test

discrepancy between the exact and the fuzzy model and as expected the value of the χ^2 test becomes high when the parameter changes are significant.

It should be noticed that the correlation of the model parameters (see Eq. (10.39)) affects the performance of the change isolation tests. Take for instance, the correlation factor $\rho_{c_1^{(1)}c_1^{(1)}}$ between $c_1^{(1)}$ and $c_1^{(3)}$ with respect to the variance υ of the fuzzy sets which is shown in Fig. 10.11 and in Table 10.2. It can be observed that the success of the change isolation tests decreases when the spread of the membership functions increases. This is because the fuzzy model becomes more redundant (a fuzzy rule success to cover all input data).

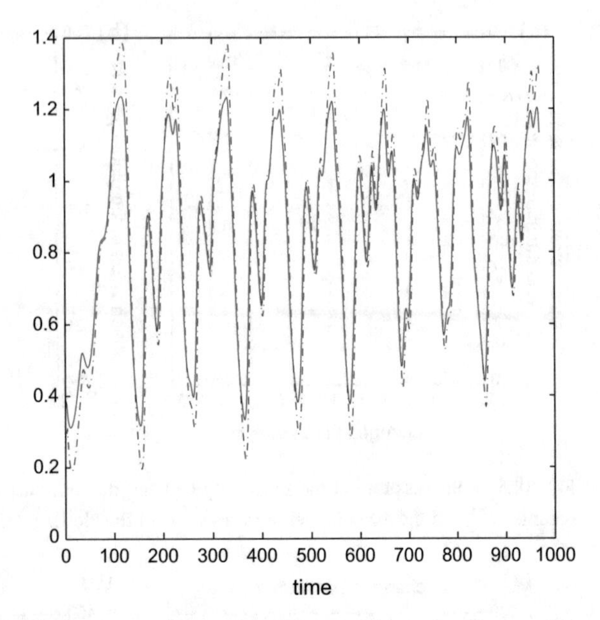

Fig. 10.10 Evolution in time of the chaotic time series

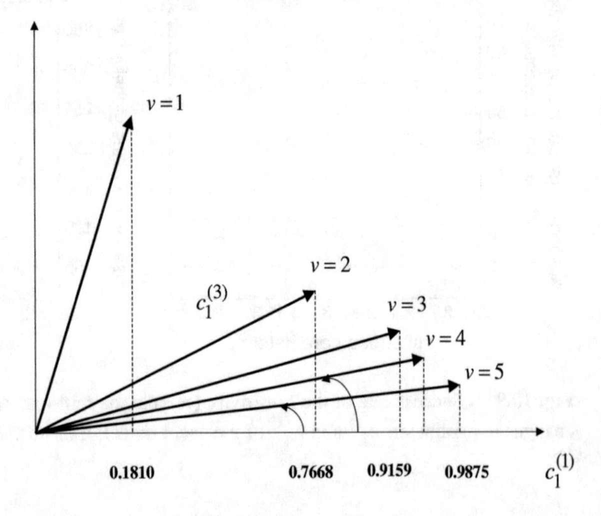

Fig. 10.11 Correlation with respect to the spread of the fuzzy sets

It is also of worth to mention that the success rate of the change isolation tests can be increased if decoupling between faults is succeeded. As already mentioned the decoupling of faults is indicated, (i) by the correlation between the different columns of the Fisher information matrix (zero correlation means that the Fisher matrix is diagonal), (ii) by the condition number of the Fisher matrix. The correlation factor and the condition number of the Fisher matrix I depend on the spread v of the membership functions. This is shown in Fig. 9 in the case of the centers $c_1^{(1)}$ and the center $c_1^{(3)}$.

Table 10.2 Condition number of Fisher Matrix

spread v	$\rho_{c_1^{(1)} c_1^{(3)}}$	*cond I*
5	0.9875	1.9742×10^{10}
4	0.9755	3.5196×10^{8}
3	0.9159	1.0457×10^{7}
2	0.7668	3.0094×10^{5}
1	0.1810	2.2321×10^{4}

10.5.2 Fuzzy Modelling with the Input Dimension Partition

Many fuzzy rule bases use the input dimension (grid) partition, i.e. the input variables are split in equal membership functions that in the 2-D case form a grid. As mentioned in Sect. 10.2.2 this partition results in comprehensible fuzzy rule bases but suffers from the curse of dimensionality when the number of dimensions and the number of fuzzy sets at each dimension increases.

The grid partition has been used in time series prediction. A fuzzy predictor is considered for a chaotic time series which is described by the following delay differential equation:

$$\frac{dx(t)}{dt} = \frac{0.2x(t-\tau)}{1+x^{10}(t-\tau)} - 0.1x(t) \tag{10.41}$$

A diagram of the time-series for sampling period $T_s = 1sec$ is depicted in Fig. 10.10.

Using the RMSE or the NRMSE to decide if there is a discrepancy between the real and the predicted output $(x - \hat{x})$ requires a threshold but there is no systematic way to define it. On the contrary in the χ^2 test the threshold is clearly defined as the dimension of the parameters vector η. The following training sets for the predictor of the chaotic time series were obtained

$$\begin{pmatrix} x(k-n) & \cdots x(k-1) \rightarrow x(k+1) \\ x(k-n-1) & \cdots x(k-2) \quad \rightarrow x(k) \\ \vdots & \vdots \quad\quad \vdots \\ x(1) & \cdots x(n-1) \rightarrow x(n+1) \end{pmatrix}$$

The rule base consists of rules of the form:

$$\begin{aligned} &R^l : IF\ X(K-3)\ is\ A_i\ AND\ X(k-2)\ is\ B_j\ AND\ X(k-1)\ is\ C_k \\ &THEN\ X(k+1)\ is\ D_m \end{aligned} \tag{10.42}$$

with $i, j, k, m = 1, 2, 3, 4$. The LMS algorithm was used for the adaptation of the linear weights $\alpha_1^{(l)}$. The rule base consists of 64 rules (3 input variables partitioned in 4 fuzzy subsets each).τ was set equal to 30. All fuzzy sets where assumed to have the same spread.

Inactive rules can be removed from the fuzzy rule base. The reduced-size rule base contained 22 rules. Therefore, the dimension of the parameters vector is 34, (22 linear weights in the antecedent part of the rules, and 12 nonlinear centers) and the value of the change threshold was set to $\eta = 34$. In the case of absence of change to the parameters of the rule base the global χ^2 test was (averaging over 10 trials) $t = 38.0319$. The condition of the Fisher Information Matrix was *cond* $(M^T \Sigma^{-1} M) = 5.5471 \times 10^5$.

Next, faults were imposed to the linear and the nonlinear parameters of the fuzzy system. Detailed diagrams on the performance of the change detection and isolation tests, in the case of a change in parameter $\alpha_1^{(2)} = 0.4092$ of the exact model, are given in Fig. 10.12. The success rate of the two tests is depicted in Fig. 10.12a. The mean value of the global χ^2 is depicted in Fig. 10.12b.

Detailed diagrams on the performance of the change detection and isolation tests, in the case of a change in parameter $c_1^{(1)} = 0.2771$ of the exact model, are given in Fig. 10.13. The parameters vector contained only the centers of the fuzzy model. The success rate of the two tests is depicted in Fig. 10.13a. It can be observed that the sensitivity test was more successful than the min-max test. The success rate was low when the parameters vector contained both the weights and the centers of the fuzzy model and this can be explained using the previous corollary.

The mean value of the global χ^2 is depicted in Fig. 10.13b. It is clear that when $c_1^{(1)}$ is close to its nominal value the χ^2 test takes a value close to the change threshold

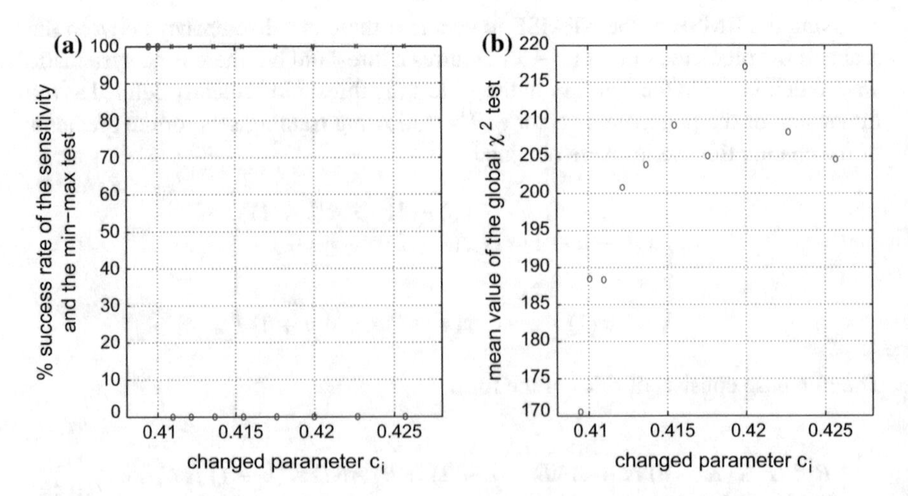

Fig. 10.12 Success rate of the sensitivity ($\times$) and the min-max (o) tests in case of a change in parameter $\alpha_1^{(2)}$ of the exact model of the Mackey-Glass time series. The parameters vector contained both weights and centers. **b** Mean value of the global χ^2 test

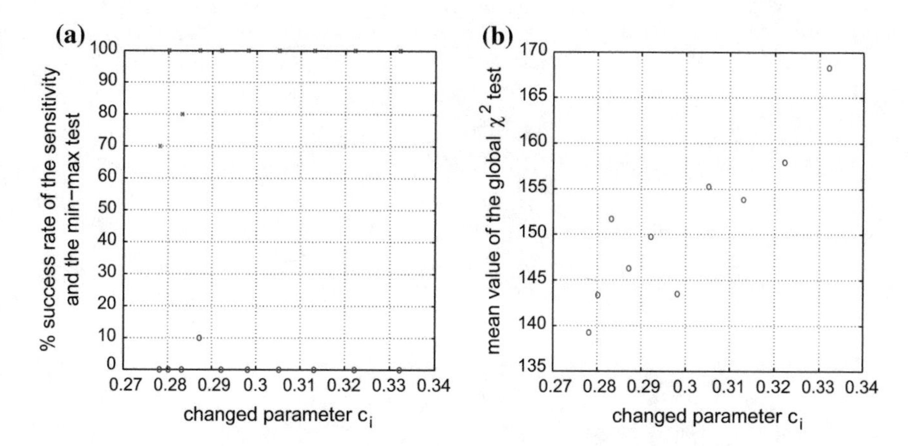

Fig. 10.13 Success rate of the sensitivity ($\times$) and the min-max (o) tests in case of a change in parameter $c_1^{(1)}$ of the exact model of the chaotic time series. The parameters vector contained only the centers of the fuzzy model. **b** Mean value of the global χ^2 test

$\eta = 34$. On the other hand when the deviation of $c_1^{(1)}$ from its nominal value increases the value of the χ^2 test grows significantly.

Moreover, it is pointed out that apart from the parameters correlation mentioned in Sect. 10.5.1 the success rate of the change isolation tests is affected by three more factors [6]: (a) the magnitude of the parameter change: a parameter change close to the nominal value results in increased success of the sensitivity test, (b) the size of the data set used for the statistical tests: if the number of data used in the statistical tests is large then the success rate is high, (c) the signal-to-noise ratio.

Finally, it can be stated that methods for local retraining can be used to correct the "changed" parameters of the fuzzy model of a finance system and these are the same with methods of global retraining. The major difference is that in global retraining, the dimension of the updated parameters vectors θ is N, while in local retraining this dimension is usual 1 (in case of single faults) or close to 1 (in case of simultaneously occurring faults). The algorithms used for local retraining are usually based on first and second order gradient techniques. These algorithms belong to: (i) batch-mode learning, where to perform parameters update the outputs of a large training set are accumulated and the mean square error is calculated (back-propagation algorithm, Gauss-Newton method, Levenberg-Marquardt method, etc.), (ii) pattern-mode learning, where training examples are run in cycles and the parameters update is carried out each time a new datum appears (Extended Kalman Filter algorithm).

Chapter 11
Distributed Validation of Option Price Forecasting Tools Using a Statistical Fault Diagnosis Approach

11.1 Overview

In this chapter the problem of validation of modelling and forecasting tools for financial systems is generalized to the distributed case. Distributed (multi-agent) modeling of financial systems' dynamics has received significant attention during the last years [76, 308, 309]. The chapter analyzes a distributed scheme for validation of option price forecasting models enabling early diagnosis of options mispricing. It is considered that N independent agents monitor and forecast the variation of option prices through locally parameterized Kalman Filters. It is also assumed that final decision about the options' price is taken through a fuzzy consensus scheme, that is the individual forecasts of the distributed agents, provided by local Kalman Filters are fused with a fuzzy weighting process. Thus forecasting is finally performed by a Fuzzy Kalman Filter.

Fuzzy Kalman Filtering is a method for distributed state estimation in which the local state estimates (in this case about options' price) produced by distributed Kalman Filters are fused into one single state estimate through weighting with fuzzy membership values. The Fuzzy Kalman Filter is designed based on the coverage of the state space by several local linear estimators according to the concept of fuzzy local linearization. It can be shown that the fuzzy Kalman Filter for distributed forecasting of options' price can be written in the form of a Takagi–Sugeno–Kung (TSK) fuzzy model, which stands for local ARMAX models weighted by fuzzy membership functions [101, 162, 225, 272]. The development of a method for validation of a distributed options' price estimation tool (Fuzzy Kalman Filter) and for detecting incipient changes in its parameters, is important for financial engineering.

The problem of Kalman Filter's validation has been studied also in past, however the developed methods are limited to linear systems and give no evidence that detection of incipient parameters change is possible. Indicative results on validation and consistency checking of Kalman Filters can be found in [89, 310]. Actually in [38, 118, 310], the χ^2 test and other statistical indexes are used for testing theaccuracy of

G.G. Rigatos, *State-Space Approaches for Modelling and Control in Financial Engineering*, Intelligent Systems Reference Library 125,
DOI 10.1007/978-3-319-52866-3_11

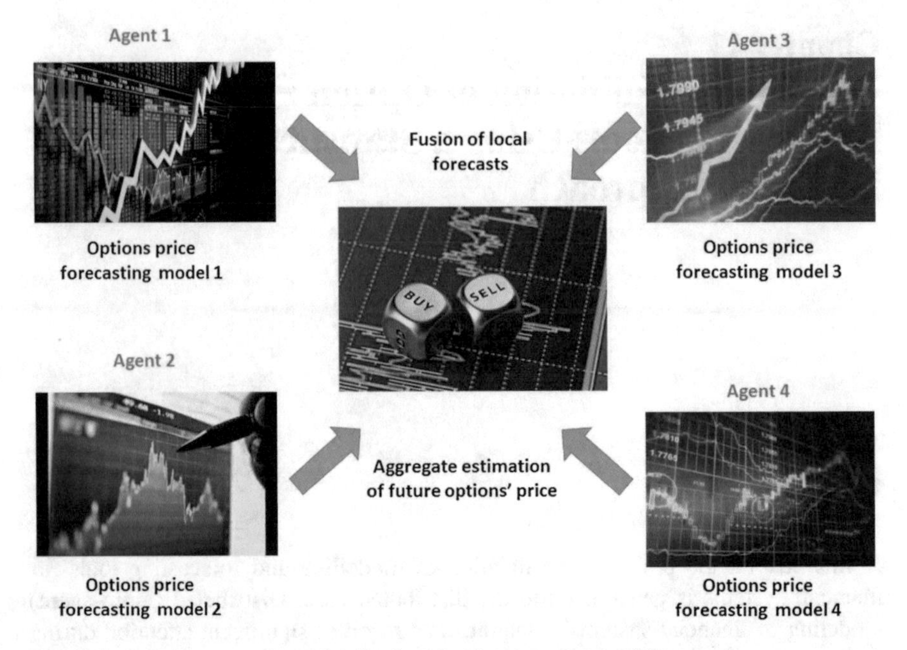

Fig. 11.1 A multi-agent scheme for distributed estimation of options' future prices

the estimation provided by the linear Kalman Filter, whereas in [138, 262, 306] these methods are applied to multiple and distributed Kalman Filters and to the Ensemble Kalman Filter. In [89] statistical methods are developed, in the form of a residuals and a smoother test, to check inconsistencies in the Kalman Filter. Other frequently met tests used in the assessment of Kalman Filter's accuracy and consistency are (i) the normalized error square test, (ii) the autocorrelation test and (iii) the normalized mean error test [9, 10, 60]. If the aforementioned tests are not satisfied then this means that the Kalman Filter is not running optimally, and the filter has to be retuned, or that the filter's design has to be reconsidered.

In this chapter the *local statistical approach to fault diagnosis* is proposed as a method for validation of distributed options' price forecasting tools, in the form of Fuzzy Kalman Filtering. This method, first appearing in [21], has been successfully tested in a wide range of fault diagnosis problems as shown in [195, 205, 290], while some of its recent applications in nonlinear and distributed parameter systems can be found in [198, 202, 216]. Here, the local statistical approach is used to check whether the models of the local Kalman Filters that constitute the Fuzzy Kalman Filter remain consistent with respect to parameters of the real system. The chapter provides one of the few approaches for testing the accuracy of distributed Kalman Filters and the only one that permits to detect parametric changes that are of magnitude of less than 1% of the nominal value of the parameter.

Comparing to the previously mentioned methods for Kalman Filter validation, the proposed approach is a significant contribution to this area. Validation of such a type of options' price estimation tools with the Local Statistical Approach has two significant advantages: (i) it provides a credible criterion (upgraded χ^2 test) to detect if the fuzzy Kalman Filter is acceptable or not, no matter what the distribution of the measurement data is. This criterion is more efficient than the previously mentioned normalized square error and mean error tests since it employs the modeling error derivative and records the tendency for change. Thus early change detection for faults in the filter's parameters becomes possible (ii) it recognizes those parameters of the Fuzzy Kalman Filter which are responsible for the deviation of the filter's estimates from the real output of the monitored dynamical system. Thus the retuning or redesign of the Fuzzy Kalman Filter can focus on a small number of its parameters.

11.2 State Estimation for the Black–Scholes PDE

11.2.1 State-Space Description of the Black–Scholes PDE

The focus of this chapter is on distributed estimation of the future prices of options and on validation methods of the commonly used forecasting tools (Fig. 11.1).

Next, the following nonlinear Black–Scholes PDE is considered:

$$\frac{\partial V(t,S)}{\partial t} + r_t S \frac{\partial V(t,S)}{\partial S} + \frac{1}{2}\sigma_t^2 S^2 \frac{\partial^2 V(t,S)}{\partial S^2} - r_t V(t,S) = 0 \tag{11.1}$$

It is considered that the volatility σ_t has a nonlinear dependence on V_{ss}, that is

$$\sigma_t = \sigma(V_{ss}) \tag{11.2}$$

where $V_{ss} = \frac{\partial^2 V(t,s)}{\partial S^2}$ and σ is a nonlinear function. The PDE model will be decomposed into an equivalent set of N ODEs using the finite differences method. Again, a grid of N points is considered, that is $\{s_1, s_2, \ldots, s_{N-1}, s_N\}$ which are placed at equal distances on the S axis. At the points of spatial discretization it holds

$$\frac{\partial V(t,s_i)}{\partial t} = r_t V(t,s_i) - r_t s_i \frac{\partial V(t,s_i)}{\partial S} - \frac{1}{2}\sigma_t^2(V_{ss}) s_i^2 \frac{\partial^2 V(t,s_i)}{\partial S^2} \tag{11.3}$$

where $i = 1, 2, \ldots, N$. The following state vector is defined

$$\begin{array}{c} \tilde{V} = [V(t,s_1), V(t,s_2), V(t,s_3), \ldots, V(t,s_{N-1}), V(t,s_N)] \text{ or} \\ \tilde{V} = [V_1, V_2, V_3, \ldots, V_{N-1}, V_N] \end{array} \tag{11.4}$$

thus one has

$$\begin{array}{c}
\frac{\partial V_1}{\partial t} = r_t V_1 - r_t s_1 \frac{V_2 - V_1}{\Delta S} - \frac{1}{2}\sigma_t^2(V_{ss}) s_1^2 \frac{V_2 - 2V_1 + V_0}{\Delta S^2} \\
\frac{\partial V_2}{\partial t} = r_t V_2 - r_t s_2 \frac{V_3 - V_2}{\Delta S} - \frac{1}{2}\sigma_t^2(V_{ss}) s_2^2 \frac{V_3 - 2V_2 + V_1}{\Delta S^2} \\
\frac{\partial V_3}{\partial t} = r_t V_3 - r_t s_3 \frac{V_4 - V_3}{\Delta S} - \frac{1}{2}\sigma_t^2(V_{ss}) s_3^2 \frac{V_4 - 2V_3 + V_2}{\Delta S^2} \\
\cdots \\
\cdots \\
\frac{\partial V_{N-1}}{\partial t} = r_t V_{N-1} - r_t s_{N-1} \frac{V_N - V_{N-1}}{\Delta S} - \frac{1}{2}\sigma_t^2(V_{ss}) s_{N-1}^2 \frac{V_N - 2V_{N-1} + V_{N-2}}{\Delta S^2} \\
\frac{\partial V_N}{\partial t} = r_t V_N - r_t s_N \frac{V_{N+1} - V_N}{\Delta S} - \frac{1}{2}\sigma_t^2(V_{ss}) s_N^2 \frac{V_{N+1} - 2V_N + V_{N-1}}{\Delta S^2}
\end{array} \tag{11.5}$$

Next, the following control inputs are defined

$$\begin{array}{c}
u_1 = -\frac{1}{2}\sigma_t^2(V_{ss}) s_1^2 \frac{V_2 - 2V_1 + V_0}{\Delta S^2} \\
u_2 = -\frac{1}{2}\sigma_t^2(V_{ss}) s_2^2 \frac{V_3 - 2V_2 + V_1}{\Delta S^2} \\
u_3 = -\frac{1}{2}\sigma_t^2(V_{ss}) s_3^2 \frac{V_4 - 2V_3 + V_2}{\Delta S^2} \\
\cdots \\
u_{N-1} = -\frac{1}{2}\sigma_t^2(V_{ss}) s_{N-1}^2 \frac{V_N - 2V_{N-1} + V_{N-2}}{\Delta S^2} \\
u_N = -\frac{1}{2}\sigma_t^2(V_{ss}) s_N^2 \frac{V_{N+1} - 2V_N + V_{N-1}}{\Delta S^2}
\end{array} \tag{11.6}$$

Therefore, the dynamics of the Black–Scholes PDE is written as

$$\dot{\tilde{V}} = \tilde{A}\tilde{V} + \tilde{B}\tilde{U} \tag{11.7}$$

where $\tilde{U} = [u_1, u_2, \ldots, u_{N-1}, u_N]^T$ and

$$\tilde{A} = \begin{pmatrix}
\tilde{a}_{11} & \tilde{a}_{12} & \tilde{a}_{13} & \tilde{a}_{14} & \cdots & \tilde{a}_{1,N-2} & \tilde{a}_{1,N-1} & \tilde{a}_{1,N} \\
\tilde{a}_{21} & \tilde{a}_{22} & \tilde{a}_{23} & \tilde{a}_{24} & \cdots & \tilde{a}_{2,N-2} & \tilde{a}_{2,N-1} & \tilde{a}_{2,N} \\
\tilde{a}_{31} & \tilde{a}_{32} & \tilde{a}_{33} & \tilde{a}_{34} & \cdots & \tilde{a}_{3,N-2} & \tilde{a}_{3,N-1} & \tilde{a}_{3,N} \\
\cdots & \cdots & \cdots & \cdots & \cdots & \cdots & \cdots & \cdots \\
\tilde{a}_{N-1,1} & \tilde{a}_{N-1,2} & \tilde{a}_{N-1,3} & \tilde{a}_{N-1,4} & \cdots & \tilde{a}_{N-1,N-2} & \tilde{a}_{N-1,N-1} & \tilde{a}_{N-1,N} \\
\tilde{a}_{N,1} & \tilde{a}_{N,2} & \tilde{a}_{N,3} & \tilde{a}_{N,4} & \cdots & \tilde{a}_{N,N-2} & \tilde{a}_{N,N-1} & \tilde{a}_{N,N}
\end{pmatrix} \tag{11.8}$$

where
$\tilde{a}_{11} = r_t + r_t \frac{s_1}{\Delta s}, \tilde{a}_{12} = -r_t \frac{s_2}{\Delta s}, \tilde{a}_{13} = 0, \tilde{a}_{14} = 0, \ldots, \tilde{a}_{1,N-2} = 0, \tilde{a}_{1,N-1} = 0, \tilde{a}_{1,N} = 0.$

$\tilde{a}_{21} = 0, \tilde{a}_{22} = r_t + r_t \frac{s_2}{\Delta s}, \tilde{a}_{23} = -r_t \frac{s_3}{\Delta s}, \tilde{a}_{24} = 0, \ldots, \tilde{a}_{2,N-2} = 0, \tilde{a}_{2,N-1} = 0, \tilde{a}_{2,N} = 0.$

$\tilde{a}_{31} = 0, \tilde{a}_{32} = 0, \tilde{a}_{33} = r_t + r_t \frac{s_3}{\Delta s}, \tilde{a}_{34} = -r_t \frac{s_4}{\Delta s}, \ldots, \tilde{a}_{3,N-2} = 0, \tilde{a}_{3,N-1} = 0, \tilde{a}_{3,N} = 0.$

$\cdots$

$\tilde{a}_{N-1,1} = 0,\ \tilde{a}_{N-1,2} = 0,\ \tilde{a}_{N-1,3} = 0,\ \tilde{a}_{N-1,4} = 0,\ \ldots,\ \tilde{a}_{N-1,N-2} = 0,\ \tilde{a}_{N-1,N-1} = r_t + r_t \frac{s_{N-1}}{\Delta s},\ \tilde{a}_{N-1,N} = -r_t \frac{s_N}{\Delta s}.$

$\tilde{a}_{N,1}=0,\ \tilde{a}_{N,2}=0,\ \tilde{a}_{N,3}=0,\ \tilde{a}_{N,4}=0,\ \ldots,\ \tilde{a}_{N,N-2}=0,\ \tilde{a}_{N,N-1}=0,\ \tilde{a}_{N,N}=r_t+r_t\frac{s_N}{\Delta s}$.

Matrix $\tilde{B}$ is defined as

$$\tilde{B}=\begin{pmatrix} 1 & 0 & 0 & 0 & \cdots & 0 & 0 & 0 \\ 0 & 1 & 0 & 0 & \cdots & 0 & 0 & 0 \\ 0 & 0 & 1 & 0 & \cdots & 0 & 0 & 0 \\ \cdots & \cdots & \cdots & \cdots & \cdots & \cdots & \cdots & \cdots \\ 0 & 0 & 0 & 0 & \cdots & 0 & 1 & 0 \\ 0 & 0 & 0 & 0 & \cdots & 0 & 0 & 1 \end{pmatrix} \tag{11.9}$$

The measurement (observation) matrix $\tilde{C}$ is taken to be the identity matrix. Therefore, one obtains again a description of the Black–Scholes PDE in the linear canonical form. For the state-space model of the Black–Scholes nonlinear PDE given in Eq. (11.7) one can perform estimation using the Derivative-free nonlinear Kalman Filter [78, 222].

As previously proven, the system of Eq. (11.5) is differentially flat, with the state vector elements $\tilde{V}=[V_1,V_2,V_3,\ldots,V_{N-1},V_N]^T$ to stand for a flat output. Actually, the auxiliary control inputs which have been defined in Eq. (11.6) can be expressed as function of the state vector elements, and this means that all variables of the state-space description of the Black–Scholes PDE are finally written as functions of the flat output and its derivatives. Therefore, the Black–Scholes PDE model is confirmed to be a differentially flat one.

11.2.2 State Estimation with Kalman Filtering

As mentioned above, for the system of Eq. (11.7), state estimation is possible by applying the standard Kalman Filter. The system is first turned into discrete-time form using common discretization methods and then the recursion of the linear Kalman Filter is applied.

To implement Kalman Filtering for the state-space model of the Black–Scholes PDE one has to substitute the previously defined matrices $\tilde{A}$, $\tilde{B}$ and $\tilde{C}$ by their discrete-time equivalents $\tilde{A}_d$, $\tilde{B}_d$ and $\tilde{C}_d$, respectively. Matrices A_d, B_d and C_d can be computed using established discretization methods. Moreover, the covariance matrices $P(k)$ and $P^-(k)$ are the ones obtained from the linear Kalman Filter update equations.

For the linear description of the system one can perform estimation using the standard Kalman Filter recursion. The discrete-time Kalman filter can be decomposed into two parts: (i) time update (prediction stage), and (ii) measurement update (correction stage).

measurement update:

$$\begin{aligned}
K(k) &= P^{-}(k)\tilde{C}_d^T[\tilde{C}_d \cdot P^{-}(k)\tilde{C}_d^T + R]^{-1} \\
\hat{V}(k) &= \hat{V}^{-}(k) + K(k)[\tilde{C}_d\tilde{V}(k) - \tilde{C}_d\hat{V}^{-}(k)] \\
P(k) &= P^{-}(k) - K(k)\tilde{C}_d P^{-}(k)
\end{aligned} \tag{11.10}$$

time update:

$$\begin{aligned}
P^{-}(k+1) &= \tilde{A}_d(k)P(k)\tilde{A}_D^T(k) + Q(k) \\
\hat{V}^{-}(k+1) &= \tilde{A}_d(k)\hat{V}(k) + \tilde{B}_d(k)\tilde{U}(k)
\end{aligned} \tag{11.11}$$

11.3 Distributed Forecasting Model

In the considered distributed forecasting model, estimates of the state vector of local models of the Black–Scholes PDE are fused through fuzzy weighting, so as to provide an aggregate estimate. Fuzzy Kalman Filtering is a distributed filtering approach in which the aggregate state estimate is provided by fuzzy weighting of the estimates generated by local and spatially distributed Kalman Filters [162]. Here, the state-space of the financial system is partitioned into local areas, each one monitored by a different Kalman Filter. The area that each Kalman Filter covers is described by fuzzy rules R^l of the form:

$$R^l \text{ IF } x_1 \text{ is } A_i \text{ AND } \cdots \; x_n \text{ is } A_j \text{ THEN } KF^l \text{ estimates } \hat{x}^l \tag{11.12}$$

where $i = 1, 2, \ldots, n, j = 1, 2, \ldots, n$ and $l = 1, 2, \ldots n \times m$.

As an example, it is assumed that the partitioning of the financial system's state-space is as depicted in Fig. 11.2. Next, the number of the fuzzy rules is denoted as $M = n \times n$. The aggregate estimate that is provided by the fuzzy Kalman Filter is of the form

$$\hat{x} = \sum_{l=1}^{M} \frac{\prod_{i=1}^{N} A_i^l}{\sum_{j=1}^{M}\prod_{i=1}^{N} A_i^j}\hat{x}_l \Rightarrow \hat{x} = \sum_{l=1}^{M} w_l \hat{x}_l \tag{11.13}$$

According to the above the financial system's model was written in the linear state-space form

$$\begin{aligned}
\dot{x} &= Ax + Bu + v \\
y &= Cx + w
\end{aligned} \tag{11.14}$$

where v and w stand for the process and measurement noise, while after discretization, the discrete-time description of the linearized financial system dynamics is obtained

$$\begin{aligned}
x(k+1) &= A_d x(k) + B_d u(k) \\
y(k) &= C_d x(x)
\end{aligned} \tag{11.15}$$

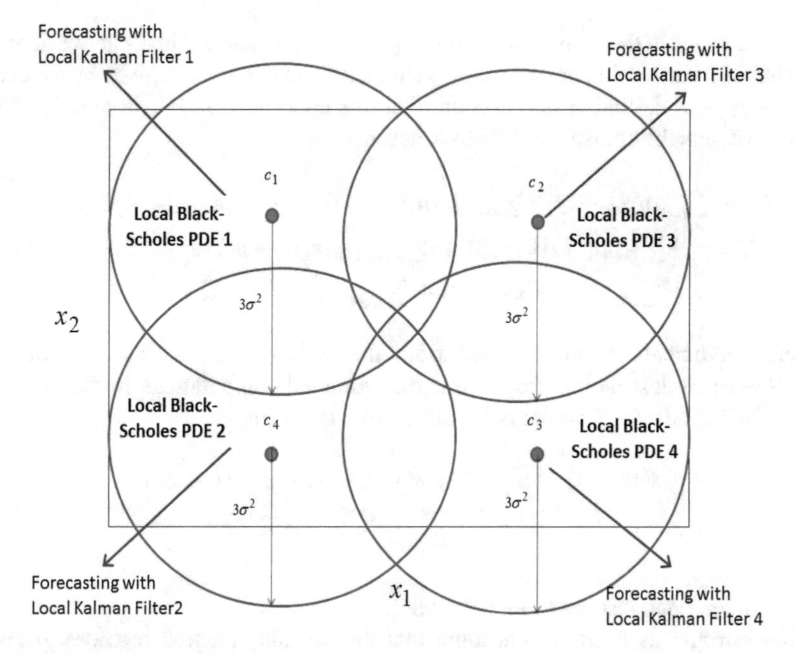

Fig. 11.2 Forecasting of option price with the use of local Black–Scholes PDE models and with the use of local Kalman filters

The i-th Kalman Filter, which is associated with the l-th fuzzy rule is given by

$$\hat{x}^l(k+1) = A_d^l \hat{x}(k) + B_d^l u(k) + K_f^l C_d^l (x^l(k)) - \hat{x}^l(k) \tag{11.16}$$

The difference $\varepsilon(k) = C_d^l(x^l(k)) - \hat{x}^l(k)$ between the real and the estimated output of the financial system's model, is the residual and follows a zero-mean Gaussian distribution. By applying the z transformation the equivalent description of the Kalman Filter in the z-frequency domain can be obtained.

Without loss of generality the case of Black–Scholes PDE state-space model with 3 inputs and 3 outputs (comprising 3 different grid points) is considered. This has a MIMO transfer function form

$$\begin{pmatrix} Y_1(z) \\ Y_2(z) \\ Y_3(z) \end{pmatrix} = \begin{pmatrix} H_{11}^A & H_{12}^A & H_{13}^A \\ H_{21}^A & H_{22}^A & H_{23}^A \\ H_{31}^A & H_{32}^A & H_{33}^A \end{pmatrix} \begin{pmatrix} U_1(z) \\ U_2(z) \\ U_3(z) \end{pmatrix} + \begin{pmatrix} H_{11}^B & H_{12}^B & H_{13}^B \\ H_{21}^B & H_{22}^B & H_{23}^B \\ H_{31}^B & H_{32}^B & H_{33}^B \end{pmatrix} \begin{pmatrix} E_1(z) \\ E_2(z) \\ E_3(z) \end{pmatrix} \tag{11.17}$$

Next, examining for instance the subsystem

$$\begin{aligned} Y_1(z) = H_{11}^A U_1(z) + H_{12}^A U_2(z) + H_{13}^A U_3(z) + \\ + H_{11}^B E_1(z) + H_{12}^B E_2(z) + H_{13}^B E_3(z) \end{aligned} \tag{11.18}$$

where each one of the transfer functions $H_{1j}\, j = 1, 2, 3$ included in the above description has in its denominator the system's characteristic polynomial given by the determinant $|zI - A_d|$. Taking into account that this characteristic polynomial is of 3rd order one gets the equivalent ARMAX description

$$\begin{aligned} y_1(k) = \textstyle\sum_{i=1}^{3} a_i y(k-i) + \sum_{j1=1}^{2} b_{j1} u_1(k-j1) + \sum_{j2=1}^{2} c_{j2} u_2(k-j2) + \\ + \textstyle\sum_{j3=1}^{2} c_{j3} u_3(k-j3) + \sum_{m1=1}^{2} p_{m1} \varepsilon_1(k-m1) + \\ + \textstyle\sum_{m2=1}^{2} q_{m2} \varepsilon_2(k-m2) + \sum_{m3=1}^{2} r_{m3} \varepsilon_3(k-m3) \end{aligned} \tag{11.19}$$

where coefficients a_i are obtained from the system's characteristic polynomial $det|zI - A|$ in descending order. To avoid extended computations in the chapter's example, the ARMAX model is simplified into the form

$$\begin{aligned} y^l(k+1) = a_1^l y^l(k) + a_2^l y^l(k-1) + a_3^l y^l(k-2) + \\ + b_1^l u_1^l(k) + b_2^l u_2^l(k) + b_3^l u_3^l(k) + \\ + c_1^l \varepsilon_1^l(k) \end{aligned} \tag{11.20}$$

where index l denotes the l-th local model.

Moreover, it is taken into account that the transfer function matrices given in Eq. (11.17), are $H^A(z) = C_d(zI - A_d)^{-1}B$ and $H^B(z) = C_d(zI - A_d)^{-1}K_f$ (where K_f is the gain of the Kalman Filter).

Considering next that the number of local models is $l = 1, 2, 3, 4$ and using the above the fuzzy Kalman Filter for the monitoring of the financial system is described by the following fuzzy rule base:

$$\begin{aligned} &\hat{y}^1(k+1) = a_1^1 \hat{y}^1(k) + a_2^1 \hat{y}^1(k-1) + a_3^1 \hat{y}^1(k-2) + \\ &+ b_1^1 u_1^1(k) + b_2^1 u_2^1(k) + b_3^1 u_3^1(k) + c_1^1 \varepsilon_1^1(k) \\ \\ &\hat{y}^2(k+1) = a_1^1 \hat{y}^2(k) + a_2^1 \hat{y}^2(k-1) + a_3^2 \hat{y}^2(k-2) + \\ &+ b_1^1 u_1^2(k) + b_2^2 u_2^2(k) + b_3^2 u_3^2(k) + c_1^2 \varepsilon_1^2(k) \\ \\ &\hat{y}^3(k+1) = a_1^3 \hat{y}^1(k) + a_2^3 \hat{y}^3(k-1) + a_3^3 \hat{y}^3(k-2) + \\ &+ b_1^3 u_1^3(k) + b_2^3 u_2^3(k) + b_3^3 u_3^3(k) + c_1^3 \varepsilon_1^3(k) \\ \\ &\hat{y}^4(k+1) = a_1^4 \hat{y}^4(k) + a_2^4 \hat{y}^4(k-1) + a_3^4 \hat{y}^4(k-2) + \\ &+ b_1^4 u_1^4(k) + b_2^4 u_2^4(k) + b_3^4 u_3^4(k) + c_1^4 \varepsilon_1^4(k) \end{aligned} \tag{11.21}$$

For a properly functioning fuzzy Kalman Filter it should hold $a_j^1 = a_j^2 = a_j^3 = a_j^4\; j = 1, 2, 3$ and similarly $b_j^1 = b_j^2 = b_j^3 = b_j^4\; j = 1, 2, 3$, and finally $c_j^1 = c_j^2 = c_j^3 = c_j^4\, j = 1$. If the above condition does not hold then for at least one local Kalman Filter the parameters of the financial system's model used in the estimation procedure are incorrect. The statistical change detection test which is proposed in this chapter is capable of detecting the inconsistent local Kalman Filter.

11.4 Consistency of the Kalman Filter

The statistical validation of the Kalman Filter after exploiting the properties of the χ^2 distribution was explained in the previous chapter. An overview of the main criteria used to check the consistency of the Kalman Filter is given next.

To obtain accurate estimates with the Kalman Filter, and consequently to forecast with precision options' values, a tuning process is required. A question that arises is about which state estimates can be considered as reliable. There is need for systematic methods showing when the Kalman Filter is not performing optimally and when its retuning, either in terms of the used model or in terms of the covariance matrices, should be performed. Several methods can be applied to test the consistency of the Kalman Filter, from the desired characteristics of the measurement residuals. These include the normalized error square (NES) test, the autocorrelation test, and the normalized mean error (NME) test and have been analyzed in [9, 60].

(i) It is assumed that a discrete error process e_k with dimension $m\times 1$ is a zero-mean Gaussian white-noise process with covariance given by E_k. This process can be the Kalman Filter's residual associated to the state estimation error or the residual associated to the measurement estimation error. Then, the following *normalized error square* (NES) is defined

$$\varepsilon_k = e_k^T E_k^{-1} e_k \tag{11.22}$$

The normalized error square follows a χ^2 distribution (Fig. 11.3). An appropriate test for the normalized error sum is to numerically show that the following condition is met within a level of confidence (according to the properties of the χ^2 distribution)

$$E\{\varepsilon_k\} = m \tag{11.23}$$

This can be succeeded using statistical hypothesis testing, which is associated with confidence intervals. A 95% confidence interval is frequently applied, which is specified using $100(1-a)$ with $a = 0.05$. Actually, a two-sided probability region is considered cutting-off two end tails of 2.5% each. For M runs of Monte-Carlo experiments the normalized error square that is obtained is given by

$$\bar{\varepsilon}_k = \frac{1}{M}\sum_{i=1}^{M}\varepsilon_k(i) = \frac{1}{M}\sum_{i=1}^{M} e_k^T(i)E_k^{-1}(i)e_k(i) \tag{11.24}$$

where ε_i stands for the i-th run at time t_k. Then $M\bar{\varepsilon}_k$ will follow a χ^2 density with Mm degrees of freedom. This condition can be checked using a χ^2 test. The hypothesis holds true if the following condition is satisfied

$$\bar{\varepsilon}_k \in [\zeta_1, \zeta_2] \tag{11.25}$$

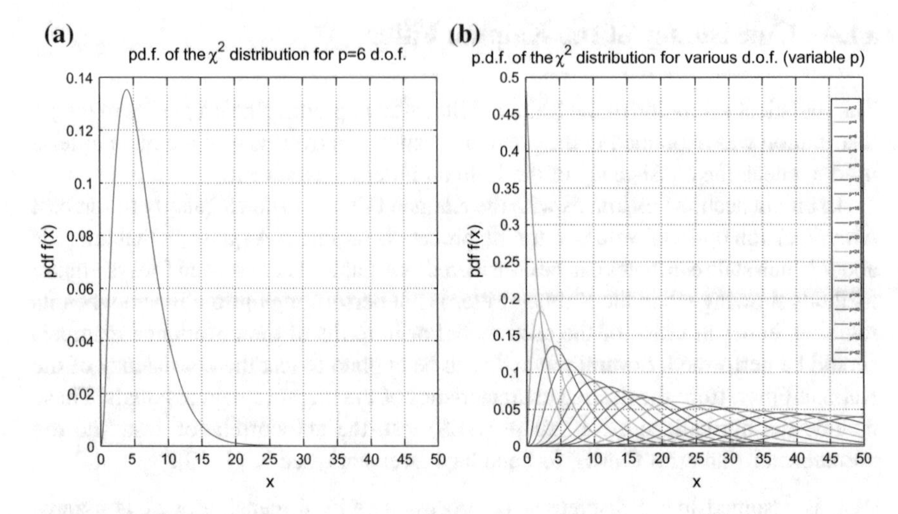

Fig. 11.3 **a** Probability density function of the χ^2 distribution for $p = 6$ degrees of freedom, **b** probability density function of the distribution for several values of the degrees of freedom (variable p)

where ζ_1 and ζ_2 are derived from the tail probabilities of the χ^2 density. For example, for $m = 2$ and $M = 100$ one has $\chi^2_{Mm}(0.025) = 162$ and $\chi^2_{Mm}(0.975) = 241$. Using that $M = 100$ one obtains $\zeta_1 = \chi^2_{Mm}(0.025)/M = 1.62$ and $\zeta_2 = \chi^2_{Mm}(0.975)/M = 2.41$.

(ii) Another consistency checking method is the *test for whiteness*. This is obtained by using the following sample autocorrelation:

$$\bar{\rho}_{k,j} = \frac{1}{\sqrt{M}} \sum_{i=1}^{M} e_k^T(i) \left[\sum_{i=1}^{M} e_k(i) e_k^T(i) = \sum_{i=1}^{M} e_j(i) e_j^T(i) \right]^{-1/2} e_j(i) \tag{11.26}$$

For a sufficiently large value of M, variable $\bar{\rho}_{j,k}$ for $k \neq j$ is zero mean with variance given by $1/M$. Next the application of the central limit theorem provides a normal approximation, and considering a 95% confidence interval one finally obtains

$$\bar{\rho}_{j,k} \in [-\tfrac{1.96}{M}, +\tfrac{1.96}{M}] \tag{11.27}$$

(iii) An additional consistency test is based on the normalized mean error (NME) for the jth element of e_k

$$[\bar{\mu}_k]_j = \frac{1}{M} \sum_{j=1}^{M} \frac{[e_k]_j}{\sqrt{[E_k]_{jj}}}, \quad j = 1, 2, \ldots, M \tag{11.28}$$

Then, since the variance of $[\bar{\mu}_k]_j$ is $\frac{1}{M}$ for a 95% acceptance interval one has

$$[\bar{\mu}_k]_j \in \left[-\tfrac{1.96}{\sqrt{M}}, +\tfrac{1.96}{\sqrt{M}}\right] \quad (11.29)$$

The hypothesis holds true, if Eq. (11.29) is satisfied. The NES, NME and autocorrelation consistency tests can be all performed with a single run using N data points. Using a time-averaging approach one obtains a low variability test statistic, which can be executed in real-time. In the latter case the time-average NES is given by

$$\bar{\varepsilon} = \frac{1}{N}\sum_{k=1}^{N} e_k^T E_k^{-1} e_k \quad (11.30)$$

Considering that e_k is a zero mean, white-noise process, then $N\bar{\varepsilon}$ follows a χ^2 density distribution with Nm degrees of freedom. Through the computation of the time-average auto-correlation the whiteness test for e_k is

$$\bar{\rho}_j = \frac{1}{\sqrt{N}}\sum_{k=1}^{N} e_k^T e_{k+j}\left[\sum_{k=1}^{N} e_k^T e_k \sum_{k=1}^{N} e_{k+j}^T e_{k+j}\right]^{\frac{-1}{2}} \quad (11.31)$$

For N sufficiently large, $\bar{\rho}_j$ has zero mean and variance given by $1/N$. With a 95% acceptance interval one has

$$\bar{\rho}_j \in \left[-\tfrac{1.96}{\sqrt{N}}, +\tfrac{1.96}{\sqrt{N}}\right] \quad (11.32)$$

The hypothesis is accepted if Eq. (11.32) is satisfied. The aforementioned tests can be applied to the residuals of the Kalman Filter or to the Kalman Filter state errors for checking the consistency of the obtained estimation and for checking the necessary consistency for filter optimality. If the tests are not satisfied then this means that the Kalman Filter is not running optimally, and the filter has to be retuned, or the filter's design has to be reconsidered.

In this chapter a systematic method, the *local statistical approach to fault diagnosis*, will be applied for checking the consistency of Fuzzy Kalman Filtering. It will be shown that the method is capable of identifying the elements responsible for the filter's failure, in the model associated with the estimation performed by the local Kalman filters. Thus, the method can assure reliable estimation and forecasting of financial systems, in applications such as options' pricing.

11.5 Equivalence Between Kalman Filters and Regressor Models

The Fuzzy Kalman Filter can be transformed into a weighted sum of local ARMAX models which finally take the form of a Takagi–Sugeno–Kung (TSK) fuzzy model. The TSK model contains both moving average part and auto-regressive parts. Actually, Kalman Filters are equivalent to ARMAX models while Fuzzy Kalman Filters are equivalent to a sum of ARMAX models with fuzzy weighting, which finally take the form of a TSK neurofuzzy model [10, 292]. The description of the Fuzzy Kalman Filter as a TSK fuzzy model will be used in the following sections for performing the model validation tests and for checking the consistency of this distributed filter.

The Fuzzy Kalman Filter can be represented as a set of fuzzy rules of the Takagi–Sugeno–Kung form. These are written as:

$$\begin{array}{c} R_l \,:\, IF\, x_1 \text{ is } A_1^l \, AND \, x_2 \text{ is } A_2^l \, AND \cdots AND \, x_n \text{ is } A_n^l \\ THEN\, \bar{y}^l = \sum_{i=1}^{n} c_{f_i}^l x_i \;\; l = 1, 2, \ldots, L \end{array} \tag{11.33}$$

where R^l is the l-th rule, $x = [x_1, x_2, \ldots, x_n]^T$ is the input (antecedent) variable, $\bar{y}^l$ is the output (consequent) variable, and w_i^l, b^l are the parameters of the local linear models. The output of the Takagi–Sugeno model is given by the weighted average of the rules consequents:

$$\hat{y} = \frac{\sum_{l=1}^{L} \bar{y}^l \prod_{i=1}^{n} \mu_{A_i^l}(x_i)}{\sum_{l=1}^{L} \prod_{i=1}^{n} \mu_{A_i^l}(x_i)} \tag{11.34}$$

where $\mu_{A_i^l}(x_i) \,:\, R \to [0, 1]$ is the membership function of the fuzzy set A_i^l in the antecedent part of the rule R^l. The output of the l-th local model is given by $\bar{y}^l = \sum_{i=1}^{n} c_{f_i}^l x_i$.

First, the residual e_i is defined as the difference between the fuzzy model output $\hat{y}_i$ and the physical system output y_i, i.e. $e_i = \hat{y}_i - y_i$. It is also acceptable to define the residual as the difference between the fuzzy model output and the exact model output, where the exact model replaces the physical system and has the same number of parameters as the fuzzy model (see Fig. 11.4). The partial derivative of the residual square is:

$$H(\theta, y_i) = \frac{1}{2}\frac{\partial e_i^2}{\partial \theta} = e_i \frac{\partial y_i}{\partial \theta} \tag{11.35}$$

The vector H having as elements the above $H(\theta, y_i)$ is called primary residual. Next, the gradients of the output with respect to the model's parameters are computed. In the case of fuzzy models the gradient of the output with respect to the consequent parameters w_i^l is given by

$$\frac{\partial \hat{y}}{\partial w_i^l} = \frac{x_i \mu_{R^l}(x)}{\sum_{l=1}^{L} \mu_{R^l}(x)} \tag{11.36}$$

The gradient with respect to the center c_i^l is

$$\frac{\partial \hat{y}}{\partial c_i^l} = \sum_{l=1}^{L} \frac{\bar{y}^l \frac{2(x_i - c_i^l)}{v_i^l} \mu_{R^l}(x_i) [\sum_{j=1}^{L} \mu_{R^j}(x_i) - \mu_{R^l}(x_i)]}{[\sum_{l=1}^{L} \mu_{R^l}(x_i)]^2} \tag{11.37}$$

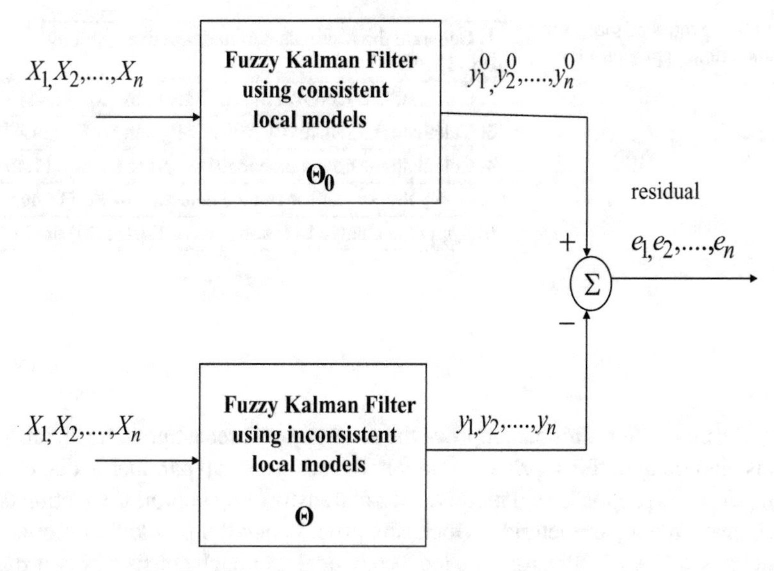

Fig. 11.4 Residual between the fuzzy Kalman filter that uses consistent local models and the fuzzy Kalman filter that uses inconsistent (distorted) local models

The gradient with respect to the spread v_i^l is

$$\frac{\partial \hat{y}}{\partial v_i^l} = \sum_{l=1}^{L} \frac{\bar{y}^l \frac{2(x_i - c_i^l)^2}{v_i^{l^3}} \mu_{R^l}(x_i)[\sum_{j=1}^{L} \mu_{R^j}(x_i) - \mu_{R^l}(x_i)]}{[\sum_{l=1}^{L} \mu_{R^l}(x_i)]^2} \tag{11.38}$$

The equivalence between the fuzzy Kalman filter and a Takagi–Sugeno neurofuzzy model enables to exploit previous results on fault detection and isolation for non-parametric estimators, such as neurofuzzy networks, by making use the *local statistical approach* to fault diagnosis. By describing the Fuzzy Kalman filter in the form of a Takagi–Sugeno neurofuzzy model it becomes easy to complete the intermediate stages for the application of the change detection method, which are described in Table 11.1 [15–18].

11.6 Change Detection of the Fuzzy Kalman Filter Using the Local Statistical Approach

11.6.1 The Global χ^2 Test for Change Detection

As previously explained, the *local statistical approach* to fault diagnosis is a statistical method of fault diagnosis which can be used for consistency checking of the

Table 11.1 Stages of the local statistical approach FKF validation

1. Generate the residuals partial derivative given by Eq. (11.39)
2. Calculate the Jacobian matrix J given by Eq. (11.41)
3. Calculate the sensitivity matrix M given by Eq. (11.44)
4. Calculate the covariance matrix S given by Eq. (11.45)
5. Apply the χ^2 test for change detection of Eq. (11.46)
6. Apply the change isolation tests of Eq. (11.49) or (11.58)

Fuzzy Kalman Filter. This can improve the accuracy of forecasting in financial applications, including options' price estimation. Based on a small parametric disturbance assumption, the proposed FDI method aims at transforming complex detection problems concerning a parameterized stochastic process into the problem of monitoring the mean of a Gaussian vector. The local statistical approach consists of two stages: (i) the global χ^2 test which indicates the existence of a change in some parameters of the fuzzy model, (ii) the diagnostics tests (sensitivity or min–max) which isolate the parameter affected by the change [15–18]. The method's stages are analyzed first, following closely the method presented in [10, 290].

As shown in Fig. 11.4 the proposed method is based on the definition of the residual e_i described as the difference between the output from the nonlinear ARMAX model of the Fuzzy Kalman Filter obtained with the use of the changed dynamics of the financial system and the output of the nonlinear ARMAX model of the Fuzzy Kalman Filter obtained with the use of the unchanged dynamics of the monitored finance entity. The nonlinear ARMAX model is actually a neuro-fuzzy model of the Takagi–Sugeno type that is based on the system's dynamics in an undistorted (fault-free) mode.

The concept of this FDI technique is as explained in the previous chapter: there is a nonlinear ARMAX model that represents the unchanged dynamics of the financial system. At each time instant the output of the aforementioned reference nonlinear ARMAX model is compared to the output of the nonlinear ARMAX model that represents the changed dynamics of the financial system. The difference between these two output measurements is called residual. The statistical processing of a sufficiently large number of residuals through a FDI method provides an index-variable that is compared against a fault threshold and which can give early indication about deviation of the model used by the Kalman Filter from the real system dynamics. Under certain conditions (detectability of changes) the proposed FDI method enables also fault isolation, i.e. to identify the source of fault within the model used by the Fuzzy Kalman Filter. In practical terms this means that the proposed change detection method can find out the i-th local Kalman Filter (out of the N local Kalman Filters that constitute the Fuzzy Kalman Filter) which makes use of an inconsistent model of the system's dynamics.

Considering the representation of the Fuzzy Kalman Filter as a neuro-fuzzy model of the Takagi–Sugeno type, the partial derivative of the residual square is:

$$H(\theta, y_i) = \frac{1}{2}\frac{\partial e_i^2}{\partial \theta} = e_i \frac{\partial \hat{y}_i}{\partial \theta} \tag{11.39}$$

where θ is the vector of model's parameters. The vector H having as elements the above $H(\theta, y_i)$ is called primary residual. Since the nonlinear ARMAX model is a neuro-fuzzy model, the gradient of the output with respect to the consequent parameters $c_{f_i}^l$ is given by

$$\frac{\partial \hat{y}}{\partial c_{f_i}^l} = \frac{x_i \mu_{R^l}(x)}{\sum_{l=1}^{L} \mu_{R^l}(x)} \tag{11.40}$$

The gradient with respect to the center c_i^l has been given in Eq. (11.37) while the gradient with respect to the spread v_i^l has been given in Eq. (11.38). Next, having calculated the partial derivatives of Eqs. (11.40), (11.37) and 11.38), the rows of the Jacobian matrix J are found by

$$J(\theta_0, y_k) = \left. \frac{\partial \hat{y}_k(\theta)}{\partial \theta} \right|_{\theta=\theta_0} \tag{11.41}$$

where θ_0 represents the nominal value of the parameters. The problem of change detection with the χ^2 test consists of monitoring a change in the mean of the Gaussian variable which for the one-dimensional parameter vector θ is formulated as

$$X = \frac{1}{\sqrt{N}} \sum_{k=1}^{N} e_k \frac{\partial \hat{y}_k}{\partial \theta} \sim N(\mu, \sigma^2) \tag{11.42}$$

where $\hat{y}_k$ is the output of the neural model generated by the input pattern x_k, e_k is the associated residual and θ is the vector of the model's parameters. It is noted that X is the monitored parameter for the FDI test, which means that when the mean value of X is 0 the system is in the fault-free condition, while when the mean value of X has moved away from 0 the system is in a faulty condition. For a multivariable parameter vector θ should hold $X \sim N(M\delta\theta, S)$, where S denotes the covariance matrix of X. In order to decide if the system (Kalman Filter) is in fault-free operating conditions, given a set of data of N measurements, let θ_* be the value of the parameters vector μ minimizing the RMSE. The notation is introduced only for the convenience of problem formulation, and its actual value does not need to be known. Then the model validation problem amounts to make a decision between the two hypotheses [10, 290]:

$$\begin{aligned} H_0 &: \theta_* = \theta_0 \\ H_1 &: \theta_* = \theta_0 + \tfrac{1}{\sqrt{N}}\delta\theta \end{aligned} \tag{11.43}$$

where $\delta\theta \neq 0$. It is known from the central limit theorem that for a large data sample, the normalized residual given by Eq. (11.42) asymptotically follows a Gaussian distribution when N→∞ [10, 12]. More specifically, the hypothesis that has to be tested is:

$$H_0 : X \sim N(0, S)$$
$$H_1 : X \sim N(M\delta\theta, S)$$

where M is the sensitivity matrix (see Eq. (11.44)), $\delta\theta$ is the change in the parameters' vector and S is the covariance matrix (see Eq. (11.45)). The product $M\delta\theta$ denotes the new center of the monitored Gaussian variable X, after a change on the system's parameter θ. The sensitivity matrix M of $\frac{1}{\sqrt{N}}X$ is defined as the mean value of the partial derivative with respect to θ of the primary residual defined in Eq. (11.39), i.e. $E\{\frac{\partial}{\partial\theta}H(\theta, y_k)\}$ and is approximated by [10, 290]:

$$M(\theta_0) \simeq \frac{\partial}{\partial\theta}\frac{1}{N}\sum_{k=1}^{N} H(\theta_0, y_k) \simeq \frac{1}{N}J^T J \tag{11.44}$$

The covariance matrix S is defined as $E\{H(\theta, y_k)H^T(\theta, y_{k+m})\}$, $m = 0, \pm 1, \cdots$ and is approximated by [12]:

$$\begin{aligned} S \simeq & \sum_{k=1}^{N}[H(\theta_0, y_k)H^T(\theta_0, y_k)] + \\ & + \sum_{m=1}^{I}\frac{1}{N-m}\sum_{k=1}^{N-m}[H(\theta_0, y_k)H^T(\theta_0, y_{k+m}) + \\ & + H(\theta_0, y_{k+m})H^T(\theta_0, y_k)] \end{aligned} \tag{11.45}$$

where an acceptable value for I is 3. The decision tool is the likelihood ratio $s(X) = ln\frac{p_{\theta_1}(x)}{p_{\theta_0}(x)}$, where $p_{\theta_1}(X) = e^{[X-\mu(X)]^T S^{-1}[X-\mu(X)]}$ and $p_{\theta_0}(X) = e^{X^T S^{-1} X}$ [294]. The center of the Gaussian distribution of the changed system is denoted as $\mu(X) = M\delta\theta$ where $\delta\theta$ is the change in the parameters vector. The *Generalized Likelihood Ratio* (GLR) is calculated by maximizing the likelihood ratio with respect to $\delta\theta$ [12]. This means that the most likely case of parameter change is taken into account. This gives the global χ^2 test t:

$$t = X^T S^{-1} M(M^T S^{-1} M)^{-1} M^T S^{-1} X \tag{11.46}$$

Since X asymptotically follows a Gaussian distribution, the statistics defined in Eq. (11.46) follows a χ^2 distribution with n degrees of freedom. Mapping the change detection problem to this χ^2 distribution enables the choice of the change threshold. Assume that the desired probability of false alarm is α then the change threshold λ should be chosen from the relation [10, 290]

$$\int_{\lambda}^{\infty} \chi_n^2(s)ds = \alpha, \tag{11.47}$$

where $\chi_n^2(s)$ is the probability density function (p.d.f.) of a variable that follows the χ^2 distribution with n degrees of freedom.

11.6.2 Isolation of Inconsistent Kalman Filter Parameters with the Sensitivity Test

As explained in the previous chapter, fault isolation is needed to identify the source of faults in the dynamic model of the system used by the Fuzzy Kalman Filter. This means that the fault diagnosis method should also be able to find out (among the N local Kalman Filters that constitute the Fuzzy Kalman Filter) which is the local Kalman Filter that makes use of an inconsistent model. A first approach to change isolation is to focus only on a subset of the parameters while considering that the rest of the parameters unchanged [12]. The parameters vector η can be written as $\eta = [\phi, \psi]^T$, where ϕ contains those parameters to be subject to the isolation test, while ψ contains those parameters to be excluded from the isolation test. M_ϕ contains the columns of the sensitivity matrix M which are associated with the parameters subject to the isolation test. Similarly M_ψ contains the columns of M that are associated with the parameters to be excluded from the sensitivity test.

Assume that among the parameters η, it is only the subset ϕ that is suspected to have undergone a change. Thus η is restricted to $\eta = [\phi, 0]^T$. The associated columns of the sensitivity matrix are given by M_ϕ and the mean of the Gaussian to be monitored is $\mu = M_\phi \phi$, i.e.

$$\mu = MA\phi, \quad A = [0, I]^T \tag{11.48}$$

Matrix A is used to select the parameters that will be subjected to the fault isolation test. The rows of A correspond to the total set of parameters while the columns of A correspond only to the parameters selected for the test. Thus the fault diagnosis (χ^2) test of Eq. (11.46) can be restated as [10, 290]:

$$t_\phi = X^T S^{-1} M_\phi (M_\phi^T S^{-1} M_\phi)^{-1} M_\phi^T S^{-1} X \tag{11.49}$$

11.6.3 Isolation of Inconsistent Kalman Filter Parameters with the Min–Max Test

In this approach the aim is to find a statistic that will be able to detect a change on the part ϕ of the parameters vector η and which will be robust to a change in the non observed part ψ [12]. Assume the vector partition $\eta = [\phi, \psi]^T$. The following notation is used:

$$M^T S^{-1} M = \begin{pmatrix} I_{\varphi\varphi} & I_{\varphi\psi} \\ I_{\psi\varphi} & I_{\psi\psi} \end{pmatrix} \tag{11.50}$$

$$\gamma = \begin{pmatrix} \varphi \\ \psi \end{pmatrix}^T \cdot \begin{pmatrix} I_{\varphi\varphi} & I_{\varphi\psi} \\ I_{\psi\varphi} & I_{\psi\psi} \end{pmatrix} \cdot \begin{pmatrix} \varphi \\ \psi \end{pmatrix} \tag{11.51}$$

where S is the previously defined covariance matrix. The min–max test aims to minimize the non-centrality parameter γ with respect to the parameters that are not suspected for change. The minimum of γ with respect to ψ is given for [10, 290]:

$$\psi^* = \arg\min_{\psi} \gamma = \varphi^T(I_{\varphi\varphi} - I_{\varphi\psi}I_{\psi\psi}^{-1}I_{\psi\varphi})\varphi \tag{11.52}$$

and is found to be

$$\begin{aligned}\gamma^* &= \min_{\psi} \gamma = \varphi^T(I_{\varphi\varphi} - I_{\varphi\psi}I_{\psi\psi}^{-1}I_{\psi\varphi})\varphi = \\ &= \begin{pmatrix}\varphi \\ -I_{\psi\psi}^{-1}I_{\psi\varphi}\varphi\end{pmatrix}^T \begin{pmatrix}I_{\varphi\varphi} & I_{\varphi\psi} \\ I_{\psi\varphi} & I_{\psi\psi}\end{pmatrix} \begin{pmatrix}\varphi \\ -I_{\psi\psi}^{-1}I_{\psi\varphi}\varphi\end{pmatrix}\end{aligned} \tag{11.53}$$

which results in

$$\gamma^* = \varphi^T\{[I, -I_{\varphi\psi}I_{\psi\psi}^{-1}]M^T\Sigma^{-1}\}\,\Sigma^{-1}\{\Sigma^{-1}M[I, -I_{\varphi\psi}I_{\psi\psi}^{-1}\}\varphi \tag{11.54}$$

The following linear transformation of the observations is considered:

$$X_{\phi}^* = [I, -I_{\varphi\psi}I_{\psi\psi}^{-1}]M^T\Sigma^{-1}X \tag{11.55}$$

The transformed variable X_{ϕ}^* follows a Gaussian distribution $N(\mu_{\phi}^*, I_{\phi}^*)$ with mean:

$$\mu_{\varphi}^* = I_{\varphi}^*\varphi \tag{11.56}$$

and with covariance:

$$I_{\varphi}^* = I_{\varphi\varphi} - I_{\varphi\psi}I_{\psi\psi}^{-1}I_{\psi\varphi} \tag{11.57}$$

The max–min test decides between the hypotheses:

$$\begin{aligned}H_0^* &: \mu^* = 0 \\ H_1^* &: \mu^* = I_{\varphi}^*\varphi\end{aligned}$$

and is described by:

$$\tau_{\varphi}^* = X_{\varphi}^{*T}I_{\varphi}^{*-1}X_{\varphi}^* \tag{11.58}$$

The stages of validation of the Fuzzy Kalman Filter (FKF) with the use of the local statistical approach are summarized in the following table:

11.7 Simulation Tests

11.7.1 Distributed State Estimation of the Black–Scholes PDE

The state-space of the Black–Scholes PDE is partitioned into sub-spaces, where each subspace is associated with a different model of the PDE having its own parametrization. For instance, the previously analyzed nonlinearities in the Black–Scholes PDE model, which are due to the dependence of the volatility variables $\sigma^2(t, S, V_{ss})$ on the option values at different points of the state-space justify the need for a multi-model representation of the Black–Scholes PDE. One can arrive at local descriptions of the Black–Scholes PDE related with specific values of the volatility.

For each local Black–Scholes PDE model the state-space description is obtained and a Kalman Filter is designed capable of performing state estimation and state forecasting. Each local state estimate is assigned with a fuzzy membership degree to the subspace described by the associated Black–Scholes PDE model.

Each discrete-time state-space model of Black–Scholes PDE is associated with a different operating point and takes also into account variations in the parameters of the finance system. Thus, one can consider that the Black–Scholes PDE dynamics is described by N local state-space models $i = 1, 2, \ldots, N$.

$$\begin{aligned} x^i(k+1) &= A^i(k)x^i(k) + B^i(k)u(k) \\ y^i(k) &= C^i(k)x^i(k) \end{aligned} \tag{11.59}$$

The Kalman Filter for the i-th local Black–Scholes PDE state-space model is

measurement update:

$$\begin{aligned} K(k)^{(i)} &= P^-(k)^{(i)}C^{T^{(i)}}[C^{(i)} \cdot P^-(k)^{(i)}C^{T^{(i)}} + R^{(i)}]^{-1} \\ \hat{x}(k)^{(i)} &= \hat{x}^-(k)^{(i)} + K(k)^{(i)}[z(k)^{(i)} - C^{(i)}\hat{x}^-(k)^{(i)}] \\ P(k)^{(i)} &= P^-(k)^{(i)} - K(k)^{(i)}C^{(i)}P^-(k)^{(i)} \end{aligned} \tag{11.60}$$

time update:

$$\begin{aligned} P^-(k+1)^{(i)} &= A(k)^{(i)}P(k)^{(i)}A^T(k)^{(i)} + Q(k)^{(i)} \\ \hat{x}^-(k+1)^{(i)} &= A(k)^{(i)}\hat{x}(k)^{(i)} + B(k)^{(i)}u(k)^{(i)} \end{aligned} \tag{11.61}$$

Using the previous state-space description with the use of local linear models, ARMA models of the financial system's dynamics can be obtained. Next, the Kalman Filter that tracks the financial system's dynamics is considered. The associated local ARMAX model is:

$$\begin{aligned} y^l(k+1) = {} & a_1^l y^l(k) + a_2^l y^l(k-1) + a_3^l y^l(k-2) + \\ & + b_1^l u_1^l(k) + b_2^l u_2^l(k) + b_3^l u_3^l(k) + c_1^l \varepsilon_1^l(k) \end{aligned} \tag{11.62}$$

Assuming $n = 7$ input variables, each one partitioned into 3 fuzzy sets and following the input dimension partitioning the input space will be partitioned into $N = 3^7 = 2187$ fuzzy regions, which also means $N = 2187$ local models. To avoid the explosion of dimensionality due to input dimension partitioning one can assume a limited number of local models $M \ll N$ according to the concept of input space partitioning.

Thus, one obtains the following nonlinear ARMAX model for the fuzzy Kalman Filter

$$y(k+1) = \frac{\sum_{l=1}^{M} \underline{c_f^l}^T \underline{x} \prod_{i=1}^{n} \mu_{A_i}^l(\underline{x})}{\sum_{l=1}^{M} \prod_{i=1}^{n} \mu_{A_i}^l(\underline{x})} \quad (11.63)$$

where $\underline{c_f^l}$ and $\underline{x}$ are vectors defined as

$$\underline{c_f^l} = \left(c_{f_1}^l \; c_{f_2}^l \; c_{f_3}^l \; c_{f_4}^l \; c_{f_5}^l \; c_{f_6}^l \; c_{f_7}^l\right) \quad (11.64)$$

$$\underline{x} = \left(y(k) \; y(k-1) \; y(k-2) \; u_1(k) \; u_2(k) \; u_3(k) \; \varepsilon_1(k)\right) \quad (11.65)$$

Residuals are generated as differences between the output of the fuzzy Kalman Filter associated with the undistorted dynamics of the system and fuzzy Kalman Filter associated with the changed dynamics of the system.

11.7.2 Simulation Results

The fuzzy Kalman Filter associated with the description of the financial system's dynamics through local state-space models of the Black–Scholes was taken to consist of the following rules:

$$\begin{aligned}
R^{(1)} &: \; IF \; x_1 \; is \; (c_1^{(1)}, v) \; AND \; x_2 \; is \; (c_2^{(1)}, v) \; AND \cdots \\
&\quad AND \; x_n \; is \; (c_n^{(1)}, v) \cdots \; THEN \; \hat{y} = \underline{c_f}^{(1)} \underline{x}^T \\
R^{(2)} &: \; IF \; x_1 \; is \; (c_1^{(2)}, v) \; AND \; x_2 \; is \; (c_2^{(2)}, v) \; AND \cdots \\
&\quad AND \; x_n \; is \; (c_n^{(2)}, v) \cdots \; THEN \; \hat{y} = \underline{c_f}^{(2)} \underline{x}^T \\
R^{(3)} &: \; IF \; x_1 \; is \; (c_1^{(3)}, v) \; AND \; x_2 \; is \; (c_2^{(3)}, v) \; AND \cdots \\
&\quad AND \; x_n \; is \; (c_n^{(3)}, v) \cdots \; THEN \; \hat{y} = \underline{c_f}^{(3)} \underline{x}^T \\
R^{(4)} &: \; IF \; x_1 \; is \; (c_1^{(4)}, v) \; AND \; x_2 \; is \; (c_2^{(4)}, v) \; AND \cdots \\
&\quad AND \; x_n \; is \; (c_n^{(4)}, v) \cdots \; THEN \; \hat{y} = \underline{c_f}^{(4)} \underline{x}^T
\end{aligned}$$

The above model implies description of the initial Black–Scholes PDE using 4 sub-models. The spread of the membership functions is denoted by v. A 2D projection of the input space partition is demonstrated in Fig. 11.5, but actually this partition should be considered taking place in a state-space of higher dimensionality.

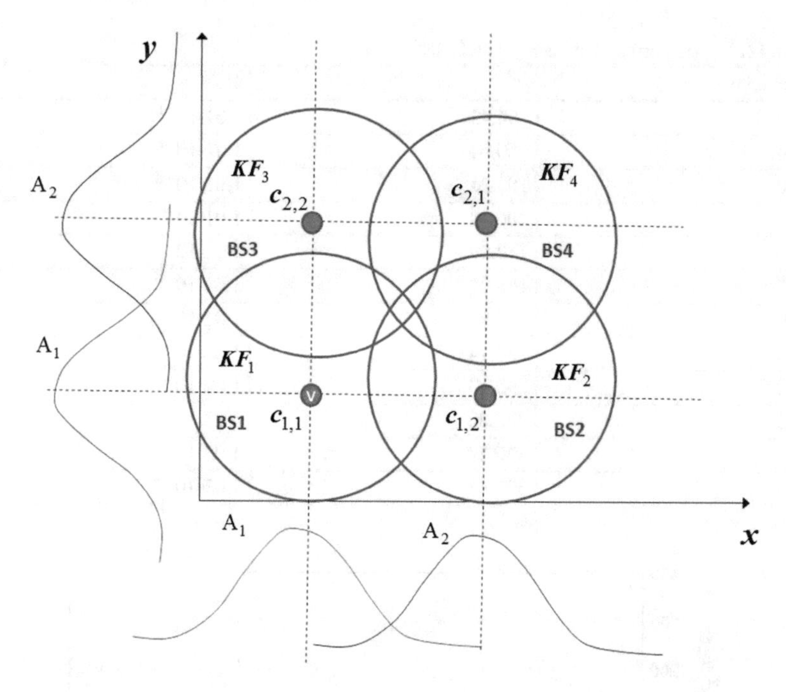

Fig. 11.5 Fuzzy Kalman filtering for financial system monitoring (distributed forecasting of option prices)

The parameters set in the new TSK fuzzy model consisted of $4 \times 7 + 4 \times 3 = 40$ parameters (28 linear parameters which were the output layer weights and 12 nonlinear parameters which were the centers of the fuzzy sets in the antecedent part of the rules). This means that by applying the local statistical approach to FDI and the χ^2 change detection test to the considered model, the fault threshold should be equal to 40.

The numerical tests confirmed theory. In case that no fault was assumed for the monitored financial system the mean value of the χ^2 test over a number of trials was found to be close to the threshold value 40. Such a value was anticipated according to the theoretical analysis of the χ^2 test. For slight deviations of the parameters of the fuzzy Kalman Filter from their nominal (fault-free) values, the global χ^2 test was capable of giving a clear indication about the existence of a fault. Thus for changes which varied between 0.1 and 1% of the nominal parameter's value the score of the χ^2 test deviated significantly from the fault threshold (which as mentioned before was set equal to 40).

A comparison between (i) the proposed χ^2 test based on the local statistical approach and the Generalized Likelihood ratio and (ii) of the mean square error (MSE) test, for detecting model inconsistencies in the distributed/fuzzy Kalman Filter is given in Table 11.2 and in Fig. 11.6. It can be clearly noticed that for small parametric changes in the local models used by the fuzzy Kalman Filter, the MSE

Table 11.2 Comparison between χ^2 and MSE tests

% change	χ^2	*MSE*
0.20	44.45	$1.01 \cdot 10^{-6}$
0.24	60.54	$1.01 \cdot 10^{-6}$
0.30	101.21	$1.01 \cdot 10^{-6}$
0.35	106.99	$1.01 \cdot 10^{-6}$
0.40	172.49	$1.01 \cdot 10^{-6}$
0.45	165.97	$1.02 \cdot 10^{-6}$
0.50	187.87	$1.02 \cdot 10^{-6}$
0.55	230.14	$1.02 \cdot 10^{-6}$
0.60	295.33	$1.02 \cdot 10^{-6}$
0.65	282.00	$1.03 \cdot 10^{-6}$
0.70	330.59	$1.03 \cdot 10^{-6}$
0.75	365.88	$1.03 \cdot 10^{-6}$

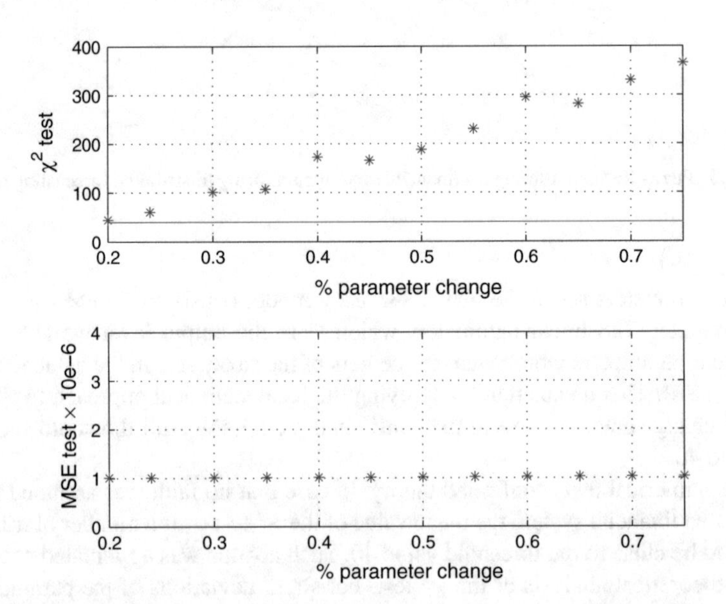

Fig. 11.6 Comparison between (i) the proposed χ^2 test based on the local statistical approach and the Generalized Likelihood ratio and (ii) of the mean square error (MSE) test, for detecting model inconsistencies in the distributed/fuzzy Kalman filter

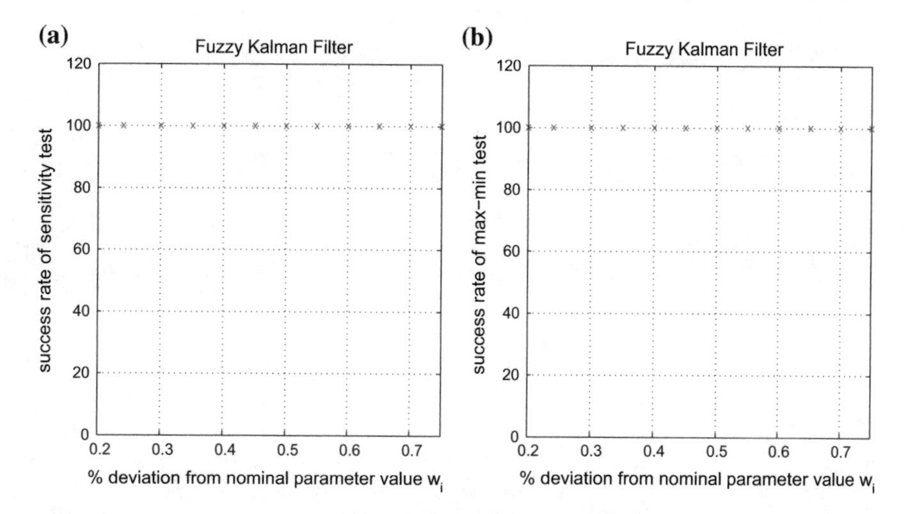

Fig. 11.7 **a** Success rate of the fault isolation test (sensitivity method) for changes in a parameter of the first local Kalman Filter, ranging between 0.1 and 1.0% of the nominal value. **b** Success rate of the fault isolation test (max–min method) for changes in a parameter of the first local Kalman filter, ranging between 0.1 and 1.0% of the nominal value

test gives the erroneous conclusion that the functioning of the Kalman Filter remains accurate. Actually it is observed that there is no change in the MSE value despite changes in the parameters of the model used by the Kalman Filter, and the MSE value remains low as in the case of fault-free operation. Besides in the MSE test the fault threshold is defined in an ad-hoc manner and this is another reason for the low credibility of this test. On the other hand the proposed χ^2 test based on the local statistical approach and the Generalized Likelihood ratio provides a clear indication about inconsistencies between the models used by the fuzzy Kalman Filter and the dynamics of the real system. Despite the small magnitude of parametric changes, the output of the χ^2 test based on the local statistical approach becomes several times larger than the fault threshold (that is 40). Thus a clear indication is provided about the need to correct the parameters of the local models used by the Fuzzy Kalman Filter.

As far as fault isolation is concerned, the numerical results showed that the sensitivity method for fault isolation was very efficient in distinguishing the parameter subject to fault among all parameters in the fuzzy Kalman Filter's model. The sensitivity fault isolation test and the min–max fault isolation test was performed for the parameters (weight w_i) of the local Kalman Filter. As it can be observed from the test's success rate depicted in Fig. 11.7 the proposed fault isolation methods can detect the local Kalman Filter, that uses an inconsistent model with reference to the real system's model. Thus correction of the parameters of this particular filter can be carried out instead of redesign of all local Kalman Filters constituting the Fuzzy Kalman Filter.

Chapter 12
Stabilization of Financial Systems Dynamics Through Feedback Control of the Black-Scholes PDE

12.1 Outline

In the previous chapters the focus has been on (i) control and stabilization of financial systems which are described by nonlinear ordinary differential equations, (ii) modelling and forecasting for financial systems including both those described by nonlinear ordinary differential equations and those described by partial differential or integro differential equations. Next, results are provided on the control and stabilization of financial systems described by partial differential equations (distributed parameter financial systems).

As noted, control and stabilization of financial systems is a difficult problem since the associated models have spatiotemporal dynamics and are either described by partial differential equations or by stochastic differential equations [184, 189]. On the one side control approaches for financial systems have been developed with the use of stochastic differential equations [177]. On the other side, control of financial dynamics through the use of the associated partial differential equations description remains an open problem for which efficient solutions have to be provided [231, 249]. To this end, in this research work a new control method is developed for the diffusion-type Black-Scholes PDE which describes the dynamics of options in financial markets. It is shown that boundary control, which is control based on the boundary conditions of the PDE, can be exerted on the Black-Scholes PDE thus modifying its dynamics and leading the options prices to convergence to specific reference values.

Controlling the dynamics of PDE financial systems with the application of exogenous inputs and feedback has been attempted in several research works [8, 80, 81, 164]. Control of financial systems is important for minimizing default risks, maximizing profits or stabilizing the dynamics of stock market indexes [26, 75, 82, 275]. The present chapter treats the problem of boundary control of the nonlinear Black-Scholes PDE, which means that the boundary conditions are used as control inputs to modify this PDE dynamics. Boundary control or lumped-input control of

G.G. Rigatos, *State-Space Approaches for Modelling and Control in Financial Engineering*, Intelligent Systems Reference Library 125,
DOI 10.1007/978-3-319-52866-3_12

nonlinear distributed parameter systems are problems of elevated difficulty and many remarkable results on this topic exist [35, 78, 170, 277]. If the PDE system is decomposed into an equivalent set of ordinary nonlinear differential equations, then controlling such a set of dynamical subsystems by the boundary conditions implies underactuation (actually one has to control the different degrees of freedom associated with the set of nonlinear ODEs by using only a couple of control inputs).

The control approach followed in this chapter is as follows. First, by implementing semi-discretization and numerical approximation of partial derivatives in the Black-Scholes PDE, a set of equivalent nonlinear ordinary differential equations is obtained [67, 87, 95, 124, 187, 276]. Next it is shown that the system of the nonlinear ODEs is a differentially flat one. This means that all its state variables and the control inputs can be written as differential functions of one single algebraic variable which is the flat output [34, 77, 133, 134, 191]. Moreover, by examining independently each nonlinear ODE it is shown that this stands again for a differentially flat system, for which a virtual control input can be computed as in the case of flatness-based control for the trivial system. The virtual control input is chosen such that the ODE subsystem is linearized and the tracking error is eliminated. The boundary condition that appears in the nonlinear ODE subsystem that comprises the last rows of the state-space description, stands for the aggregate control input. The computation of the boundary control input uses recursively all virtual control inputs mentioned above, moving from the last ODE system to the first one. Thus, by tracing the rows of the state-space model backwards, at each iteration of the control algorithm, one can finally obtain the control input that should be applied to the Black-Scholes PDE system so as to assure that all its state vector elements will converge to the desirable setpoints. By analyzing the dynamics of the closed loop system that results from the application of the aforementioned control method, asymptotic stability is confirmed.

12.2 Transformation of the Black-Scholes PDE into Nonlinear ODEs

12.2.1 Decomposition of the PDE Model into Equivalent ODEs

The decomposition of the 1D diffusion PDE model into an equivalent set of ODEs was first presented in Chap. 6 and is overviewed here. The following nonlinear diffusion equation is considered again

$$\frac{\partial \phi}{\partial t} = k\frac{\partial^2 \phi}{\partial x^2} + f(\phi) \tag{12.1}$$

Using the approximation for the partial derivative

$$\frac{\partial^2\phi}{\partial x^2} \simeq \frac{\phi_{i+1}-2\phi_i+\phi_{i-1}}{\Delta x^2} \tag{12.2}$$

and considering spatial measurements of variable ϕ along axis x at points $x_0 + i\Delta x,\ i = 1, 2, \cdots, N$ one has

$$\frac{\partial \phi_i}{\partial t} = \frac{K}{\Delta x^2}\phi_{i+1} - \frac{2K}{\Delta x^2}\phi_i + \frac{K}{\Delta x^2}\phi_{i-1} + f(\phi_i) \tag{12.3}$$

By considering the associated samples of ϕ given by $\phi_0, \phi_1, \cdots, \phi_N, \phi_{N+1}$ one has

$$\begin{aligned}
\frac{\partial \phi_1}{\partial t} &= \frac{K}{\Delta x^2}\phi_2 - \frac{2K}{\Delta x^2}\phi_1 + \frac{K}{\Delta x^2}\phi_0 + f(\phi_1)\\
\frac{\partial \phi_2}{\partial t} &= \frac{K}{\Delta x^2}\phi_3 - \frac{2K}{\Delta x^2}\phi_2 + \frac{K}{\Delta x^2}\phi_1 + f(\phi_2)\\
\frac{\partial \phi_3}{\partial t} &= \frac{K}{\Delta x^2}\phi_4 - \frac{2K}{\Delta x^2}\phi_3 + \frac{K}{\Delta x^2}\phi_2 + f(\phi_3)\\
&\cdots\\
\frac{\partial \phi_{N-1}}{\partial t} &= \frac{K}{\Delta x^2}\phi_N - \frac{2K}{\Delta x^2}\phi_{N-1} + \frac{K}{\Delta x^2}\phi_{N-2} + f(\phi_{N-1})\\
\frac{\partial \phi_N}{\partial t} &= \frac{K}{\Delta x^2}\phi_{N+1} - \frac{2K}{\Delta x^2}\phi_N + \frac{K}{\Delta x^2}\phi_{N-1} + f(\phi_N)
\end{aligned} \tag{12.4}$$

By defining the following state vector

$$x^T = (\phi_1, \phi_2, \cdots, \phi_N) \tag{12.5}$$

one obtains the following state-space description

$$\begin{aligned}
\dot{x}_1 &= \frac{K}{\Delta x^2}x_2 - \frac{2K}{\Delta x^2}x_1 + \frac{K}{\Delta x^2}\phi_0 + f(x_1)\\
\dot{x}_2 &= \frac{K}{\Delta x^2}x_3 - \frac{2K}{\Delta x^2}x_2 + \frac{K}{\Delta x^2}x_1 + f(x_2)\\
\dot{x}_3 &= \frac{K}{\Delta x^2}x_4 - \frac{2K}{\Delta x^2}x_3 + \frac{K}{\Delta x^2}x_2 + f(x_3)\\
&\cdots\\
\dot{x}_{N-1} &= \frac{K}{\Delta x^2}x_N - \frac{2K}{\Delta x^2}x_{N-1} + \frac{K}{\Delta x^2}x_{N-2} + f(x_{N-1})\\
\dot{x}_N &= \frac{K}{\Delta x^2}\phi_{N+1} - \frac{2K}{\Delta x^2}x_N + \frac{K}{\Delta x^2}x_{N-1} + f(x_N)
\end{aligned} \tag{12.6}$$

Next, the following state variables are defined

$$y_{1,i} = x_i \tag{12.7}$$

and the state-space description of the system becomes as follows

$$
\begin{aligned}
\dot{y}_{1,1} &= \frac{K}{\Delta x^2}y_{1,2} - \frac{2K}{\Delta x^2}y_{1,1} + \frac{K}{\Delta x^2}\phi_0 + f(y_{1,1}) \\
\dot{y}_{1,2} &= \frac{K}{\Delta x^2}y_{1,3} - \frac{2K}{\Delta x^2}y_{1,2} + \frac{K}{\Delta x^2}y_{1,1} + f(y_{1,2}) \\
\dot{y}_{1,3} &= \frac{K}{\Delta x^2}y_{1,4} - \frac{2K}{\Delta x^2}y_{1,3} + \frac{K}{\Delta x^2}y_{1,2} + f(y_{1,3}) \\
&\cdots \\
&\cdots \\
\dot{y}_{1,N-1} &= \frac{K}{\Delta x^2}y_{1,N} - \frac{2K}{\Delta x^2}y_{1,N-1} + \frac{K}{\Delta x^2}y_{1,N-2} + f(y_{1,N-1}) \\
\dot{y}_{1,N} &= \frac{K}{\Delta x^2}\phi_{N+1} - \frac{2K}{\Delta x^2}y_{1,N} + \frac{K}{\Delta x}y_{1,N-1} + f(y_{1,N})
\end{aligned}
\tag{12.8}
$$

The dynamical system described in Eq. (12.8) is a differentially flat one with flat output defined as the vector $\tilde{y} = [y_{1,1}, y_{1,2}, \cdots, y_{1,N}]$. Indeed all state variables can be written as functions of the flat output and its derivatives.

Denoting $a = \frac{K}{Dx^2}$ and $b = -\frac{2K}{Dx^2}$, the initial description of the system is rewritten as follows

$$
\begin{pmatrix} \dot{y}_{1,1} \\ \dot{y}_{1,2} \\ \cdots \\ \dot{y}_{1,N-1} \\ \dot{y}_{1,N} \end{pmatrix} = \begin{pmatrix} b\,0\,a\,0\,0\,0\,0 \cdots 0\,0\,0\,0\,0\,0 \\ a\,0\,b\,0\,a\,0\,0 \cdots 0\,0\,0\,0\,0\,0 \\ 0\,0\,a\,0\,b\,0\,a \cdots 0\,0\,0\,0\,0\,0 \\ \cdots \cdots \cdots \cdots \cdots \cdots \cdots \\ 0\,0\,0\,0\,0\,0\,0 \cdots a\,0\,b\,0\,a\,0 \\ 0\,0\,0\,0\,0\,0\,0 \cdots 0\,0\,a\,0\,b\,0 \end{pmatrix} \begin{pmatrix} y_{1,1} \\ y_{1,2} \\ \cdots \\ y_{1,N-1} \\ y_{1,N} \end{pmatrix} +
+ \begin{pmatrix} 1\,0\,0 \cdots 0\,0 \\ 0\,1\,0 \cdots 0\,0 \\ 0\,0\,1 \cdots 0\,0 \\ \cdots \cdots \cdots \cdots \\ 0\,0\,0 \cdots 1\,0 \\ 0\,0\,0 \cdots 0\,1 \end{pmatrix} \begin{pmatrix} v_1 \\ v_2 \\ v_3 \\ \cdots \\ v_{N-1} \\ v_N \end{pmatrix}
\tag{12.9}
$$

The associated control inputs are defined as

$$
\begin{aligned}
v_1 &= \frac{K}{\Delta x^2}\phi_0 + f(y_{1,1}) \\
v_2 &= f(y_{1,2}) \\
v_3 &= f(y_{1,3}) \\
&\cdots \\
v_{N-1} &= f(y_{1,N-1}) \\
v_N &= \frac{K}{\Delta x^2}\phi_{N+1} + f(y_{1,N})
\end{aligned}
\tag{12.10}
$$

By selecting measurements from a subset of points x_j $j \in [1, 2, \cdots, m]$, the associated observation (measurement) equation becomes

$$
\begin{pmatrix} z_1 \\ z_2 \\ \cdots \\ z_m \end{pmatrix} = \begin{pmatrix} 1\,0\,0 \cdots 0\,0 \\ \cdots \cdots \cdots \cdots \\ 0\,0\,0 \cdots 0\,1 \end{pmatrix} \begin{pmatrix} y_{1,1} \\ y_{1,2} \\ \cdots \\ y_{1,N} \end{pmatrix}
\tag{12.11}
$$

Thus, in matrix form one has the following state-space description of the system

$$\dot{\tilde{y}} = A\tilde{y} + Bv$$
$$\tilde{z} = C\tilde{y} \tag{12.12}$$

12.2.2 Modeling in State-Space Form of the Black-Scholes PDE

As first presented in Chap. 6, modelling of the Black-Scholes PDE will be performed with semi-discretization and the finite differences approach. The following nonlinear Black-Scholes PDE is considered:

$$\frac{\partial V(t,S)}{\partial t} + r_t S\frac{\partial V(t,S)}{\partial S} + \frac{1}{2}\sigma_t^2 S^2 \frac{\partial^2 V(t,S)}{\partial S^2} - r_t V(t,S) = 0 \tag{12.13}$$

It is considered that the volatility σ_t has a nonlinear dependence on V_{ss}, that is

$$\sigma_t = \sigma(V_{ss}) \tag{12.14}$$

where $V_{ss} = \frac{\partial^2 V(t,s)}{\partial S^2}$ and σ is a nonlinear function. Again, a grid of N points is used, that is $\{s_1, s_2, \cdots, s_{N-1}, s_N\}$ which are placed at equal distances on the S axis. At the points of spatial discretization it holds

$$\frac{\partial V(t,s_i)}{\partial t} = r_t V(t,s_i) - r_t s_i \frac{\partial V(t,s_i)}{\partial S} - \frac{1}{2}\sigma_t^2(V_{ss})\frac{\partial^2 V(t,s_i)}{\partial S^2} \tag{12.15}$$

where $i = 1, 2, \cdots, N$. The following state vector is defined

$$\tilde{V} = [V(t,s_1), V(t,s_2), V(t,s_3), \cdots, V(t,s_{N-1}), V(t,s_N)] \text{ or}$$
$$\tilde{V} = [V_1, V_2, V_3, \cdots, V_{N-1}, V_N] \tag{12.16}$$

thus one has

$$\begin{aligned}
\frac{\partial V_1}{\partial t} &= r_t V_1 - r_t s_1 \frac{V_2 - V_1}{\Delta S} - \frac{1}{2}\sigma_t^2(V_{ss})\frac{V_2 - 2V_1 + V_0}{\Delta S^2} \\
\frac{\partial V_2}{\partial t} &= r_t V_2 - r_t s_2 \frac{V_3 - V_2}{\Delta S} - \frac{1}{2}\sigma_t^2(V_{ss})\frac{V_3 - 2V_2 + V_1}{\Delta S^2} \\
\frac{\partial V_3}{\partial t} &= r_t V_3 - r_t s_3 \frac{V_4 - V_3}{\Delta S} - \frac{1}{2}\sigma_t^2(V_{ss})\frac{V_4 - 2V_3 + V_2}{\Delta S^2} \\
&\cdots \\
&\cdots \\
\frac{\partial V_{N-1}}{\partial t} &= r_t V_{N-1} - r_t s_{N-1} \frac{V_N - V_{N-1}}{\Delta S} - \frac{1}{2}\sigma_t^2(V_{ss})\frac{V_N - 2V_{N-1} + V_{N-2}}{\Delta S^2} \\
\frac{\partial V_N}{\partial t} &= r_t V_N - r_t s_N \frac{V_{N+1} - V_N}{\Delta S} - \frac{1}{2}\sigma_t^2(V_{ss})\frac{V_{N+1} - 2V_N + V_{N-1}}{\Delta S^2}
\end{aligned} \tag{12.17}$$

For the i-th ODE one has at sampling point S_i along the S axis one has

$$\frac{\partial V_i}{\partial t} = r_t V_i - r_t S_i \frac{V_{i+1}-V_i}{\Delta S} - \frac{1}{2}\sigma_t^2(V_{ss})\frac{V_{i+1}}{\Delta S^2} + \sigma_t^2(V_{ss})\frac{V_i}{\Delta S^2} + \frac{1}{2}\sigma_t^2(V_{ss})\frac{V_{i-1}}{\Delta S^2} \quad (12.18)$$

Eq. (12.18) is also written in the form

$$\frac{\partial V_i}{\partial t} = \frac{1}{2}\frac{\sigma_t^2(V_{ss})}{\Delta S^2}V_{i-1} + \frac{\sigma_t^2(V_{ss})}{\Delta S^2}V_i - \left[\frac{r_t S_i}{\Delta S} + \frac{1}{2}\frac{\sigma_t^2(V_{ss})}{\Delta S^2}\right]V_{i+1} + \left[r_t + \frac{r_t S_i}{\Delta S}\right]V_i \quad (12.19)$$

By denoting $K_1 = \sigma_t^2(V_{ss})$, $K_2 = -[2r_t S_i \Delta S + \sigma_t^2(V_{ss})]$ and $f(V_i) = [r_t + \frac{r_t S_i}{\Delta S}]V_i$ one obtains the following description for Eq. (12.19)

$$\frac{\partial V_i}{\partial t} = \frac{K_1}{2\Delta S^2}V_{i-1} + \frac{K_1}{\Delta S^2}V_i + \frac{K_2}{\Delta S^2}V_{i+1} + f(V_i) \quad (12.20)$$

12.3 Differential Flatness of the Black-Scholes PDE Model

Next, the state vector $\tilde{Y} = [y_{1,1}, y_{1,2}, \cdots, y1, i, \cdots, y_{1,N-1}, y_{1,N}]$ is defined for the PDE model, where $y_{1,1} = V_1$, $y_{1,2} = V_2, \cdots, y_{1,i} = V_i, \cdots, y_{1,N-1} = V_{N-1}$ and $y_{1,N} = V_N$. It will be shown that the state-space description of the nonlinear PDE dynamics, which has as control input only the boundary condition ϕ_0 is a differentially flat one. One has

$$\dot{y}_{1,1} = \frac{K_1}{\Delta S^2}y_{1,1} + \frac{K_1}{\Delta S^2}\phi_0 + f(y_{1,1}) + \frac{K_2}{2\Delta S^2}y_{1,2} \quad (12.21)$$

$$\dot{y}_{1,2} = \frac{K_1}{\Delta S^2}y_{1,2} + \frac{K_1}{\Delta S^2}y_{1,1} + f(y_{1,2}) + \frac{K_2}{2\Delta S^2}y_{1,3} \quad (12.22)$$

$$\dot{y}_{1,3} = \frac{K_1}{\Delta S^2}y_{1,3} + \frac{K_1}{\Delta S^2}y_{1,2} + f(y_{1,3}) + \frac{K_2}{2\Delta S^2}y_{1,4} \quad (12.23)$$

$$\cdots\ \cdots$$

$$\dot{y}_{1,i} = \frac{K_1}{\Delta S^2}y_{1,i} + \frac{K_1}{\Delta S^2}y_{1,i-1} + f(y_{1,i}) + \frac{K_2}{2\Delta S^2}y_{1,i+1} \quad (12.24)$$

$$\cdots\ \cdots$$

$$\dot{y}_{1,N-1} = \frac{K_1}{\Delta S^2}y_{1,N-1} + \frac{K_1}{\Delta S^2}y_{1,N-2} + f(y_{1,N-1}) + \frac{K_2}{2\Delta S^2}y_{1,N} \quad (12.25)$$

$$\dot{y}_{1,N} = \frac{K_1}{\Delta S^2}y_{1,N} + \frac{K_1}{\Delta S^2}y_{1,N-1} + f(y_{1,N}) + \frac{K_2}{2\Delta S^2}\phi_{N+1} \quad (12.26)$$

The flat output is considered to be the state variable $y_{1,N}$, which is denoted as $y = y_{1,N}$. It is shown that all state variables which stand also for virtual control inputs of the system $\alpha_i = y_{1,N-i}$, can be written as functions of the flat output $y = y_{1,N}$.

From Eq. (12.26) and considering that ϕ_{N+1} is constant one obtains

$$\begin{aligned} y_{1,N-1} = \alpha_1 = \tfrac{1}{K/2\Delta S^2}\left[\dot{y}_{1,N} - \tfrac{K_1}{\Delta S^2}y_{1,N} - \tfrac{K}{2\Delta S^2}\phi_{1,N+1} - f(y_{1,N})\right] \\ \Rightarrow y_{1,N-1} = h_1(y, \dot{y}, \cdots) \end{aligned} \quad (12.27)$$

and following a similar procedure, from Eq. (12.25) one gets

$$\begin{aligned} y_{1,N-2} = \alpha_2 = \tfrac{1}{K/2\Delta S^2}\left[\dot{y}_{1,N-1} - \tfrac{K_1}{\Delta S^2}y_{1,N-1} - \tfrac{K}{2\Delta S^2}y_{1,N} - f(y_{1,N-1})\right] \\ \Rightarrow y_{1,N-2} = h_2(y, \dot{y}, \cdots) \end{aligned} \quad (12.28)$$

Continuing in a similar manner, from Eq. (12.24) one obtains

$$\begin{aligned} y_{1,i-1} = \alpha_{N-i+1} = \tfrac{1}{K/2\Delta S^2}\left[\dot{y}_{1,i} - \tfrac{K_1}{\Delta S^2}y_{1,i} - \tfrac{K}{2\Delta S^2}y_{1,i+1} - f(y_{1,i})\right] \\ \Rightarrow y_{1,i-1} = h_{N-i+1}(y, \dot{y}, \cdots) \end{aligned} \quad (12.29)$$

From Eq. (12.23) one obtains

$$\begin{aligned} y_{1,2} = \alpha_{N-2} = \tfrac{1}{K/2\Delta S^2}\left[\dot{y}_{1,3} - \tfrac{K_1}{\Delta S^2}y_{1,3} - \tfrac{K}{\Delta S^2}y_{1,4} - f(y_{1,3})\right] \\ \Rightarrow y_{1,2} = h_{N-2}(y, \dot{y}, \cdots) \end{aligned} \quad (12.30)$$

From Eq. (12.22) one obtains

$$\begin{aligned} y_{1,1} = \alpha_{N-1} = \tfrac{1}{K/2\Delta S^2}\left[\dot{y}_{1,2} - \tfrac{K_1}{\Delta S^2}y_{1,2} - \tfrac{K}{\Delta S^2}y_{1,3} - f(y_{1,2})\right] \\ \Rightarrow y_{1,1} = h_{N-1}(y, \dot{y}, \cdots) \end{aligned} \quad (12.31)$$

Finally, From Eq. (12.21) one obtains

$$\begin{aligned} \phi_0 = \alpha_N = \tfrac{1}{K/2\Delta S^2}\left[\dot{y}_{1,1} - \tfrac{K_1}{\Delta S^2}y_{1,1} - \tfrac{K}{\Delta S^2}y_2 - f(y_{1,1})\right] \\ \Rightarrow \phi_0 = h_N(y, \dot{y}, \cdots) \end{aligned} \quad (12.32)$$

The above procedure confirms that all state variables of the model

$$\begin{matrix} y_{1,1} & y_{1,2} & y_{1,3} & \cdots \\ y_{1,i} & \cdots & y_{1,N-1} & y_{1,N} \end{matrix} \quad (12.33)$$

and the control input which is the boundary condition ϕ_0 can be written as functions of the flat output $y = y_{1,N}$ and of the flat output's derivatives. Consequently, the state-space model of the PDE dynamics is differentially flat.

Additionally, one can consider decomposition of the PDE state-space equation into submodels, where at each submodel the virtual control input is $\alpha_i = y_{1,i+1}$

$$\dot{y}_{1,i} = -\frac{2K}{\Delta x^2} y_{1,i} + \frac{K}{\Delta x^2} y_{1,i-1} + f(y_{1,i}) + \frac{K}{\Delta x^2} y_{1,i+1} \qquad (12.34)$$
$$\cdots\cdots$$

and with local flat output $y_{1,i}$. Then, one can confirm that all such subsystems $i = 1, 2, \cdots, N$ are differentially flat. The aggregate flat output of the PDE system is the vector of grid points $Y = [y_{1,1}, y_{1,2}, \cdots, y_{1,i}, \cdots, y_{1,N}]$. Next, one can compute the virtual inputs which are applied to each subsystem.

12.4 Computation of a Boundary Conditions-Based Feedback Control Law

To implement boundary feedback control in the system of Eqs. (12.21–12.26), the nonlinear diffusion-PDE model is rewritten as follows:

$$\dot{y}_{1,N} = \frac{K_1}{\Delta S^2} y_{1,N} + f(y_{1,N}) + \frac{K_2}{2\Delta S^2} \phi_{N+1} + \frac{K_1}{\Delta S^2} y_{1,N-1} \qquad (12.35)$$

$$\dot{y}_{1,N-1} = \frac{K_1}{\Delta S^2} y_{1,N-1} + f(y_{1,N-1}) + \frac{K_2}{2\Delta S^2} y_{1,N} + \frac{K_1}{\Delta S^2} y_{1,N-2} \qquad (12.36)$$

$$\dot{y}_{1,N-2} = \frac{K_1}{\Delta S^2} y_{1,N-2} + f(y_{1,N-2}) + \frac{K_2}{2\Delta S^2} y_{1,N-1} + \frac{K_1}{\Delta S^2} y_{1,N-1} \qquad (12.37)$$
$$\cdots\cdots$$

$$\dot{y}_{1,i} = \frac{K_1}{\Delta S^2} y_{1,i} + f(y_{1,i}) + \frac{K_2}{2\Delta S^2} y_{1,i+1} + \frac{K_1}{\Delta S^2} y_{1,i-1} \qquad (12.38)$$
$$\cdots\cdots$$

$$\dot{y}_{1,3} = \frac{K_1}{\Delta S^2} y_{1,3} + f(y_{1,3}) + \frac{K_2}{2\Delta S^2} y_{1,4} + \frac{K_1}{\Delta S^2} y_{1,2} \qquad (12.39)$$
$$\cdots\cdots$$

$$\dot{y}_{1,2} = \frac{K_1}{\Delta S^2} y_{1,2} + f(y_{1,2}) + \frac{K_2}{2\Delta S^2} y_{1,3} + \frac{K_1}{\Delta S^2} y_{1,1} \qquad (12.40)$$

$$\dot{y}_{1,1} = \frac{K_1}{\Delta S^2} y_{1,1} + f(y_{1,1}) + \frac{K_2}{2\Delta S^2} y_{1,2} + \frac{K_1}{\Delta S^2} \phi_0 \qquad (12.41)$$

The boundary condition ϕ_{N+1} is assumed to be known and remains steady. The boundary condition ϕ_0 stands for the control input (for example, it can be dependent on the interest rate). The feedback control law is designed as follows:

$$\alpha_1 = y^*_{1,N-1} = \frac{1}{(K_1/\Delta S^2)} \left[\dot{y}^d_{1,N} - k_{p,1}(y_{1,N} - y^d_{1,N}) - \frac{K_1}{\Delta S^2} y_{1,N} - f(y_{1,N}) - \frac{K_2}{2\Delta S^2} \phi_{N+1}\right] \qquad (12.42)$$

$$\alpha_2 = y^*_{1,N-2} = \frac{1}{(K_1/\Delta S^2)}\left[\dot{y}^d_{1,N-1} - k_{p,2}(y_{1,N-1} - y^d_{1,N-1}) - \frac{K_1}{\Delta S^2}y_{1,N-1} - f(y_{1,N-1}) - \frac{K_2}{2\Delta S^2}y_N\right] \Rightarrow$$
$$\alpha_2 = \frac{1}{(K_1/\Delta S^2)}\left[\dot{\alpha}_1 - k_{p,2}(y_{1,N-1} - \alpha_1) - \frac{K_1}{\Delta S^2}y_{1,N-1} - f(y_{1,N-1}) - \frac{K_2}{2\Delta S^2}y_N\right] \tag{12.43}$$

$$\alpha_3 = y^*_{1,N-3} = \frac{1}{(K_1/\Delta S^2)}\left[\dot{y}^d_{1,N-2} - k_{p,3}(y_{1,N-2} - y^d_{1,N-2}) - \frac{K_1}{\Delta S^2}y_{1,N-2} - f(y_{1,N-2}) - \frac{K_2}{2\Delta S^2}y_{N-1}\right] \Rightarrow$$
$$\alpha_3 = \frac{1}{(K_1/\Delta S^2)}\left[\dot{\alpha}_2 - k_{p,3}(y_{1,N-2} - \alpha_2) - \frac{K_1}{\Delta S^2}y_{1,N-2} - f(y_{1,N-2}) - \frac{K_2}{2\Delta S^2}y_{N-1}\right] \tag{12.44}$$

and continuing in a similar manner

$$\begin{aligned}\alpha_i = y^*_{1,N-i} &= \frac{1}{(K_1/\Delta S^2)}\left[\dot{y}^d_{1,N-i+1} - k_{p,i}(y_{1,N-i+1} - y^d_{1,N-i+1}) - \frac{K_1}{\Delta S^2}y_{1,N-i+1} - f(y_{1,N-i+1}) - \frac{K_2}{2\Delta S^2}y_{N-i+2}\right] \Rightarrow \\ \alpha_i &= \frac{1}{(K_1/\Delta S^2)}\left[\dot{\alpha}_{i-1} - k_{p,i}(y_{1,N-i+1} - \alpha_{i-1}) - \frac{K_1}{\Delta S^2}y_{1,N-i+1} - f(y_{1,N-i+1}) - \frac{K_2}{2\Delta S^2}y_{N-i+2}\right]\end{aligned} \tag{12.45}$$

Following this procedure one arrives to compute the control inputs which are associated with the last two rows of the state-space model

$$\alpha_{N-2} = y^*_{1,2} = \frac{1}{(K_1/\Delta S^2)}\left[\dot{y}^d_{1,3} - k_{p,N-1}(y_{1,3} - y^d_{1,3}) - \frac{K_1}{\Delta S^2}y_{1,3} - f(y_{1,3}) - \frac{K_2}{2\Delta S^2}y_{1,4}\right] \Rightarrow$$
$$\alpha_{N-2} = \frac{1}{(K_1/\Delta S^2)}\left[\dot{\alpha}_{N-3} - k_{p,N-2}(y_{1,3} - \alpha_{N-3}) - \frac{K_1}{\Delta S^2}y_{1,3} - f(y_{1,3}) - \frac{K_2}{2\Delta S^2}y_{1,4}\right] \tag{12.46}$$

$$\alpha_{N-1} = y^*_{1,1} = \frac{1}{(K_1/\Delta S^2)}\left[\dot{y}^d_{1,2} - k_{p,N-1}(y_{1,2} - y^d_{1,2}) - \frac{K_1}{\Delta S^2}y_{1,2} - f(y_{1,2}) - \frac{K_2}{2\Delta S^2}y_{1,3}\right] \Rightarrow$$
$$\alpha_{N-1} = \frac{1}{(K_1/\Delta S^2)}\left[\dot{\alpha}_{N-2} - k_{p,N-1}(y_{1,2} - \alpha_{N-2}) - \frac{K_1}{\Delta S^2}y_{1,2} - f(y_{1,2}) - \frac{K_2}{2\Delta S^2}y_{1,3}\right] \tag{12.47}$$

and finally

$$\alpha_N = \phi_0 = \frac{1}{(K_1/\Delta S^2)}\left[\dot{y}^d_{1,1} - k_{p,N}(y_{1,1} - y^d_{1,1}) - \frac{K_1}{\Delta S^2}y_{1,1} - f(y_{1,1}) - \frac{K_2}{2\Delta S^2}y_{1,2}\right] \Rightarrow$$
$$\alpha_N = \frac{1}{(K_1/\Delta S^2)}\left[\dot{\alpha}_{N-1} - k_{p,N}(y_{1,1} - \alpha_{N-1}) - \frac{K_1}{\Delta S^2}y_{1,1} - f(y_{1,1}) - \frac{K_2}{2\Delta S^2}y_{1,2}\right] \tag{12.48}$$

Consequently, the computation of the aggregate control input $\alpha_N = \phi_0$ which is exerted on the PDE model is performed backwards, by substituting recursively into ϕ_0 the virtual control inputs $a_{N-1}, a_{N-2}, \cdots, a_i s, \cdots, a_2, a_1$.

It is noted that implementation of flatness-based control in cascaded loops has been studied in the case of lumped parameter systems, as for example electric machines [62, 213, 214]. The approach followed in this chapter extends this concept to a problem of elevated difficulty, and with a state-space description of higher dimensionality, that is control of distributed parameter systems and in particular control of the Black-Scholes PDE.

12.5 Closed Loop Dynamics

By substituting Eq. (12.48) into Eq. (12.41) of the state-space model of the PDE dynamics, and using the definition $y_{1,1} - y_{1,1}^d = z_1$ one has

$$\begin{gathered} \dot{y}_{1,1} = \dot{a}_{N-1} - k_{p,1}(y_{1,1} - a_{N-1}) \Rightarrow \\ (\dot{y}_{1,1} - \dot{a}_{N-1}) + k_{p,1}(y_{1,1} - a_{N-1}) = 0 \Rightarrow \\ \dot{z}_1 + k_{p,1} z_1 = 0 \end{gathered} \tag{12.49}$$

Equivalently, by substituting Eq. (12.47) into Eq. (12.40), and using the definition $y_{1,2} - \alpha_{N-2} = z_2$ one has

$$\begin{gathered} \dot{y}_{1,2} = \dot{\alpha}_{N-2} - k_{p,2}(y_{1,2} - \alpha_{N-2}) \Rightarrow \\ (\dot{y}_{1,2} - \dot{\alpha}_{N-2}) + k_{p,2}(y_{1,2} - \alpha_{N-2}) = 0 \Rightarrow \\ \dot{z}_2 + k_{d,2}\dot{z}_2 + k_{p,2} z_2 = 0 \end{gathered} \tag{12.50}$$

Similarly, continuing with the rest of the equations of the state-space model and by substituting Eq. (12.46) into Eq. (12.39), while also using the definition $y_{1,3} - \alpha_{N-3} = z_3$ one has

$$\begin{gathered} \dot{y}_{1,3} = \dot{\alpha}_{N-3} - k_{p,3}(y_{1,3} - \alpha_{N-3}) \Rightarrow \\ (\dot{y}_{1,3} - \dot{\alpha}_{N-3}) + k_{p,3}(y_{1,3} - \alpha_{N-3}) = 0 \Rightarrow \\ \ddot{z}_3 + k_{p,3} z_3 = 0 \end{gathered} \tag{12.51}$$

Moving backwards, and by substituting Eq. (12.45) into Eq. (12.38) of the state-space model of the PDE dynamics, and using the definition $y_{1,i} - \alpha_{i-1} = z_i$ one has

$$\begin{gathered} \dot{y}_{1,i} = \dot{\alpha}_{i-1} - k_{p,i}(y_{1,i} - \alpha_{i-1}) \Rightarrow \\ (\dot{y}_{1,i} - \dot{\alpha}_{i-1}) + k_{p,i}(y_{1,i} - \alpha_{i-1}) = 0 \Rightarrow \\ \dot{z}_i + k_{p,i} z_i = 0 \end{gathered} \tag{12.52}$$

By substituting Eq. (12.43) into Eq. (12.36) of the state-space model of the PDE dynamics, and using the definition $y_{1,N-1} - \alpha_1 = z_{N-1}$ one has

$$\begin{array}{c} \dot{y}_{1,N-1} = \dot{\alpha}_1 - k_{p,N-1}(y_{1,N-1} - \alpha_1) \Rightarrow \\ (\dot{y}_{1,N-1} - \dot{\alpha}_1) + k_{p,N-1}(y_{1,N-1} - \alpha_1) = 0 \Rightarrow \\ \dot{z}_{N-1} + k_{p,N-1} z_{N-1} = 0 \end{array} \tag{12.53}$$

Finally, by substituting Eq. (12.42) into Eq. (12.35), and using the definition $y_{1,N} - y_{1,N}^d = z_N$ one obtains

$$\begin{array}{c} \dot{y}_{1,N} = \dot{y}_{1,N}^d - k_{p,N}(y_{1,N} - y_{1,N}^d) \Rightarrow \\ (\dot{y}_{1,N} - \dot{y}_{1,N}^d) + k_{p,N}(y_{1,N} - y_{1,N}^d) = 0 \Rightarrow \\ \dot{z}_N + k_{p,N} z_N = 0 \end{array} \tag{12.54}$$

Thus, the dynamics of the closed-loop system becomes

$$\begin{array}{c} \ddot{z}_1 + k_{p,1} z_1 = 0 \\ \ddot{z}_2 + k_{p,2} z_2 = 0 \\ \ddot{z}_3 + k_{p,3} z_3 = 0 \\ \cdots \cdots \\ \ddot{z}_i + k_{p,i} z_i = 0 \\ \cdots \cdots \\ \ddot{z}_{N-1} + k_{p,N-1} z_{N-1} = 0 \\ \ddot{z}_N + k_{p,N} z_N = 0 \end{array} \tag{12.55}$$

The dynamics of the closed-loop system can be also written in matrix form

$$\ddot{\tilde{Z}} + K_P \tilde{Z} = 0 \tag{12.56}$$

where

$$\begin{array}{c} \tilde{Z} = [z_1, z_2, z_3, \cdots, z_i, \cdots, z_{N-1}, z_N] \\ K_P = \text{diag}[k_{p,1}, k_{p,2}, k_{p,3}, \cdots, k_{p,i}, \cdots, k_{p,N-1}, k_{p,N}] \end{array} \tag{12.57}$$

After suitable selection of the coefficients $k_{P,i}$, $i = 1, 2, \cdots, N$ such that the monomials $p(s) = s + k_{p,i}$ to have negative roots, it can be assured that $lim_{t \to \infty} z_i(t) = 0$ and that the closed-loop system is asymptotically stable.

Moreover, to prove asymptotic stability for the closed-loop system, the following Lyapunov function can be used

$$V = \sum_{i=1}^{N} \frac{1}{2} \dot{z}_i^2 \tag{12.58}$$

The derivative of this Lyapunov function with respect to time is

$$\begin{array}{c}\dot{V} = \sum_{i=1}^{N} \frac{1}{2}\dot{z}_i^2 \Rightarrow \dot{V} = \sum_{i=1}^{N} z_i(-k_{p_i} z_i) \Rightarrow \\ \dot{V} = -\sum_{i=1}^{N} k_{p,i} z_i^2 < 0\end{array} \tag{12.59}$$

Thus, it is proven again that the closed-loop system is globally asymptotically stable.

12.6 Simulation Tests

Simulation examples about the proposed control method for distributed parameter systems are provided for the Black-Scholes PDE that was given in Eq. (12.13). Results have been obtained for two different test cases. Those of the first test case are depicted in Figs. 12.1, 12.2, 12.3, 12.4, 12.5. Those of the second test case are depicted in Figs. 12.6, 12.7, 12.8, 12.9, 12.10. The spatial discretization of the PDE model consisted of $N = 25$ points. The boundary condition ϕ_{N+1} of the PDE was taken to be known and constant. The boundary condition ϕ_0 served as the control input and was computed as each iteration of the control algorithm according to the procedure described in Sect. 12.4.

The numerical simulation experiments have confirmed the theoretical findings of this chapter. It has been shown that by applying the proposed control method the Black-Scholes PDE dynamics can be modified so as to converge to the desirable reference profile. The control input that succeeds this, changes smoothly and has a moderate range of variation. The accuracy of tracking of the reference setpoints was

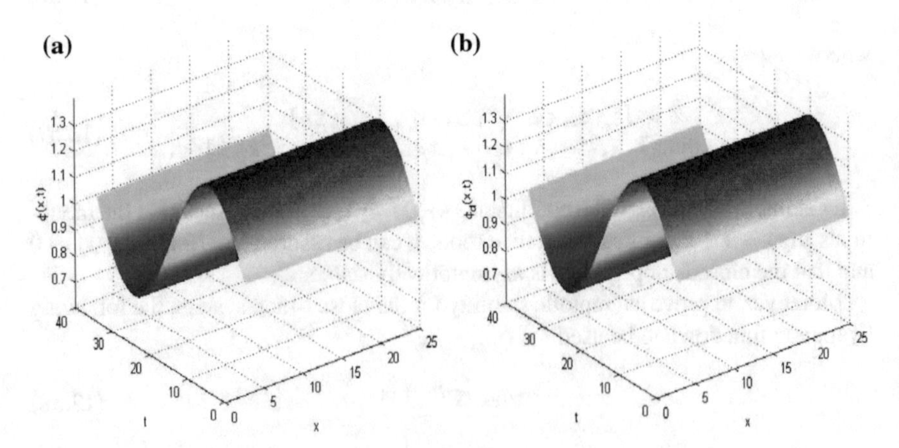

Fig. 12.1 Test case 1: **a** plot of the 2D-BSE function after applying boundary control **b** plot of the reference 2D-BSE function

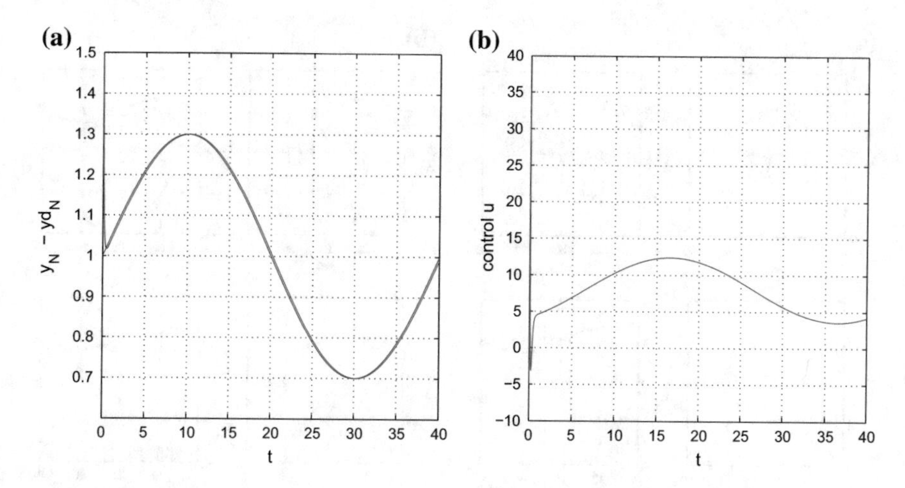

Fig. 12.2 Test case 1: **a** variation in time of the solution of the PDE at grid point x_N (*blue* line) and associated reference setpoint (*red* line) **b** control input exerted to the PDE system through the boundary condition ϕ_0

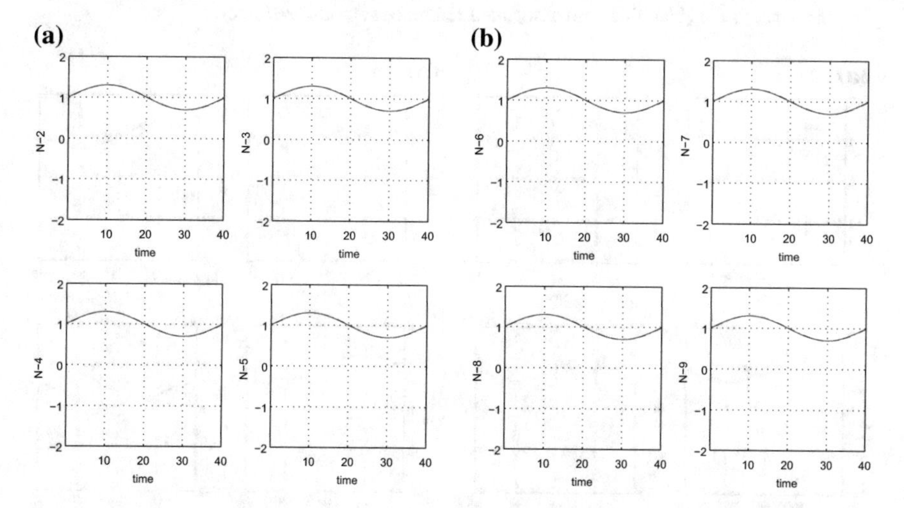

Fig. 12.3 Test case 1: **a** variation in time of the solution of the PDE at grid points x_2 to x_5 (*blue* line) and associated reference setpoint (*red* line) **b** variation in time of the solution of the PDE at grid points x_6 to x_9 (*blue* line) and associated reference setpoint (*red* line)

indeed satisfactory. It is noted that the application of the proposed Black-Scholes PDE feedback control method is not limited to the case of option pricing but can be also extended to macroeconomic features and indexes with a spatiotemporal dynamics.

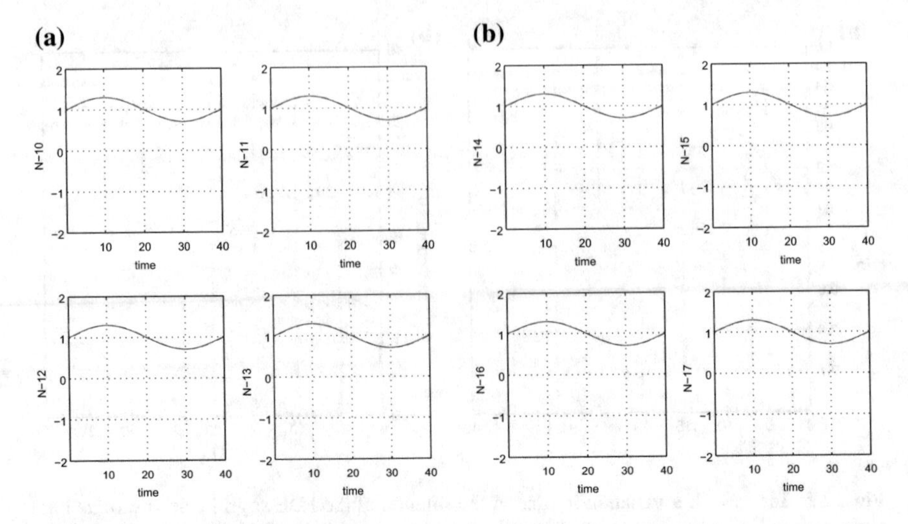

Fig. 12.4 Test case 1: **a** variation in time of the solution of the PDE at grid points x_{10} to x_{13} (*blue* line) and associated reference setpoint (*red* line) **b** variation in time of the solution of the PDE at grid points x_{14} to x_{17} (*blue* line) and associated reference setpoint (*red* line)

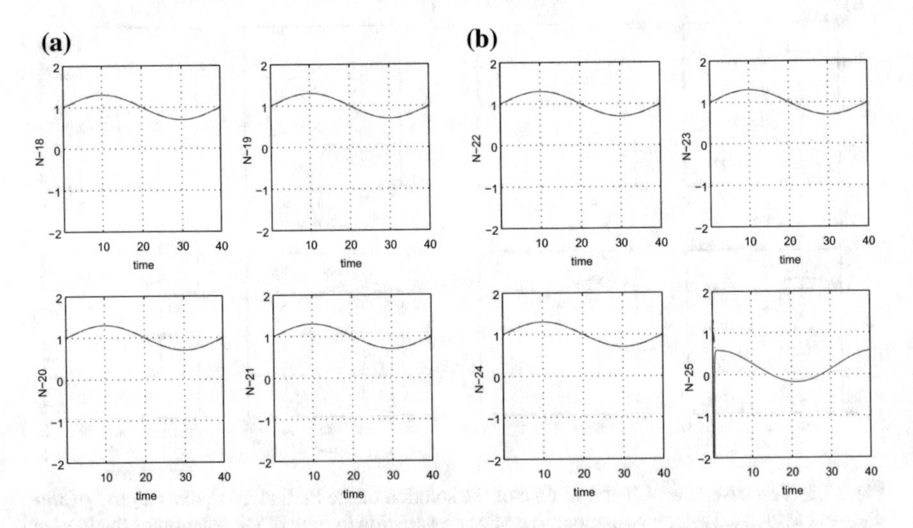

Fig. 12.5 Test case 1: **a** variation in time of the solution of the PDE at grid points x_{18} to x_{21} (*blue* line) and associated reference setpoint (*red* line) **b** variation in time of the solution of the PDE at grid points x_{22} to x_{25} (*blue* line) and associated reference setpoint (*red* line)

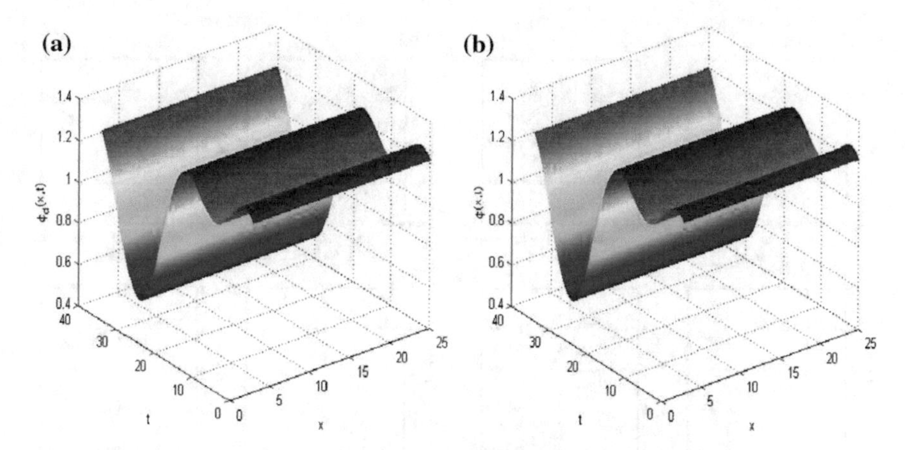

Fig. 12.6 Test case 2: **a** plot of the 2D-BSE function after applying boundary control **b** plot of the reference 2D-BSE function

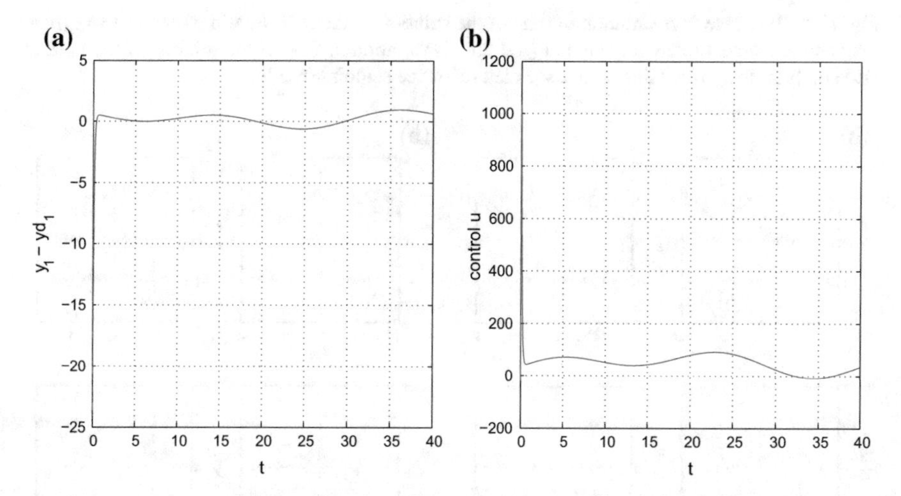

Fig. 12.7 Test case 2: **a** variation in time of the solution of the PDE at grid point x_N (*blue* line) and associated reference setpoint (*red* line) **b** control input exerted to the PDE system through the boundary condition ϕ_0

Regarding robustness to uncertainty of the control scheme, it can be noted that: (i) each local control loop for the subsystems into which the PDE model is decomposed, is of the proportional-derivative type and has its own robustness. This means that according to the values of the feedback gains in it a specific phase and gain margin can be exhibited, (ii) elevated robustness of each local control loop associated with an individual subsystem of the PDE model can be achieved, if the associated virtual

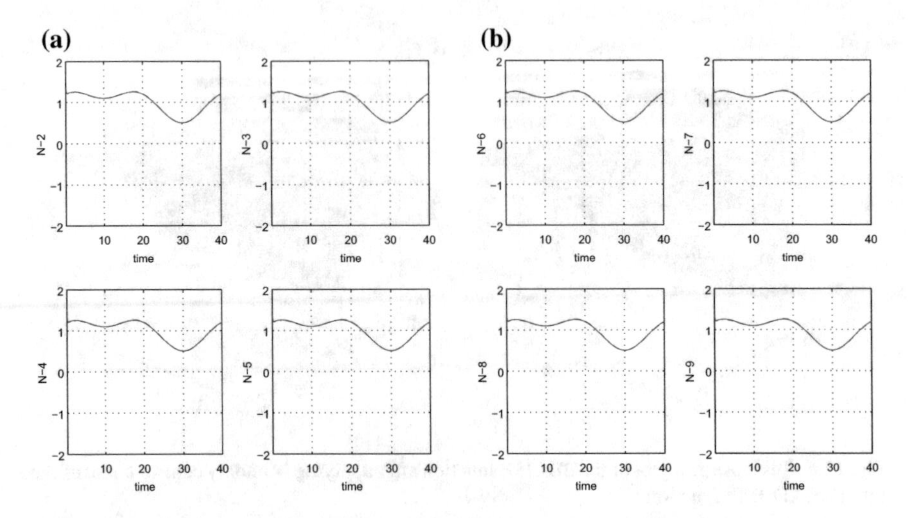

Fig. 12.8 Test case 2: **a** variation in time of the solution of the PDE at grid points x_2 to x_5 (*blue* line) and associated reference setpoint (*red* line) **b** variation in time of the solution of the PDE at grid points x_6 to x_9 (*blue* line) and associated reference setpoint (*red* line)

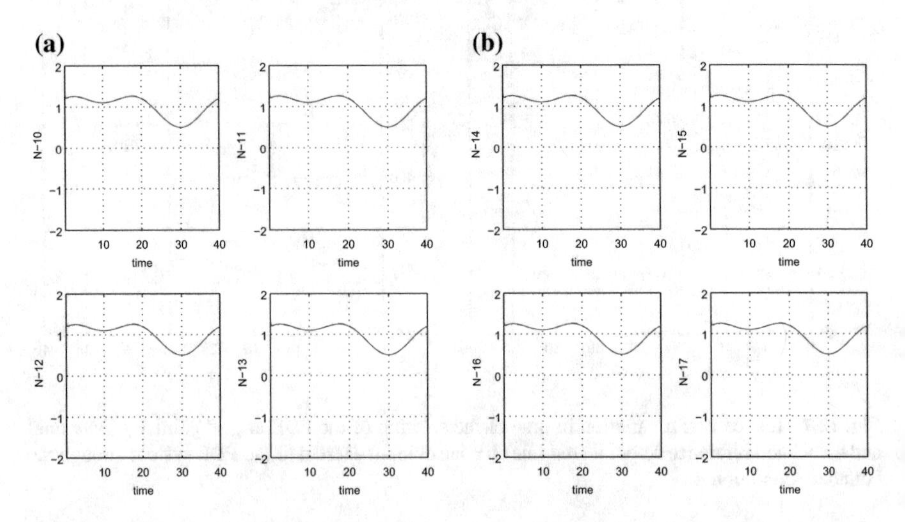

Fig. 12.9 Test case 2: **a** variation in time of the solution of the PDE at grid points x_{10} to x_{13} (*blue* line) and associated reference setpoint (*red* line) **b** variation in time of the solution of the PDE at grid points x_{14} to x_{17} (*blue* line) and associated reference setpoint (*red* line)

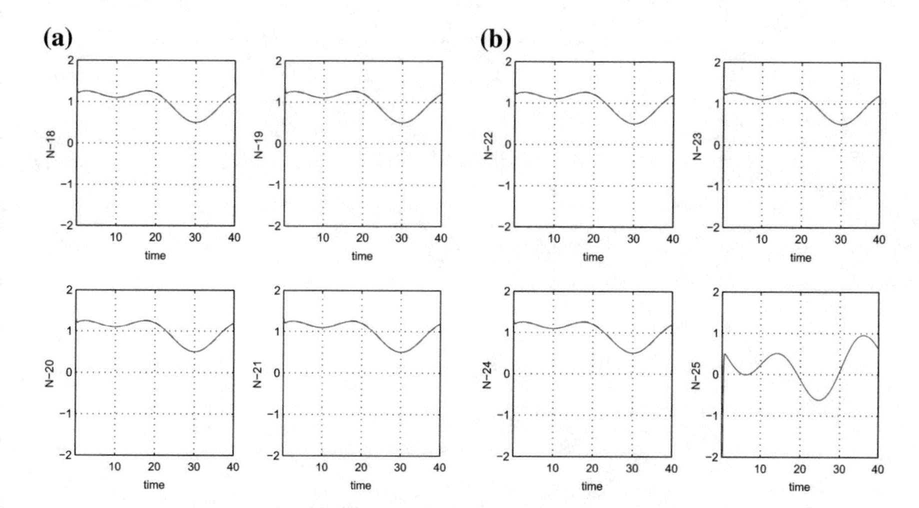

Fig. 12.10 Test case 2: **a** variation in time of the solution of the PDE at grid points x_{18} to x_{21} (*blue* line) and associated reference setpoint (*red* line) **b** variation in time of the solution of the PDE at grid points x_{22} to x_{25} (*blue* line) and associated reference setpoint (*red* line)

control input is complemented by an additional term which will aim at the compensation of the disturbances and model uncertainty effects. This will finally result in an additional control term that will complement the control input u that is actually exerted on the PDE model.

Chapter 13
Stabilization of the Multi-asset Black–Scholes PDE Using Differential Flatness Theory

13.1 Outline

The use of PDE models, such as the Black–Scholes PDE, for modelling options' price dynamics, has been previously explained. Next, the multi-asset Black–Scholes PDE will be considered as an advanced modelling tool in option pricing, with potential applications to electricity markets. It will be shown that even for such a high-dimensional PDE model, feedback control and stabilization can be achieved.

Since electricity markets move progressively to more elaborated schemes for the trading of electric power (based on long term contracts and options' bargain), optimization policies and stabilization measures for such financial systems become necessary. It is considered that the value of the electric power that is offered to the grid by a power corporation is defined by options which in turn depend on multiple assets. In such a case, the dynamics of the options' value is described by the multi-asset Black–Scholes PDE [143, 157, 165, 166, 266]. The chapter treats the problem of feedback control and stabilization of the aforemention PDE. Methods for feedback stabilization of systems with nonlinear PDE dynamics have been a flourishing research subject in the last years [7, 10, 20, 33, 35, 139, 249]. In particular, feedback control of diffusion-type (parabolic) PDEs has been a subject of extensive research and several remarkable results have been produced [151, 170, 277, 311]. For the control of diffusion PDEs, boundary and distributed control methods have been developed [78, 276].

By showing the feasibility of control of the multi-asset Black–Scholes PDE it is also proven that through selected purchases and sales during the trading procedure, the price of the negotiated electric power can be made to converge and stabilize at specific reference values. By applying semi-discretization and the finite differences method, the multi-asset Black–Scholes PDE model is written in a state-space form [124, 179]. It is proven that this state-space description stands for a differentially flat system which means that all its state variables and control inputs can be written as differential functions of the flat output vector [34, 133, 134]. One can note

G.G. Rigatos, *State-Space Approaches for Modelling and Control in Financial Engineering*, Intelligent Systems Reference Library 125,
DOI 10.1007/978-3-319-52866-3_13

again several results on the use of differential flatness theory in the control of PDEs [222, 231]. As in the case of the single-asset Black–Scholes PDE, the first stage in the proposed control scheme is to decompose the state-space description of the PDE into an equivalent set of nonlinear ODEs. Next by examining independently each nonlinear ODE it is shown that this stands again for a differentially flat system, for which a virtual control input can be computed as in the case of flatness-based control for the trivial system. The virtual control input is chosen such that the ODE subsystem dynamics is linearized and the tracking error is eliminated. The boundary condition that appears in the nonlinear ODE subsystem that comprises the last rows of the state-space description, stands for the aggregate control input [95, 127, 187, 261].

The computation of the boundary control input uses recursively all virtual control inputs mentioned above, moving from the last ODE system to the first one. This stands for implementation of flatness-based control in successive (cascading) loops. Thus, by tracing the rows of the state-space model backwards, at each iteration of the control algorithm, one can finally obtain the control input that should be applied to the multi-asset Black–Scholes PDE system so as to assure that all its state vector elements will converge to the desirable setpoints. The stability of the control loop is proven in two manners. First, convergence to zero is proven for the tracking error of all subsystems into which the PDE's state-space model is decomposed. Next, with the use of Lyapunov analysis it is reconfirmed that this control scheme is asymptotically stable.

13.2 Boundary Control of the Multi-asset Black–Scholes PDE

First, the single-asset nonlinear Black–Scholes PDE is used to describe the financial system's (option price) dynamics:

$$\frac{\partial V(t,S)}{\partial t} + r_t S \frac{\partial V(t,S)}{\partial S} + \frac{1}{2}\sigma_t^2 S^2 \frac{\partial^2 V(t,S)}{\partial S^2} - r_t V(t,S) = 0 \tag{13.1}$$

Variable σ_t stands for volatility and variable r_t stands for an interest rate. Next, the multi-asset Black–Scholes PDE is introduced [143, 157, 165, 166, 266]:

$$\frac{\partial V}{\partial t} = \sum_{i=1}^{N}\sum_{j=1}^{N} \rho\sigma_i\sigma_j S_i S_j \frac{\partial^2 V}{\partial S_i \partial S_j} + \sum_{i=1}^{N} r S_i \frac{\partial V}{\partial S_i} - rV \tag{13.2}$$

Moreover, without loss of generality the two-asset Black–Scholes PDE is considered

$$\frac{\partial V}{\partial t} = \frac{1}{2}\sigma_1^2 S_1^2 \frac{\partial V}{\partial^2 S_1^2} + \frac{1}{2}\sigma_2^2 S_2^2 \frac{\partial V}{\partial^2 S_2^2} + \rho\sigma_1\sigma_2 S_1 S_2 \frac{\partial^2 V}{\partial S_1 \partial S_2} + rS_1\frac{\partial V}{\partial S_1} + rS_2 \frac{\partial V}{\partial S_2} - rV \tag{13.3}$$

Semi-discretization and the finite differences method is applied. To this end the partial derivatives appearing in Eq. (13.3) are computed as follows:

$$\frac{\partial V}{\partial S_1} = \frac{V(S_{1,i+1}, S_{2,j}) - V(S_{1,i}, S_{2,j})}{\Delta S_1} \tag{13.4}$$

$$\frac{\partial^2 V}{\partial^2 S_1} = \frac{V(S_{1,i+1},S_{2,j})-2V(S_{1,i},S_{2,j})+V(S_{1,i-1},S_{2,j})}{{\Delta S_1}^2} \tag{13.5}$$

$$\frac{\partial V}{\partial S_2} = \frac{V(S_{1,i},S_{2,j+1})-V(S_{1,i},S_{2,j})}{\Delta S_2} \tag{13.6}$$

$$\frac{\partial^2 V}{\partial^2 S_2} = \frac{V(S_{1,i},S_{2,j+1})-2V(S_{1,i},S_{2,j})+V(S_{1,i},S_{2,j-1})}{{\Delta S_2}^2} \tag{13.7}$$

$$\frac{\partial^2 V}{\partial S_1 \partial S_2} = \frac{V(S_{1,i+i},S_{2,j+i})-V(S_{1,i+1},S_{2,j})-V(S_{1,i},S_{2,j+1})+V(S_{1,i},S_{2,j})}{\Delta S_1 \Delta S_2} \tag{13.8}$$

Using the previous semi-discretization, for grid point $(i,\ j)$ it holds

$$\begin{aligned}
\frac{\partial V(S_{1,i},S_{2,j})}{\partial t} &= \tfrac{1}{2}\sigma_1^2 S_{1,i}^2 \left[\frac{V(S_{1,i+1},S_{2,j})-2V(S_{1,i},S_{2,j})+V(S_{1,i-1},S_{2,j})}{\Delta S_1^2}\right]+ \\
&\tfrac{1}{2}\sigma_2^2 S_{2,j}^2 \left[\frac{V(S_{1,i},S_{2,j+1})-2V(S_{1,i},S_{2,j})+V(S_{1,i},S_{2,j-1})}{\Delta S_1^2}\right]+ \\
&\rho\sigma_1\sigma_2 S_{1,i} S_{2,j}\left[\frac{V(S_{1,i+1},S_{2,j+1})-V(S_{1,i+1},S_{2,j})-V(S_{1,i},S_{2,j+1})+V(S_{1,i},S_{2,j})}{\Delta S_1 \Delta S_2}\right]+ \\
&r S_{1,i}\left[\frac{V(S_{1,i+1},S_{2,j})-V(S_{1,i},S_{2,j})}{\Delta S_1}\right]+r S_{2,j}\left[\frac{V(S_{1,i},S_{2,j+1})-V(S_{1,i},S_{2,j})}{\Delta S_2}\right]-rV(S_{1,i},\ S_{2,j})
\end{aligned} \tag{13.9}$$

The boundary conditions of the PDE are taken to be $V_{i,0}{\neq}0$ only if $i = 1$, $V_{0,j}{\neq}0$ only if $j = 1$ and $V(i,\ j) = ct$ (constant) if $i > N$ or $j > N$.

Considering that $i = 1, 2, \ldots, N$ and $j = 1, 2, \ldots, N$ the option's values at the grid points $(i,\ j)$ are denoted as $V_{i,j}$. Using this notation, the semi-discretized model of the PDE takes the following form:
At grid point $i = 1$ and $j = 1$

$$\begin{aligned}
\frac{\partial V_{1,1}}{\partial t} &= \tfrac{1}{2}\sigma_1^2 S_{1,1}^2\left[\frac{V_{2,1}-2V_{1,1}+V_{0,1}}{\Delta S_1^2}\right]+ \\
&\tfrac{1}{2}\sigma_2^2 S_{2,1}^2\left[\frac{V_{1,2}-2V_{1,1}+V_{1,0}}{\Delta S_2^2}\right]+\rho\sigma_1\sigma_2 S_{1,1}S_{2,1}\left[\frac{V_{2,2}-V_{2,1}-V_{1,2}+V+1,1}{\Delta S_1 \Delta S_2}\right]+ \\
&r S_{1,1}\left[\frac{V_{2,1}-V_{1,1}}{\Delta S_1}\right]+r S_{2,1}\left[\frac{V_{1,2}-V_{1,1}}{\Delta S_2}\right]-rV_{1,1}
\end{aligned} \tag{13.10}$$

At grid point $i > 1$ and $j > 1$ it holds

$$\begin{aligned}
\frac{\partial V_{i,j}}{\partial t} &= \tfrac{1}{2}\sigma_1^2 S_{1,i}^2\left[\frac{V_{i+1,j}-2V_{i,j}+V_{i-1,j}}{\Delta S_1^2}\right]+ \\
&\tfrac{1}{2}\sigma_2^2 S_{2,j}^2\left[\frac{V_{i,j+1}-2V_{i,j}+V_{i,j-1}}{\Delta S_2^2}\right]+\rho\sigma_1\sigma_2 S_{1,i}S_{2,j}\left[\frac{V_{i+1,j+1}-V_{i+1,j+1}-V_{i,j+1}+V+i,i}{\Delta S_1 \Delta S_2}\right]+ \\
&r S_{1,i}\left[\frac{V_{i+1,j}-V_{i,j}}{\Delta S_1}\right]+r S_{2,j}\left[\frac{V_{i,j+1}-V_{i,j}}{\Delta S_2}\right]-rV_{i,j}
\end{aligned} \tag{13.11}$$

Next, the following state vector variables are defined $x_{(i-1)N+j} = V_{i,j}$, $i = 1,2,\dots,N$ and $j = 1,2,\dots,N$. Using this notation of state variables Eq. (13.10) becomes

$$\begin{aligned}\dot{x}_1 = \tfrac{1}{2}\sigma_1^2 S_{1,1}^2\left[\tfrac{x_{N+1}-2x_1}{\Delta S_1^2}\right] + \tfrac{1}{2}\sigma_2^2 S_{2,1}^2\left[\tfrac{x_2-2x_1}{\Delta S_1^2}\right] + \\ \rho\sigma_1\sigma_2 S_{1,1}S_{1,2}\left[\tfrac{x_{N+2}-x_2-x_{N+1}+x_1}{\Delta S_1 \Delta S_2}\right] + rS_{1,1}\left[\tfrac{x_{N+1}-x_1}{\Delta S_1}\right] + rS_{2,1}\left[\tfrac{x_2-x_1}{\Delta S_2}\right] + \\ \left[\tfrac{1}{2}\sigma_1^2 S_{1,1}^2\tfrac{V_{0,1}}{\Delta S_1^2} + \tfrac{1}{2}\sigma_2^2 S_{2,1}^2\tfrac{V_{1,0}}{\Delta S_1^2}\right]\end{aligned} \tag{13.12}$$

Thus, by defining the control input associated with the boundary conditions as $u = [V_{0.1}, V_{1,0}]^T$ and $c_1 u = [\frac{1}{2}\sigma_1^2 S_{1,1}^2\frac{V_{0,1}}{\Delta S_1^2} + \frac{1}{2}\sigma_2^2 S_{2,1}^2\frac{V_{1,0}}{\Delta S_1^2}]$ one obtains a description for Eq. (13.12) in the form

$$\dot{x}_1 = f_1(x) + c_1 u \tag{13.13}$$

Equivalently for Eq. (13.11) one obtains

$$\begin{aligned}\dot{x}_{(i-1)N+j} = \tfrac{1}{2}\sigma_1^2 S_{1,i}^2\left[\tfrac{x_{iN+j}-2x_{(i-1)N+j}+x_{(i-2)N+j}}{\Delta S_1^2}\right] + \\ \tfrac{1}{2}\sigma_2^2 S_{2,j}^2\left[\tfrac{x_{(i-1)N+(j+1)}-2x_{(i-1)N+j}+x_{(i-1)N+(j-1)}}{\Delta S_2^2}\right] + \\ \rho\sigma_1\sigma_2 S_{1,i}S_{2,j}\left[\tfrac{x_{(i)N+(j+1)}-x_{(i-1)N+(j+1)}-x_{iN+j}+x_{(i-1)N+j}}{\Delta S_1 \Delta S_2}\right] + \\ rS_{1,i}\left[\tfrac{x_{iN+j}-x_{(i-1)N+j}}{\Delta S_1}\right] + rS_{2,j}\left[\tfrac{x_{(i-1)N+(j+1)}-x_{(i-1)N+j}}{\Delta S_2}\right] - rx_{(i-1)N+j}\end{aligned} \tag{13.14}$$

Equation (13.14) can be also written in the form

$$\begin{aligned}\dot{x}_{(i-1)N+j} = \tfrac{1}{2}\sigma_1^2 S_{1,i}^2\left[\tfrac{x_{iN+j}-2x_{(i-1)N+j}+x_{(i-2)N+j}}{\Delta S_1^2}\right] + \\ \tfrac{1}{2}\sigma_2^2 S_{2,j}^2\left[\tfrac{x_{(i-1)N+(j+1)}-2x_{(i-1)N+j}}{\Delta S_2^2}\right] + \\ \rho\sigma_1\sigma_2 S_{1,i}S_{2,j}\left[\tfrac{x_{(i)N+(j+1)}-x_{(i-1)N+(j+1)}-x_{iN+j}+x_{(i-1)N+j}}{\Delta S_1 \Delta S_2}\right] + \\ rS_{1,i}\left[\tfrac{x_{iN+j}-x_{(i-1)N+j}}{\Delta S_1}\right] + rS_{2,j}\left[\tfrac{x_{(i-1)N+(j+1)}-x_{(i-1)N+j}}{\Delta S_2}\right] - rx_{(i-1)N+j} + \\ \left[\tfrac{1}{2}\sigma_2^2 S_{2,j}^2\tfrac{1}{\Delta S_2^2}\right] x_{(i-1)N+(j-1)}\end{aligned} \tag{13.15}$$

Equation (13.15) is finally written in the form

$$\dot{x}_{(i-1)N+j} = f_{(i-1)N+j}(x) + c_{(i-1)N+j}x_{(i-1)N+(j-1)} \tag{13.16}$$

Considering that $i = 1, 2, \ldots, N$ and $j = 1, 2, \ldots, N$ there are N^2 state-space equations. Thus, the dynamics of the PDE model is written as

$$\begin{aligned} \dot{x}_{N^2} &= f_{N^2}(x) + c_{N^2}x_{N^2-1} \\ \dot{x}_{N^2-1} &= f_{N^2-1}(x) + c_{N^2-1}x_{N^2-2} \\ &\cdots\ \cdots \\ \dot{x}_{(i-1)N+j} &= f_{(i-1)N+j} + c_{(i-1)N+j}x_{(i-1)N+(j-1)} \\ &\cdots\ \cdots \\ \dot{x}_2 &= f_2(x) + c_2x_1 \\ \dot{x}_1 &= f_1(x) + c_1u \end{aligned} \tag{13.17}$$

13.3 Flatness-Based Control of the Multi-asset Black–Scholes PDE

First, it can be proven that the system of Eq. (13.17) is a differentially flat one, with flat output

$$Y = [x_1, x_2, \ldots, x_{(i-1)N+j}, \ldots, x_{N^2-1}, x_{N^2}] \tag{13.18}$$

Next, the following virtual control inputs are defined

$$\begin{aligned} &\alpha_1 = x_{N^2-1}, \qquad \alpha_2 = x_{N^2-2}, \qquad \cdots \\ &\alpha_{N^2-(i-1)N-(j-1)} = x_{(i-1)N+(j-1)}, \qquad \cdots \qquad \alpha_{N^2-1} = x_1 \end{aligned} \tag{13.19}$$

Obviously, from the last row of Eq. (13.17) it holds that u is a function of the flat output and its derivatives. Consequently, all state variables of the PDE model are differential functions of the flat output Y. Thus, it is confirmed that the system is a differentially flat one.

Using the virtual control inputs of Eq. (13.19) in the state-space model of Eq. (13.17) one gets

$$\begin{aligned} \dot{x}_{N^2} &= f_{N^2}(x) + c_{N^2}\alpha_1 \\ \dot{x}_{N^2-1} &= f_{N^2-1}(x) + c_{N^2-1}\alpha_2 \\ &\cdots\ \cdots \\ \dot{x}_{(i-1)N+j} &= f_{(i-1)N+j} + c_{(i-1)N+j}\alpha_{N^2-(i-1)N-(j-1)} \\ &\cdots\ \cdots \\ \dot{x}_2 &= f_2(x) + c_2\alpha_{N^2-1} \\ \dot{x}_1 &= f_1(x) + c_1u \end{aligned} \tag{13.20}$$

By examining independently each nonlinear ODE of the previous state-space description of Eq. (13.20) it can been shown that this stands again for a differentially flat system, for which the virtual control input can be computed as in the case of flatness-based control for the trivial system. The virtual control input is chosen such that the ODE subsystem dynamics is linearized and the tracking error is eliminated.

The boundary condition that appears in the nonlinear ODE subsystem that comprises the last rows of the state-space description, was used as the aggregate control input.

One can find the values α_1^* that the virtual control inputs should have, so as to eliminate the tracking error for each one of the subsystems that are obtained from the per-row decomposition of Eq. (13.20)

$$\alpha_1^* = \frac{1}{c_{N^2}}\left[\dot{x}_N^* - f_{N^2}(x) - k_{p_1}\left(x_{N^2} - x_{N^2}^*\right)\right] \tag{13.21}$$

with $k_{p_1} > 0$, while it also holds that $\alpha_1^* = x_{N^2-1}^*$. Continuing with the rest of the rows of Eq. (13.20), the associated virtual control input can be obtained:

$$\alpha_2^* = \frac{1}{c_{N^2-1}}\left[\dot{x}_{N^2-1}^* - f_{N^2-1}(x) - k_{p_2}\left(x_{N^2-1} - x_{N^2-1}^*\right)\right] \tag{13.22}$$

with $k_{p_2} > 0$, while it also holds that $\alpha_2^* = x_{N^2-2}^*$. Continuing in a similar manner, for the i-th row of Eq. (13.20), the associated virtual control input is

$$\begin{aligned}\alpha_{N^2-(i-1)N-(j-1)} = \frac{1}{c_{(i-1)N+j}}\Big[\dot{x}_{(i-1)N+j}^* - f_{(i-1)N+j}(x) - \\ -k_{p\,N^2-(i-1)N-(j-1)}\left(x_{(i-1)N+j} - x_{(i-1)N+j}^*\right)\Big]\end{aligned} \tag{13.23}$$

with $k_{p\,N^2-(i-1)N-(j-1)} > 0$, where $\alpha_{N^2-(i-1)N-(j-1)}^*$. By applying the same procedure, the virtual control input for the N^2-1 row of Eq. (13.20) is found

$$\alpha_{N^2-1}^* = \frac{1}{c_2}\left[\dot{x}_2^* - f_2(x) - k_{p\,N^2-1}\left(x_2 - x_2^*\right)\right] \tag{13.24}$$

where $\alpha_{N^2-1}^* = x_1^*$. Finally, from the N^2-th row of Eq. (13.20) one computes the boundary control input that is really exerted on the system

$$u = \frac{1}{c_1}\left[\dot{x}_1^* - f_1(x) - k_{p\,N^2}\left(x_1 - x_1^*\right)\right] \tag{13.25}$$

where $\alpha_{N^2}^* = u$. Using the previous definitions, the virtual control inputs can be written as

$$\alpha_1^* = \frac{1}{c_{N^2}}\left[\dot{x}_N^* - f_{N^2}(x) - k_{p_1}\left(x_{N^2} - x_{N^2}^*\right)\right] \tag{13.26}$$

$$\alpha_2^* = \frac{1}{c_{N^2-1}}\left[\dot{\alpha}_1^* - f_{N^2-1}(x) - k_{p_2}\left(x_{N^2-1} - \alpha_1^*\right)\right] \tag{13.27}$$

$$\begin{aligned}\alpha_{N^2-(i-1)N-(j-1)} = \frac{1}{c_{(i-1)N+j}}\Big[\dot{\alpha}_{N^2-(i-1)N-(j-1)-1}^* - f_{(i-1)N+j}(x) - \\ -k_{p\,N^2-(i-1)N-(j-1)}\left(x_{(i-1)N+j} - \alpha_{N^2-(i-1)N-(j-1)-1}^*\right)\Big]\end{aligned} \tag{13.28}$$

$$\alpha_{N^2-1}^* = \frac{1}{c_2}\left[\dot{\alpha}_{N^2-2}^* - f_2(x) - k_{p\,N^2-1}\left(x_2 - \alpha_{N^2-2}^*\right)\right] \tag{13.29}$$

$$u = \frac{1}{c_1}\left[\dot{\alpha}_{N^2-1}^* - f_1(x) - k_{p\,N^2}\left(x_1 - \alpha_{N^2-1}^*\right)\right] \tag{13.30}$$

13.4 Stability Analysis of the Control Loop

The dynamics of the PDE system has been shown to be

$$\dot{x}_{N^2} = f_{N^2}(x) + c_{N^2}\alpha_1 \tag{13.31}$$

$$\dot{x}_{N^2-1} = f_{N^2-1}(x) + c_{N^2-1}\alpha_2 \\ \cdots\cdots \tag{13.32}$$

$$\dot{x}_{(i-1)N+j} = f_{(i-1)N+j} + c_{(i-1)N+j}\alpha_{N^2-(i-1)N-(j-1)} \\ \cdots\ \cdots \tag{13.33}$$

$$\dot{x}_2 = f_2(x) + c_2\alpha_{N^2-1} \tag{13.34}$$

$$\dot{x}_1 = f_1(x) + c_1 u \tag{13.35}$$

From Eqs. (13.30) and (13.35) one gets

$$\begin{aligned} \dot{x}_1 = f_1(x) + c_1\tfrac{1}{c_1}\left[\dot{\alpha}^*_{N^2-1} - f_1(x) - k_{p_{N^2}}\left(x_1 - \alpha^*_{N^2-1}\right)\right] \Rightarrow \\ \left(\dot{x}_1 - \dot{\alpha}^*_{N^2-1}\right) + k_{p_{N^2}}\left(x_1 - \alpha^*_{N^2-1}\right) = 0 \end{aligned} \tag{13.36}$$

By defining $z_1 = x_1 - \alpha^*_{N^2-1}$ and taking $k_{p_{N^2}} > 0$ one obtains

$$\begin{aligned} \dot{z}_1 + k_{p_{N^2}} z_1 = 0 \Rightarrow lim_{t\to\infty} z_1 = 0 \\ \Rightarrow lim_{t\to\infty} x_1 = \alpha^*_{N^2-1} \Rightarrow lim_{t\to\infty} x_1 = x_1^* \end{aligned} \tag{13.37}$$

From Eqs. (13.29) and (13.34), and using that $\alpha_{N^2-1} \to \alpha^*_{N^2-1}$ one gets

$$\begin{aligned} \dot{x}_2 = f_2(x) + c_2\tfrac{1}{c_2}\left[\dot{\alpha}^*_{N^2-2} - f_2(x) - k_{p_{N^2-1}}\left(x_2 - \alpha^*_{N^2-2}\right)\right] \Rightarrow \\ \left(\dot{x}_2 - \dot{\alpha}^*_{N^2-2}\right) + k_{p_{N^2-1}}\left(x_2 - \alpha^*_{N^2-2}\right) = 0 \end{aligned} \tag{13.38}$$

By defining $z_2 = x_2 - \alpha^*_{N^2-1}$ and taking $k_{p_{N^2}} > 0$ one obtains

$$\begin{aligned} \dot{z}_2 + k_{p_{N^2}} z_2 = 0 \Rightarrow lim_{t\to\infty} z_2 = 0 \\ \Rightarrow lim_{t\to\infty} x_2 = \alpha^*_{N^2-2} \Rightarrow lim_{t\to\infty} x_2 = x_2^* \end{aligned} \tag{13.39}$$

This procedure is also applied to the rest of the rows of the PDE's state-space description. From Eqs. (13.28) and (13.33) one gets

$$\begin{aligned}\dot{x}_{(i-1)N+j} &= f_{(i-1)N+j}(x) + c_{(i-1)N+j}\frac{1}{c_{(i-1)N+j}}\Big[\dot{\alpha}^*_{N^2-(i-1)N-(j-1)-1} - f_{(i-1)N+j}(x) - \\ &k_{pN^2-(i-1)N-(j-1)}\left(x_{(i-1)N+j} - \alpha^*_{N^2-(i-1)N-(j-1)-1}\right)\Big] \Rightarrow \\ &\left(\dot{x}_{(i-1)N+j} - \dot{\alpha}^*_{N^2-(i-1)N-(j-1)-1}\right) + \\ &+k_{pN^2-(i-1)N-(j-1)}\left(x_{(i-1)N+j} - \alpha^*_{N^2-(i-1)N-(j-1)-1}\right) = 0\end{aligned} \tag{13.40}$$

By defining $z_{N^2-(i-1)N-(j-1)} = x_{(i-1)N+j} - \alpha^*_{N^2-(i-1)N-(j-1)-1} - 1$ and taking that the gain $k_{pN^2-(i-1)N-(j-1)} > 0$ one obtains

$$\begin{aligned}\dot{z}_{N^2-(i-1)N-(j-1)} + k_{pN^2-(i-1)N-(j-1)} z_{N^2-(i-1)N-(j-1)} &= 0 \Rightarrow \\ lim_{t\to\infty} z_{N^2-(i-1)N-(j-1)} &= 0 \Rightarrow \\ lim_{t\to\infty} x_{(i-1)N+j} &= \alpha^*_{N^2-(i-1)N-(j-1)-1} \Rightarrow \\ lim_{t\to\infty} x_{(i-1)N+j} &= x^*_{(i-1)N+j}\end{aligned} \tag{13.41}$$

One processes in a similar manner the rest of the rows of the state-space description of the PDE given in Eq. (13.20). From Eqs. (13.27) and (13.32), and using that $\alpha_2 \to \alpha_2^*$ one gets

$$\begin{aligned}\dot{x}_{N^2-1} &= f_{N^2-1}(x) + c_{N^2-1}\frac{1}{c_{N^2-1}}[\dot{\alpha}_1^* - f_{N^2-1}(x) - k_{p_2}(x_{N^2-1} - \alpha_1^*)] \Rightarrow \\ &(\dot{x}_{N^2-1} - \dot{\alpha}_1^*) + k_{p_2}(x_{N^2-1} - \dot{\alpha}_1^*)) = 0\end{aligned} \tag{13.42}$$

By defining $z_{N^2-1} = x_{N^2-1} - \alpha_1$ and taking $k_{p_2} > 0$ one obtains

$$\begin{aligned}\dot{z}_{N^2-1} + k_{p_2} z_{N^2-1} = 0 \Rightarrow lim_{t\to\infty} z_{N^2-1} = 0 \Rightarrow \\ lim_{t\to\infty} x_{N^2-1} = \alpha_1^* \Rightarrow lim_{t\to\infty} x_{N^2-1} = x^*_{N^2-1}\end{aligned} \tag{13.43}$$

Finally, from Eqs. (13.26) and (13.31), and using that $\alpha_1 \to \alpha_1^*$ one gets

$$\begin{aligned}\dot{x}_{N^2} &= f_{N^2}(x) + c_{N^2}\frac{1}{c_{N^2}}[\dot{x}_N^* - f_{N^2}(x) - k_{p_1}(x_{N^2} - x^*_{N^2})] \Rightarrow \\ &(\dot{x}_{N^2} - \dot{x}^*_{N^2}) + k_{p_1}(x_{N^2} - \dot{x}^*_{N^2})) = 0\end{aligned} \tag{13.44}$$

By defining $z_{N^2} = x_{N^2} - x^*_{N^2}$ and taking $k_{p_1} > 0$ one obtains

$$\begin{aligned}\dot{z}_{N^2} + k_{p_1} z_{N^2} = 0 \Rightarrow lim_{t\to\infty} z_{N^2} = 0 \Rightarrow \\ lim_{t\to\infty} x_{N^2} = x^*_{N^2}\end{aligned} \tag{13.45}$$

Through this procedure, it is proven that the tracking error for the individual control loops into which the PDE model is decomposed converges asymptotically to 0.

From the previous analysis one can also demonstrate the stability of the control loop by applying the Lyapunov method. It holds that

$$\begin{gathered}
\dot{z}_{N^2} + k_{p_1} z_{N^2} = 0 \\
\dot{z}_{N^2-1} + k_{p_2} z_{N^2-1} = 0 \\
\cdots\ \cdots \\
\dot{z}_{N^2-(i-1)N-(j-1)} + k_{p_{N^2-(i-1)N-(j-1)}} z_{N^2-(i-1)N-(j-1)} = 0 \\
\cdots\ \cdots \\
\dot{z}_2 + k_{p_{N^2-1}} z_2 = 0 \\
\dot{z}_1 + k_{p_{N^2}} z_1 = 0
\end{gathered} \tag{13.46}$$

The following Lyapunov function is defined

$$V = \frac{1}{2}\sum_{i=1}^{N^2} z_i^2 \tag{13.47}$$

Setting k_{p_i} for $i = 1, 2, \ldots, N^2$, the first derivative of this Lyapunov function is

$$\begin{gathered}
\dot{V} = \frac{1}{2}\sum_{i=1}^{N^2} 2z_i\dot{z}_i \Rightarrow \dot{V} = \sum_{i=1}^{N^2} z_i(-k_{p_i} z_i) \Rightarrow \\
\dot{V} = -\sum_{i=1}^{N^2} k_{p_i} z_i^2 \Rightarrow \dot{V} < 0
\end{gathered} \tag{13.48}$$

The above result confirms the asymptotic stability of the multi-asset Black–Scholes PDE control loop, that has been based on differential flatness theory.

13.5 Simulation Tests

Simulation examples about the proposed control method for distributed parameter systems are provided for the case of the multi-asset Black–Scholes PDE that was given in Eq. (13.3). The obtained results are depicted in Figs. 13.1, 13.2 and 13.3. The spatial discretization of the PDE model consisted of $N^2 = 16$ points. The boundary condition $V_{0,1}$ served as the control input, while the boundary condition $V_{1,0}$ was set equal to zero.

The numerical simulation experiments have confirmed the theoretical findings of this chapter. It has been shown that by applying the proposed control method, the multi-asset Black–Scholes PDE dynamics can be modified so as to converge to the desirable reference profile. The control input that succeeds this changes smoothly and has a moderate range of variation. The accuracy of tracking of the reference setpoints was quite satisfactory. The proposed method shows that stabilization of financial systems dynamics is possible through feedback control.

Through the previous theoretical analysis and through the evaluation tests it has been shown that by applying semi-discretization and the finite differences method, the multi-asset Black–Scholes PDE can be written in a state-space form for which differential flatness properties hold. Next, the state-space description of the system

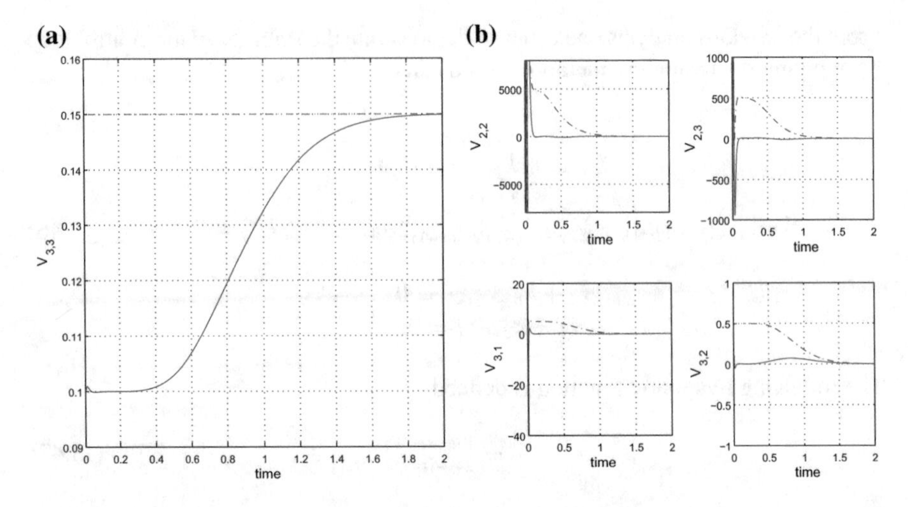

Fig. 13.1 Tracking of reference setpoint No 1 (*dashed red line*) by the value of the multi-asset Black–Scholes PDE (*blue line*) **a** at the final grid point $V_{N,N}$, **b** at preceding points of the spatial grid

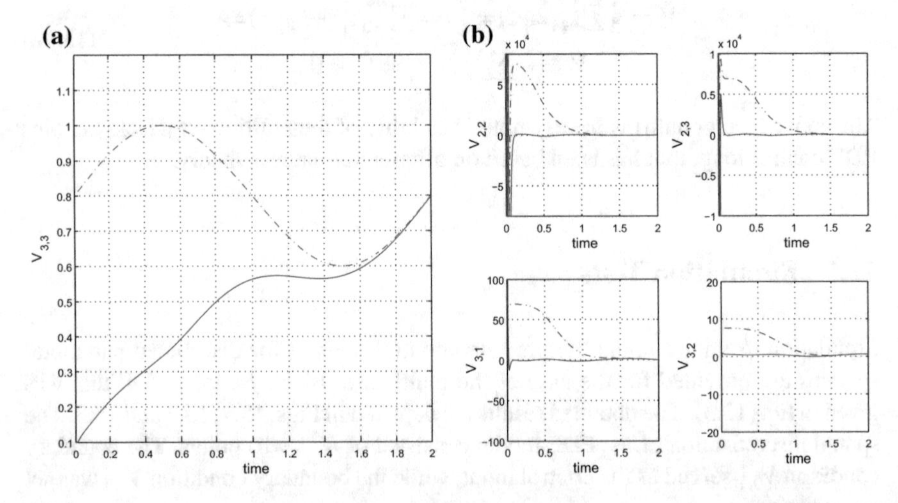

Fig. 13.2 Tracking of reference setpoint No 2 (*dashed red line*) by the value of the multi-asset Black–Scholes PDE (*blue line*) **a** at the final grid point $V_{N,N}$, **b** at preceding points of the spatial grid

was decomposed into an equivalent set of nonlinear ODEs and control algorithm consisting of successive (cascading) loops was developed.

Actually, by examining independently each nonlinear ODE it has been shown that this stands again for a differentially flat system, for which a virtual control input can be computed as in the case of flatness-based control for the trivial system. The virtual control input was chosen such that the ODE subsystem dynamics is linearized and the tracking error is eliminated. The boundary condition that appears in the nonlinear ODE subsystem that comprises the last rows of the state-space description, was used

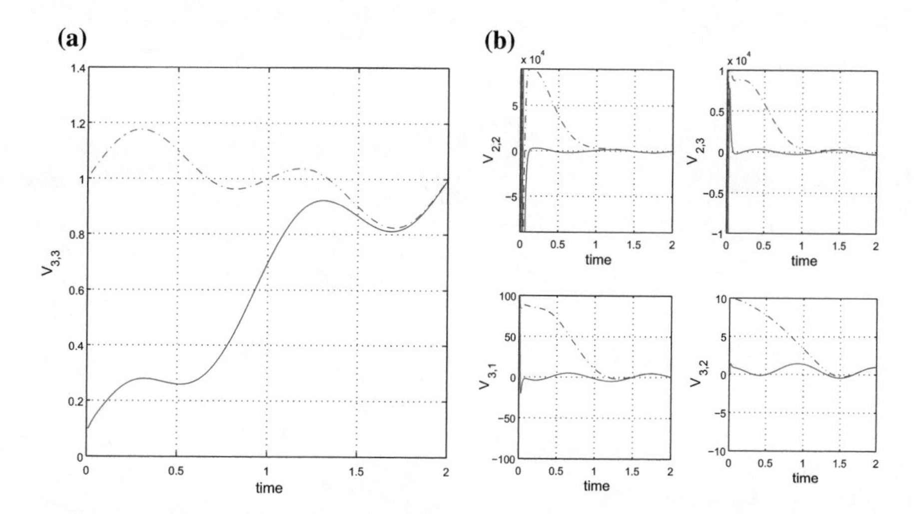

Fig. 13.3 Tracking of reference setpoint No 3 (*dashed red line*) by the value of the multi-asset Black–Scholes PDE (*blue line*) **a** at the final grid point $V_{N,N}$, **b** at preceding points of the spatial grid

as the aggregate control input. This control input contained recursively all virtual control inputs which were computed for the individual ODE subsystems associated with the previous rows of the state-space equation. Thus, by tracing the rows of the state-space model backwards, at each iteration of the control algorithm, one could find the control input that should be applied to the Black–Scholes PDE system so as to assure that all its state variables will converge to the desirable setpoints.

The method can be of interest for trading of options in open electricity markets. By showing the feasibility of such a control method it is also proven that through selected purchases and sales during the trading procedure the price of the negotiated electric power can be made to converge and stabilize at specific reference values.

Chapter 14
Stabilization of Commodities Pricing PDE Using Differential Flatness Theory

14.1 Outline

The problem of feedback control of the multi-asset Black–Scholes PDE can find application to stabilization of commodities prices. This is because the commodities price dynamics is described by a PDE that is equivalent to the multi-asset Black–Scholes PDE. Actually, in this PDE model, the first asset stands for the spot price while the second asset is the convenience yield.

Commodities trade has expanded rapidly during the last years and the need for elaborated methods for commodities pricing has emerged. Advanced pricing models for commodities are not only based on the spot pricing approach but reflect the dynamics of prices within long-term contracts [59, 146, 254, 255]. This dynamics can be expressed either in the form of stochastic differential equations or equivalently in the form of partial differential equations (PDEs) [147, 188]. In this chapter a multi-factor PDE commodities price model is considered. It is shown that, the dynamics of the commodities' price is described by a multi-dimensional PDE which is equivalent to the multi-asset Black–Scholes PDE [80, 81, 143, 224]. The problem of feedback control and stabilization of the aforemention PDE is treated. As already noted, methods for feedback stabilization of systems with nonlinear PDE dynamics have been a flourishing research subject in the last years [157, 164–166, 177, 275]. In particular, feedback control of diffusion-type (parabolic) PDEs has been a topic of extensive research and several remarkable results have been produced [10, 20, 26, 82, 249].

By showing the feasibility of control of the multi-factor commodities price PDE it is also proven that through selected purchases and sales during the trading procedure, the price of the negotiated commodity can be made to converge and stabilize at specific reference values. By applying semi-discretization and the finite differences method, the multi-factor commodities PDE model is written in a state-space form [179, 184, 189]. It is proven that this state-space description stands for a differentially flat system which means that all its state variables and control

G.G. Rigatos, *State-Space Approaches for Modelling and Control in Financial Engineering*, Intelligent Systems Reference Library 125,
DOI 10.1007/978-3-319-52866-3_14

inputs can be written as differential functions of the flat output vector of the system [133, 191, 216, 225, 231]. One can note again several results on the use of differential flatness theory in the control of PDEs [78, 170, 222, 277]. The first stage in the proposed control scheme for the multi-factor commodities PDE is to decompose its state-space description into an equivalent set of nonlinear ODEs. Next by examining independently each nonlinear ODE it is shown that this stands again for a differentially flat system, for which a virtual control input can be computed as in the case of flatness-based control for the trivial system. The virtual control input is chosen such that the ODE subsystem dynamics is linearized and the tracking error is eliminated. The boundary condition that appears in the nonlinear ODE subsystem that comprises the last row of the state-space description, stands for the aggregate control input.

The computation of the boundary control input uses recursively all virtual control inputs mentioned above, moving from the last ODE system to the first one [222, 224]. This stands for implementation of flatness-based control in successive (cascading) loops. Thus, by tracing the rows of the state-space model backwards, at each iteration of the control algorithm, one can finally obtain the control input that should be applied to the multi-factor commodities PDE system so as to assure that all its state vector elements will converge to the desirable setpoints. The stability of the control loop is proven in two manners. First, convergence to zero is proven for the tracking error of all subsystems into which the PDE's state-space model is decomposed. Next, with the use of Lyapunov analysis it is reconfirmed that this control scheme, for the commodities price PDE, is asymptotically stable.

14.2 Models for Commodities Pricing

14.2.1 Elaborated Schemes for Trading Electric Power

Electric power is a major commodity and models of commodities PDE pricing can find application to its trade. In the last years the role of the electricity market has significantly changed, now allowing trading between generators, retailers and other financial intermediaries. Wholesale transactions in the physical commodity are typically settled by a supervisory authority which is the grid operator. In most European countries, a power exchange is now based on a voluntary marketplace. A power exchange provides a spot market (mainly day-ahead) for transactions in the electricity grid to be carried out on the next day. Participants submit their purchase or sale orders electronically. The supply and demand are aggregated and compared in the power exchange to decide the market price for each hour of the following day.

Apart from spot pricing in the electricity market, options, futures and derivatives pricing have been established as a mean to manage more efficiently and profitably the financial risk associated with electricity price volatility (which in turn can be caused by variations in the production rates, condition of the transmission and distribution system, pay-off capability etc.). Long-term contracts traded in power exchanges include the so-called forward contracts, futures, swaps and options. These new pricing schemes need to be supported by forecasting and value assessment tools, that will be computationally efficient and accurate.

As previously analyzed, modeling tools for pricing in energy markets can be based on the Black–Scholes partial differential equation and its variants [184, 189]. The underlying asset S evolves in time according to the following stochastic differential equation

$$dS = \mu S dt + \sigma S dX \tag{14.1}$$

while the Black–Scholes PDE pricing model, providing finally the option; s value $V(S,t)$ is described by:

$$\frac{\partial V}{\partial t} + rS\frac{\partial V}{\partial S} + \frac{\sigma^2 S^2}{2}\frac{\partial^2 V}{\partial S^2} = rV \tag{14.2}$$

14.2.2 Commodities Pricing with the Single-Factor PDE Model

Another pricing approach in long-term contracts is based on the use of the commodities price PDE [59, 146, 147, 188, 254, 255]. In this pricing model it is first assumed that the commodity's spot price S is expressed by a stochastic process [59, 146, 147, 188, 254, 255]

$$dS = \kappa(\mu - lnS)Sdt + \sigma S dz \tag{14.3}$$

By defining the new variable $X = lnS$ the previous equation is modified into

$$\begin{aligned} dX &= \kappa(\alpha - X)dt + \sigma dz \\ \alpha &= \mu - \tfrac{\sigma^2}{2\kappa} \end{aligned} \tag{14.4}$$

An equivalent description for variable X is an Ornstein–Uhlenbeck process

$$\begin{aligned} dX &= \kappa(\alpha^* - X)dt + \sigma dz^* \\ \alpha^* &= \alpha - \lambda = \mu - \tfrac{\sigma^2}{2\kappa} - \lambda \end{aligned} \tag{14.5}$$

where λ is a constant expressing the market's price of risk. From Eq. (14.5), the conditional distribution of X at time t under the equivalent time measure is normal with mean and variance

$$E_0[X(t)] = e^{\kappa T}X(0) + (1 - e^{-\kappa t})a^* \tag{14.6}$$

$$Var_0[X(t)] = \frac{\sigma^2}{2\kappa}(1 - e^{-2\kappa t}) \tag{14.7}$$

Since $X = ln(S)$, the spot price of the commodity is log-normally distributed. Considering next that the interest rate is constant the futures (or forward) of the commodity with maturity t is the expected price of the commodity at time t, under the equivalent martingale measure. Then, from the properties of the log-normal distribution one has [254, 255]

$$F(S,t) = E(S,t) = exp(E_0[X(t)] + \frac{1}{2}Var_0[X(t)]) \tag{14.8}$$

or equivalently

$$F(S,t) = E(S,t) = exp(e^{-\kappa t}lnS + (1-e^{-\kappa t})a^* + \frac{\sigma^2}{4\kappa}(1-e^{-2\kappa t})) \tag{14.9}$$

or in log form

$$lnF(S,t) = lnE(S,t) = (e^{-\kappa t}lnS + (1-e^{-\kappa t})a^* + \frac{\sigma^2}{4\kappa}(1-e^{-2\kappa t})) \tag{14.10}$$

This last equation is the one that appears frequently in empirical tests. Eq. (14.10) is also the solution of partial differential equation [254, 255]

$$\frac{1}{2}\sigma^2 S^2\frac{\partial^2 F}{\partial S^2} + \kappa(\mu - \lambda - lnS)S\frac{\partial F}{\partial S} - \frac{\partial F}{\partial t} = 0 \tag{14.11}$$

with terminal boundary condition $F(S,0) = S$, where $F(S,t)$ is the value of the commodity's price at the maturity time t and S and S is the underlying spot price of the commodity. Consequently, to analyze commodities pricing one can work equivalently with the PDE model of Eq. (14.11).

14.2.3 Commodities Pricing with the Two-Factor PDE Model

In the two-factor model the distribution of the commodity's price $F(S,\delta,t)$ is now dependent on two variables, where the first one is the sport price of the commodity S and the second is the so-called convenience yield δ or long-term price. Actually, the convenience yield is interpreted as the benefit for holding (storing) the commodity [59, 146, 147, 188, 254, 255]. Now the variation of S and δ is described by the stochastic processes

$$dS = (\mu - \delta)Sdt + \sigma_1 Sdz_1 \tag{14.12}$$

$$d\delta = \kappa(a - \delta)dt + \sigma_2 Sdz_2 \tag{14.13}$$

where the increments to standard Brownian motion which are correlated with

$$dz_1 dz_2 = \rho dt \tag{14.14}$$

It is noted that if the convenience yield δ instead of being stochastic is a deterministic function of the spot price S and $\delta(S) = \kappa lnS$, then the two factor model coincides with the one factor model [254, 255]. Defining again $X = lnS$ and applying Ito's Lemma the process for the log price takes the form

$$dX = (\mu - \delta - \frac{1}{2}\sigma_1^2)dt + \sigma_1 dz_1 \tag{14.15}$$

The stochastic processes for the underlying factors, that is the spot price and the convenience yield can be also written as

$$dS = (r - \delta)Sdt + \sigma_1 S dz_1^* \tag{14.16}$$

$$d\delta = [\kappa(a - \delta) - \lambda]dt + \sigma_2 dz_2^* \tag{14.17}$$

$$dz_1^* dz_2^* = \rho dt \tag{14.18}$$

where λ is the market price of the convenience yield risk. Futures prices then can be equivalently computed from the solution of the following partial differential equation, which stands for the 2-factor PDE model of the commodities price

$$\begin{array}{c} \frac{1}{2}\sigma^2 S^2 \frac{\partial^2 F}{\partial S^2} + \sigma_1\sigma_2\rho S \frac{\partial^2 F}{\partial S \partial \delta} + \frac{1}{2}\delta^2 S^2 \frac{\partial^2 F}{\partial \delta^2} + (r - \delta)S\frac{\partial F}{\partial S} + \\ +(\kappa(a - \delta) - \lambda)\frac{\partial F}{\partial \delta} - \frac{\partial F}{\partial t} = 0 \end{array} \tag{14.19}$$

The analytical solution of the PDE model of Eq. (14.19) is given by [254, 255]

$$F(S, \delta, t) = Sexp[-\delta\frac{1 - e^{\kappa t}}{\kappa} + A(t)] \tag{14.20}$$

Equivalently, in logarithmic form one obtains

$$lnF(S, \delta, t) = lnS - \delta\frac{1 - e^{\kappa t}}{\kappa} + A(t) \tag{14.21}$$

where by denoting $\tilde{a} = a - \frac{\lambda}{\kappa}$

$$\begin{array}{c} A(t) = (r - \tilde{a} + \frac{1}{2}\frac{\sigma_2^2}{\kappa^2} - \frac{\sigma_1\sigma_2\rho}{\kappa}t) + \frac{1}{4}\sigma_2^2\frac{1-e^{-2\kappa t}}{\kappa^2} + \\ +(\tilde{a}\kappa + \sigma_1\sigma_2\rho - \frac{\sigma_2^2}{\kappa})\frac{1-e^{-\kappa t}}{\kappa^2} \end{array} \tag{14.22}$$

14.2.4 Commodities Pricing with the Three-Factor PDE Model

Next, a three-factor commodity PDE model is presented. These factors are the spot price S, the convenience yield δ and the interest rate r [59, 146, 147, 188, 254, 255]. The underlying stochastic processes that describe this model are

$$dS = (r - \delta)Sdt + \sigma_1 S_1 dz_1^* \tag{14.23}$$

$$d\delta = \kappa(\tilde{a} - \delta)dt + \sigma_2 dz_2^* \tag{14.24}$$

$$d\tau = \alpha(m^* - r)dt + \sigma_3 dz_3^* \tag{14.25}$$

$$dz_1^* dz_2^* = \rho_1 dt \cdots dz_2^* dz_3^* = \rho_2 dt \cdots dz_3^* dz_1^* = \rho_3 dt \tag{14.26}$$

where a and m^* are respectively, the speed of adjustment coefficient and the risk adjusted mean short rate of the interest rate process. Equivalently, the dynamics of the three-factor commodities pricing model can be described by the PDE [254, 255]

$$\begin{gathered}\frac{1}{2}\sigma_1^2 S^2 \frac{\partial^2 F}{\partial S^2} + \frac{1}{2}\sigma_2^2 \frac{\partial^2 F}{\partial \delta^2} + \frac{1}{2}\sigma_3^2 \frac{\partial^2 F}{\partial r^2} + \\ +\sigma_1\sigma_2\rho_1 S\frac{\partial^2 F}{\partial S \partial \delta} + \sigma_2\sigma_3\rho_2 S\frac{\partial^2 F}{\partial \delta \partial r} + \sigma_3\sigma_1\rho_3 S\frac{\partial^2 F}{\partial S \partial r} + \\ +(r - \delta)S\frac{\partial F}{\partial S} + \kappa(\tilde{a} - \delta)\frac{\partial F}{\partial \delta} + a(m^* - r)\frac{\partial F}{\partial r} - \frac{\partial F}{\partial t} = 0\end{gathered} \tag{14.27}$$

14.3 Boundary Control of the Multi-factor Commodities Price PDE

Next, the multi-asset Black–Scholes PDE is introduced [157, 166]:

$$\frac{\partial V}{\partial t} = \sum_{i=1}^{N} \sum_{j=1}^{N} \rho\sigma_i\sigma_j S_i S_j \frac{\partial^2 V}{\partial S_i \partial S_j} + \sum_{i=1}^{N} r S_i \frac{\partial V}{\partial S_i} - rV \tag{14.28}$$

Moreover, without loss of generality the two-asset Black–Scholes PDE is considered [224]

$$\frac{\partial V}{\partial t} = \frac{1}{2}\sigma_1^2 S_1^2 \frac{\partial V}{\partial^2 S_1^2} + \frac{1}{2}\sigma_2^2 S_2^2 \frac{\partial V}{\partial^2 S_2^2} + \rho\sigma_1\sigma_2 S_1 S_2 \frac{\partial^2 V}{\partial S_1 \partial S_2} + rS_1 \frac{\partial V}{\partial S_1} + rS_2 \frac{\partial V}{\partial S_2} - rV \tag{14.29}$$

The above 2-asset Black–Scholes PDE is shown to be equivalent to the 2-factor commodities price PDE that was described in Eq. (14.19). This is demonstrated through the change of variable $S_1 = S$ that is S_1 is equal to the spot price, $S_2 = \delta$

that is S_2 is equal to the convenience yield and after the coefficients of the three last partial derivative terms appearing in the right of Eq. (14.29) are suitably modified to arrive at the form:

$$\begin{array}{c}\frac{\partial V}{\partial t}=\frac{1}{2}\sigma_1^2 S_1^2\frac{\partial V}{\partial^2 S_1^2}+\frac{1}{2}\sigma_2^2 S_2^2\frac{\partial V}{\partial^2 S_2^2}+\rho\sigma_1\sigma_2 S_1 S_2\frac{\partial^2 V}{\partial S_1\partial S_2}+\\ +(r-S_2)\frac{\partial V}{\partial S_1}+(\kappa(a-S_2)-\lambda)\frac{\partial V}{\partial S_2}\end{array} \tag{14.30}$$

Semi-discretization and the finite differences method is applied to the PDE model of Eq. (14.30). To this end the partial derivatives appearing in Eq. (14.29) are computed as follows:

$$\frac{\partial V}{\partial S_1}=\frac{V(S_{1,i+1,S_{2,j}})-V(S_{1,i,S_{2,j}})}{\Delta S_1} \tag{14.31}$$

$$\frac{\partial^2 V}{\partial^2 S_1}=\frac{V(S_{1,i+1},S_{2,j})-2V(S_{1,i},S_{2,j})+V(S_{1,i-1},S_{2,j})}{\Delta {S_1}^2} \tag{14.32}$$

$$\frac{\partial V}{\partial S_2}=\frac{V(S_{1,i,S_{2,j+1}})-V(S_{1,i,S_{2,j}})}{\Delta S_2} \tag{14.33}$$

$$\frac{\partial^2 V}{\partial^2 S_2}=\frac{V(S_{1,i},S_{2,j+1})-2V(S_{1,i},S_{2,j})+V(S_{1,i},S_{2,j-1})}{\Delta {S_1}^2} \tag{14.34}$$

$$\frac{\partial^2 V}{\partial S_1\partial S_2}=\frac{V(S_{1,i+i},S_{2,j+i})-V(S_{1,i+1},S_{2,j})-V(S_{1,i},S_{2,j+1})+V(S_{1,i},S_{2,j})}{\Delta S_1\Delta S_2} \tag{14.35}$$

Using the previous semi-discretization, for grid point (i, j) it holds

$$\begin{array}{c}\frac{\partial V(S_{1,i},S_{2,j})}{\partial t}=\frac{1}{2}\sigma_1^2 S_{1,i}^2[\frac{V(S_{1,i+1},S_{2,j})-2V(S_{1,i},S_{2,j})+V(S_{1,i-1},S_{2,j})}{\Delta S_1^2}]+\\ \frac{1}{2}\sigma_2^2 S_{2,j}^2[\frac{V(S_{1,i},S_{2,j+1})-2V(S_{1,i},S_{2,j})+V(S_{1,i},S_{2,j-1})}{\Delta S_1^2}]+\\ \rho\sigma_1\sigma_2 S_{1,i}S_{2,j}[\frac{V(S_{1,i+1},S_{2,j+1})-V(S_{1,i+1},S_{2,j})-V(S_{1,i},S_{2,j+1})+V(S_{1,i},S_{2,j})}{\Delta S_1\Delta S_2}]+\\ (r-S_{2,j})[\frac{V(S_{1,i+1},S_{2,j})-V(S_{1,i},S_{2,j})}{\Delta S_1}]+(\kappa(a-S_{2,j})-\lambda)[\frac{V(S_{1,i},S_{2,j+1})-V(S_{1,i},S_{2,j})}{\Delta S_2}]\end{array} \tag{14.36}$$

The boundary conditions of the PDE are taken to be $V_{i,0}\neq 0$ only if $i=1$, $V_{0,j}\neq 0$ only if $j=1$ and $V(i,j)=ct$ (constant) if $i>N$ or $j>N$.

Considering that $i=1,2,\cdots,N$ and $j=1,2,\cdots,N$ the commodity's value at the grid points (i,j) are denoted as $V_{i,j}$. Using this notation, the semi-discretized model of the PDE takes the following form:

At grid point $i = 1$ and $j = 1$

$$\begin{array}{c}\frac{\partial V_{1,1}}{\partial t} = \frac{1}{2}\sigma_1^2 S_{1,1}^2[\frac{V_{2,1}-2V_{1,1}+V_{0,1}}{\Delta S_1^2}]+ \\ \frac{1}{2}\sigma_2^2 S_{2,1}^2[\frac{V_{1,2}-2V_{1,1}+V_{1,0}}{\Delta S_2^2}] + \rho\sigma_1\sigma_2 S_{1,1}S_{2,1}[\frac{V_{2,2}-V_{2,1}-V_{1,2}+V+1,1}{\Delta S_1 \Delta S_2}]+ \\ (r - S_{2,1})[\frac{V_{2,1}-V_{1,1}}{\Delta S_1}] + (\kappa(a - S_{2,1}) - \lambda)[\frac{V_{1,2}-V_{1,1}}{\Delta S_2}]\end{array} \tag{14.37}$$

At grid point $i > 1$ and $j > 1$ it holds

$$\begin{array}{c}\frac{\partial V_{i,j}}{\partial t} = \frac{1}{2}\sigma_1^2 S_{1,i}^2[\frac{V_{i+1,j}-2V_{i,j}+V_{i-1,j}}{\Delta S_1^2}]+ \\ \frac{1}{2}\sigma_2^2 S_{2,j}^2[\frac{V_{i,j+1}-2V_{i,j}+V_{i,j-1}}{\Delta S_2^2}] + \rho\sigma_1\sigma_2 S_{1,i}S_{2,j}[\frac{V_{i+1,j+1}-V_{i+1,j+1}-V_{i,j+1}+V+i,i}{\Delta S_1 \Delta S_2}]+ \\ (r - S_{2,j})[\frac{V_{i+1,j}-V_{i,j}}{\Delta S_1}] + (\kappa(a - S_{2,j}) - \lambda)[\frac{V_{i,j+1}-V_{i,j}}{\Delta S_2}]\end{array} \tag{14.38}$$

Next, the following state vector variables are defined $x_{(i-1)N+j} = V_{i,j}, i = 1, 2, \cdots,$ N and $j = 1, 2, \cdots, N$. Using this notation of state variables Eq. (14.37) becomes

$$\begin{array}{c}\dot{x}_1 = \frac{1}{2}\sigma_1^2 S_{1,1}^2[\frac{x_{N+1}-2x_1}{\Delta S_1^2}] + \frac{1}{2}\sigma_2^2 S_{2,1}^2[\frac{x_2-2x_1}{\Delta S_1^2}]+ \\ \rho\sigma_1\sigma_2 S_{1,1}S_{1,2}[\frac{x_{N+2}-x_2-x_{N+1}+x_1}{\Delta S_1 \Delta S_2}] + (r - S_{2,1})[\frac{x_{N+1}-x_1}{\Delta S_1}] + (\kappa(a - S_{2,1}) - \lambda)[\frac{x_2-x_1}{\Delta S_2}]+ \\ [\frac{1}{2}\sigma_1^2 S_{1,1}^2 \frac{V_{0,1}}{\Delta S_1^2} + \frac{1}{2}\sigma_2^2 S_{2,1}^2 \frac{V_{1,0}}{\Delta S_1^2}]\end{array} \tag{14.39}$$

Thus, by defining the control input associated with the boundary conditions as $u = [V_{0.1}, V_{1,0}]^T$ and $c_1 u = [\frac{1}{2}\sigma_1^2 S_{1,1}^2 \frac{V_{0,1}}{\Delta S_1^2} + \frac{1}{2}\sigma_2^2 S_{2,1}^2 \frac{V_{1,0}}{\Delta S_1^2}]$ one obtains a description for Eq. (14.39) in the form

$$\dot{x}_1 = f_1(x) + c_1 u \tag{14.40}$$

Equivalently for Eq. (14.38) one obtains

$$\begin{array}{c}\dot{x}_{(i-1)N+j} = \frac{1}{2}\sigma_1^2 S_{1,i}^2[\frac{x_{iN+j}-2x_{(i-1)N+j}+x_{(i-2)N+j}}{\Delta S_1^2}]+ \\ \frac{1}{2}\sigma_2^2 S_{2,j}^2[\frac{x_{(i-1)N+(j+1)}-2x_{(i-1)N+j}+x_{(i-1)N+(j-1)}}{\Delta S_2^2}]+ \\ \rho\sigma_1\sigma_2 S_{1,i}S_{2,j}[\frac{x_{(i)N+(j+1)}-x_{(i-1)N+(j+1)}-x_{iN+j}+x_{(i-1)N+j}}{\Delta S_1 \Delta S_2}]+ \\ (r - S_{2,j})[\frac{x_{iN+j}-x_{(i-1)N+j}}{\Delta S_1}] + (\kappa(a - S_{2,j}) - \lambda)[\frac{x_{(i-1)N+(j+1)}-x_{(i-1)N+j}}{\Delta S_2}]\end{array} \tag{14.41}$$

Eq. (14.41) can be also written in the form

$$\begin{array}{c}\dot{x}_{(i-1)N+j} = \frac{1}{2}\sigma_1^2 S_{1,i}^2[\frac{x_{iN+j}-2x_{(i-1)N+j}+x_{(i-2)N+j}}{\Delta S_1^2}]+ \\ \frac{1}{2}\sigma_2^2 S_{2,j}^2[\frac{x_{(i-1)N+(j+1)}-2x_{(i-1)N+j}}{\Delta S_2^2}]+ \\ \rho\sigma_1\sigma_2 S_{1,i}S_{2,j}[\frac{x_{(i)N+(j+1)}-x_{(i-1)N+(j+1)}-x_{iN+j}+x_{(i-1)N+j}}{\Delta S_1 \Delta S_2}]+ \\ (r - S_{2,j})[\frac{x_{iN+j}-x_{(i-1)N+j}}{\Delta S_1}] + (\kappa(a - S_{2,j} - \lambda)[\frac{x_{(i-1)N+(j+1)}-x_{(i-1)N+j}}{\Delta S_2}]+ \\ [\frac{1}{2}\sigma_2^2 S_{2,j}^2 \frac{1}{\Delta S_2^2}]x_{(i-1)N+(j-1)}\end{array} \tag{14.42}$$

Eq. (14.42) is finally written in the form

$$\dot{x}_{(i-1)N+j} = f_{(i-1)N+j}(x) + c_{(i-1)N+j}x_{(i-1)N+(j-1)} \quad (14.43)$$

where functions $f_{(i-1)N+j}(x)$ and $c_{(i-1)N+j}$ are given by

$$\begin{array}{c} f_{(i-1)N+j}(x) = \frac{1}{2}\sigma_1^2 S_{1,i}^2 [\frac{x_{iN+j}-2x_{(i-1)N+j}+x_{(i-2)N+j}}{\Delta S_1^2}]+ \\ \frac{1}{2}\sigma_2^2 S_{2,j}^2 [\frac{x_{(i-1)N+(j+1)}-2x_{(i-1)N+j}}{\Delta S_2^2}]+ \\ \rho\sigma_1\sigma_2 S_{1,i} S_{2,j} [\frac{x_{(i)N+(j+1)}-x_{(i-1)N+(j+1)}-x_{iN+j}+x_{(i-1)N+j}}{\Delta S_1 \Delta S_2}]+ \\ (r - S_{2,j})[\frac{x_{iN+j}-x_{(i-1)N+j}}{\Delta S_1}] + (\kappa(a - S_{2,j} - \lambda)[\frac{x_{(i-1)N+(j+1)}-x_{(i-1)N+j}}{\Delta S_2}] \end{array} \quad (14.44)$$

$$c_{(i-1)N+j} = [\frac{1}{2}\sigma_2^2 S_{2,j}^2 \frac{1}{\Delta S_2^2}] \quad (14.45)$$

Considering that $i = 1, 2, \cdots, N$ and $j = 1, 2, \cdots, N$ there are N^2 state-space equations. Thus, the dynamics of the PDE model is written as [224]

$$\begin{array}{c} \dot{x}_{N^2} = f_{N^2}(x) + c_{N^2}x_{N^2-1} \\ \dot{x}_{N^2-1} = f_{N^2-1}(x) + c_{N^2-1}x_{N^2-2} \\ \cdots\cdots \\ \dot{x}_{(i-1)N+j} = f_{(i-1)N+j} + c_{(i-1)N+j}x_{(i-1)N+(j-1)} \\ \cdots\cdots \\ \dot{x}_2 = f_2(x) + c_2x_1 \\ \dot{x}_1 = f_1(x) + c_1u \end{array} \quad (14.46)$$

14.4 Flatness-Based Control of the Multi-factor Commodities Price PDE

First, it can be proven that the state-space description of the commodities price PDE, given in Eq. (14.46), is a differentially flat one, with flat output $y = x_{N^2}$ [224]. Solving the i-th row of the state space model with respect to x_{i+1} one finds that state variables x_{i+1} is a differential function of the flat output y. Moreover, from the last row of Eq. (14.46) it holds that u is a function of the flat output and its derivatives. Next, the following virtual control inputs are defined

$$\begin{array}{ccc} \alpha_1 = x_{N^2-1}, & \alpha_2 = x_{N^2-2}, & \cdots \\ \alpha_{N^2-(i-1)N-(j-1)} = x_{(i-1)N+(j-1)}, & \cdots & \alpha_{N^2-1} = x_1 \end{array} \quad (14.47)$$

Using the virtual control inputs of Eq. (14.47) in the state-space model of Eq. (14.46) one gets

$$\begin{aligned}
\dot{x}_{N^2} &= f_{N^2}(x) + c_{N^2}\alpha_1 \\
\dot{x}_{N^2-1} &= f_{N^2-1}(x) + c_{N^2-1}\alpha_2 \\
&\cdots\ \cdots \\
\dot{x}_{(i-1)N+j} &= f_{(i-1)N+j} + c_{(i-1)N+j}\alpha_{N^2-(i-1)N-(j-1)} \\
&\cdots\ \cdots \\
\dot{x}_2 &= f_2(x) + c_2\alpha_{N^2-1} \\
\dot{x}_1 &= f_1(x) + c_1 u
\end{aligned} \tag{14.48}$$

By examining independently each nonlinear ODE of the previous state-space description of Eq. (14.48) and by defining as local flat output for the i-th ODE the state variable x_i it can been shown that the i-th row of the state-space description stands again for a differentially flat system. Actually, one has now N^2 subsystems, each one of them related to a row of the state-space model and the local flat outputs for these subsystems are

$$Y = [x_1, x_2, \cdots, x_{(i-1)N+j}, \cdots, x_{N^2-1}, x_{N^2}] \tag{14.49}$$

From the i-th row of the state-space model it can be seen that the virtual control input α_i is a differential function of the local flat output x_i, which shows again that the i-th subsystem, if independently examined, is also differentially flat.

The virtual control input for the i-th row of the state space model is chosen such that the ODE subsystem dynamics is linearized and the tracking error is eliminated. The boundary condition that appears in the nonlinear ODE subsystem that comprises the last rows of the state-space description, was used as the aggregate control input [224].

One can find the values α_1^* that the virtual control inputs should have, so as to eliminate the tracking error for each one of the subsystems that are obtained from the per-row decomposition of Eq. (14.48)

$$\alpha_1^* = \frac{1}{c_{N^2}}[\dot{x}_N^* - f_{N^2}(x) - k_{p_1}(x_{N^2} - x_{N^2}^*)] \tag{14.50}$$

with $k_{p_1} > 0$, while it also holds that $\alpha_1^* = x_{N^2-1}^*$. Continuing with the rest of the rows of Eq. (14.48), the associated virtual control input can be obtained:

$$\alpha_2^* = \frac{1}{c_{N^2-1}}[\dot{x}_{N^2-1}^* - f_{N^2-1}(x) - k_{p_2}(x_{N^2-1} - x_{N^2-1}^*)] \tag{14.51}$$

with $k_{p_2} > 0$, while it also holds that $\alpha_2^* = x_{N^2-2}^*$. Continuing in a similar manner, for the i-th row of Eq. (14.48), the associated virtual control input is

$$\begin{aligned}
\alpha_{N^2-(i-1)N-(j-1)} = \tfrac{1}{c_{(i-1)N+j}}[\dot{x}_{(i-1)N+j}^* - f_{(i-1)N+j}(x) - \\
-k_{p\,N^2-(i-1)N-(j-1)}(x_{(i-1)N+j} - x_{(i-1)N+j}^*)]
\end{aligned} \tag{14.52}$$

with $k_{pN^2-(i-1)N-(j-1)} > 0$, where $\alpha^*_{N^2-(i-1)N-(j-1)}$. By applying the same procedure, the virtual control input for the N^2-1 row of Eq. (14.48) is found

$$\alpha^*_{N^2-1} = \frac{1}{c_2}[\dot{x}^*_2 - f_2(x) - k_{pN^2-1}(x_2 - x^*_2)] \tag{14.53}$$

where $\alpha^*_{N^2-1} = x^*_1$. Finally, from the N^2-th row of Eq. (14.48) one computes the boundary control input that is really exerted on the system

$$u = \frac{1}{c_1}[\dot{x}^*_1 - f_1(x) - k_{pN^2}(x_1 - x^*_1)] \tag{14.54}$$

where $\alpha^*_{N^2} = u$. Using the previous definitions, the virtual control inputs can be written as

$$\alpha^*_1 = \frac{1}{c_{N^2}}[\dot{x}^*_N - f_{N^2}(x) - k_{p_1}(x_{N^2} - x^*_{N^2})] \tag{14.55}$$

$$\alpha^*_2 = \frac{1}{c_{N^2-1}}[\dot{\alpha}^*_1 - f_{N^2-1}(x) - k_{p_2}(x_{N^2-1} - \alpha^*_1)] \tag{14.56}$$

$$\cdots\ \cdots$$

$$\alpha_{N^2-(i-1)N-(j-1)} = \frac{1}{c_{(i-1)N+j}}[\dot{\alpha}^*_{N^2-(i-1)N-(j-1)-1} - f_{(i-1)N+j}(x) - k_{pN^2-(i-1)N-(j-1)}(x_{(i-1)N+j} - \alpha^*_{N^2-(i-1)N-(j-1)-1})] \tag{14.57}$$

$$\alpha^*_{N^2-1} = \frac{1}{c_2}[\dot{\alpha}^*_{N^2-2} - f_2(x) - k_{pN^2-1}(x_2 - \alpha^*_{N^2-2})] \tag{14.58}$$

$$\cdots\ \cdots$$

$$u = \frac{1}{c_1}[\dot{\alpha}^*_{N^2-1} - f_1(x) - k_{pN^2}(x_1 - \alpha^*_{N^2-1})] \tag{14.59}$$

14.5 Stability Analysis of the Control Loop of the Multi-factor Commodities Price PDE

The dynamics of the multi-factor commodities PDE system has been shown to be

$$\dot{x}_{N^2} = f_{N^2}(x) + c_{N^2}\alpha_1 \tag{14.60}$$

$$\dot{x}_{N^2-1} = f_{N^2-1}(x) + c_{N^2-1}\alpha_2 \tag{14.61}$$

$$\cdots\ \cdots$$

$$\dot{x}_{(i-1)N+j} = f_{(i-1)N+j} + c_{(i-1)N+j}\alpha_{N^2-(i-1)N-(j-1)} \tag{14.62}$$

$$\cdots\cdots$$

$$\dot{x}_2 = f_2(x) + c_2\alpha_{N^2-1} \tag{14.63}$$

$$\dot{x}_1 = f_1(x) + c_1 u \tag{14.64}$$

From Eq. (14.59) and Eq. (14.64) one gets

$$\begin{aligned}\dot{x}_1 = f_1(x) + c_1\tfrac{1}{c_1}[\dot{\alpha}^*_{N^2-1} - f_1(x) - k_{p_{N^2}}(x_1 - \alpha^*_{N^2-1})] \Rightarrow \\ (\dot{x}_1 - \dot{\alpha}^*_{N^2-1}) + k_{p_{N^2}}(x_1 - \alpha^*_{N^2-1})] = 0\end{aligned} \tag{14.65}$$

By defining $z_1 = x_1 - \alpha^*_{N^2-1}$ and taking $k_{p_{N^2}} > 0$ one obtains

$$\begin{aligned}\dot{z}_1 + k_{p_{N^2}} z_1 = 0 \Rightarrow lim_{t\to\infty} z_1 = 0 \\ \Rightarrow lim_{t\to\infty} x_1 = \alpha^*_{N^2-1} \Rightarrow lim_{t\to\infty} x_1 = x_1^*\end{aligned} \tag{14.66}$$

From Eqs. (14.58) and (14.63), and using that $\alpha_{N^2-1} \rightarrow \alpha^*_{N^2-1}$ one gets

$$\begin{aligned}\dot{x}_2 = f_2(x) + c_2\tfrac{1}{c_2}[\dot{\alpha}^*_{N^2-2} - f_2(x) - k_{p_{N^2-1}}(x_2 - \alpha^*_{N^2-2})] \Rightarrow \\ (\dot{x}_2 - \dot{\alpha}^*_{N^2-2}) + k_{p_{N^2-1}}(x_2 - \alpha^*_{N^2-2}) = 0\end{aligned} \tag{14.67}$$

By defining $z_2 = x_2 - \alpha^*_{N^2-1}$ and taking $k_{p_{N^2}} > 0$ one obtains

$$\begin{aligned}\dot{z}_2 + k_{p_{N^2}} z_2 = 0 \Rightarrow lim_{t\to\infty} z_2 = 0 \\ \Rightarrow lim_{t\to\infty} x_2 = \alpha^*_{N^2-2} \Rightarrow lim_{t\to\infty} x_2 = x_2^*\end{aligned} \tag{14.68}$$

This procedure is also applied to the rest of the rows of the PDE's state-space description. From Eqs. (14.57) and (14.62) one gets

$$\begin{aligned}\dot{x}_{(i-1)N+j} = f_{(i-1)N+j}(x) + c_{(i-1)N+j}\tfrac{1}{c_{(i-1)N+j}}[\dot{\alpha}^*_{N^2-(i-1)N-(j-1)-1} - f_{(i-1)N+j}(x) - \\ k_{p_{N^2-(i-1)N-(j-1)}}(x_{(i-1)N+j} - \alpha^*_{N^2-(i-1)N-(j-1)-1})] \Rightarrow \\ (\dot{x}_{(i-1)N+j} - \dot{\alpha}^*_{N^2-(i-1)N-(j-1)-1}) + \\ + k_{p_{N^2-(i-1)N-(j-1)}}(x_{(i-1)N+j} - \alpha^*_{N^2-(i-1)N-(j-1)-1}) = 0\end{aligned} \tag{14.69}$$

By defining $z_{N^2-(i-1)N-(j-1)} = x_{(i-1)N+j} - \alpha^*_{N^2-(i-1)N-(j-1)-1} - 1$ and taking that the gain $k_{p_{N^2-(i-1)N-(j-1)}} > 0$ one obtains

$$\begin{array}{c} \dot{z}_{N^2-(i-1)N-(j-1)} + k_{p_{N^2-(i-1)N-(j-1)}} z_{N^2-(i-1)N-(j-1)} = 0 \Rightarrow \\ lim_{t \to \infty} z_{N^2-(i-1)N-(j-1)} = 0 \Rightarrow \\ lim_{t \to \infty} x_{(i-1)N+j} = \alpha^*_{N^2-(i-1)N-(j-1)-1} \Rightarrow \\ lim_{t \to \infty} x_{(i-1)N+j} = x^*_{(i-1)N+j} \end{array} \tag{14.70}$$

One processes in a similar manner the rest of the rows of the state-space description of the PDE given in Eq. (14.48). From Eqs. (14.56) and (14.61), and using that $\alpha_2 \to \alpha_2^*$ one gets

$$\begin{array}{c} \dot{x}_{N^2-1} = f_{N^2-1}(x) + c_{N^2-1} \frac{1}{c_{N^2-1}} [\dot{\alpha}_1^* - f_{N^2-1}(x) - k_{p_2}(x_{N^2-1} - \alpha_1^*)] \Rightarrow \\ (\dot{x}_{N^2-1} - \dot{\alpha}_1^*) + k_{p_2}(x_{N^2-1} - \dot{\alpha}_1^*) = 0 \end{array} \tag{14.71}$$

By defining $z_{N^2-1} = x_{N^2-1} - \alpha_1$ and taking $k_{p_2} > 0$ one obtains

$$\begin{array}{c} \dot{z}_{N^2-1} + k_{p_2} z_{N^2-1} = 0 \Rightarrow lim_{t \to \infty} z_{N^2-1} = 0 \Rightarrow \\ lim_{t \to \infty} x_{N^2-1} = \alpha_1^* \Rightarrow lim_{t \to \infty} x_{N^2-1} = x^*_{N^2-1} \end{array} \tag{14.72}$$

Finally, from Eqs. (14.55) and (14.60), and using that $\alpha_1 \to \alpha_1^*$ one gets

$$\begin{array}{c} \dot{x}_{N^2} = f_{N^2}(x) + c_{N^2} \frac{1}{c_{N^2}} [\dot{x}_N^* - f_{N^2}(x) - k_{p_1}(x_{N^2} - x^*_{N^2})] \Rightarrow \\ (\dot{x}_{N^2} - \dot{x}^*_{N^2}) + k_{p_1}(x_{N^2} - \dot{x}^*_{N^2}) = 0 \end{array} \tag{14.73}$$

By defining $z_{N^2} = x_{N^2} - x^*_{N^2}$ and taking $k_{p_1} > 0$ one obtains

$$\begin{array}{c} \dot{z}_{N^2} + k_{p_1} z_{N^2} = 0 \Rightarrow lim_{t \to \infty} z_{N^2} = 0 \Rightarrow \\ lim_{t \to \infty} x_{N^2} = x^*_{N^2} \end{array} \tag{14.74}$$

Through this procedure, it is proven that the tracking error for the individual control loops into which the PDE model is decomposed converges asymptotically to 0. From the previous analysis one can also demonstrate the stability of the control loop by applying the Lyapunov method. It holds that

$$\begin{array}{c} \dot{z}_{N^2} + k_{p_1} z_{N^2} = 0 \\ \dot{z}_{N^2-1} + k_{p_2} z_{N^2-1} = 0 \\ \cdots \cdots \\ \dot{z}_{N^2-(i-1)N-(j-1)} + k_{p_{N^2-(i-1)N-(j-1)}} z_{N^2-(i-1)N-(j-1)} = 0 \\ \cdots \cdots \\ \dot{z}_2 + k_{p_{N^2-1}} z_2 = 0 \\ \dot{z}_1 + k_{p_{N^2}} z_1 = 0 \end{array} \tag{14.75}$$

The following Lyapunov function is defined

$$V = \sum_{i=1}^{N^2} \frac{1}{2} z_i^2 \tag{14.76}$$

Setting $k_{p_i} > 0$ for $i = 1, 2, \cdots, N^2$, the first derivative of this Lyapunov function is

$$\begin{matrix} \dot{V} = \sum_{i=1}^{N^2} \frac{1}{2} 2 z_i \dot{z}_i \Rightarrow \dot{V} = \sum_{i=1}^{N^2} z_i (-k_{p_i} z_i) \Rightarrow \\ \dot{V} = -\sum_{i=1}^{N^2} k_{p_i} z_i^2 \Rightarrow \dot{V} < 0 \end{matrix} \tag{14.77}$$

The above result confirms the asymptotic stability of the multi-factor commodities price PDE control loop, that has been based on differential flatness theory.

14.6 Simulation Tests

Simulation examples about the proposed control method for distributed parameter systems are provided for the case of the multi-factor commodities price PDE that was given in Eq. (14.29). The boundary condition served as the control input and was computed as each iteration of the control algorithm according to the procedure described in Sect. 14.4. The obtained results are depicted in Figs. 14.1, 14.2 and 14.3. The spatial discretization of the PDE model consisted of $N^2 = 16$ points. The boundary condition $V_{0,1}$ served as the control input, while the boundary condition $V_{1,0}$ was set equal to zero.

The numerical simulation experiments have confirmed the theoretical findings of this chapter. It has been shown that by applying the proposed control method, the multi-factor commodities PDE dynamics can be modified so as to converge to the desirable reference profile. The control input that succeeds this changes smoothly

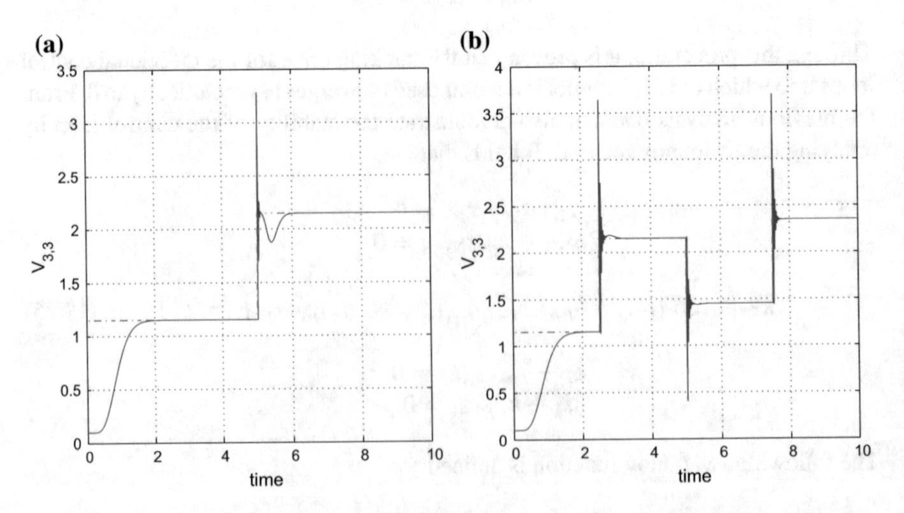

Fig. 14.1 2-factors commodities price PDE: tracking of reference setpoint (*dashed red line*) by the value of the PDE system (*blue line*) at the final grid point V_{NN} (**a**) Setpoint 1 (**b**) Setpoint 2

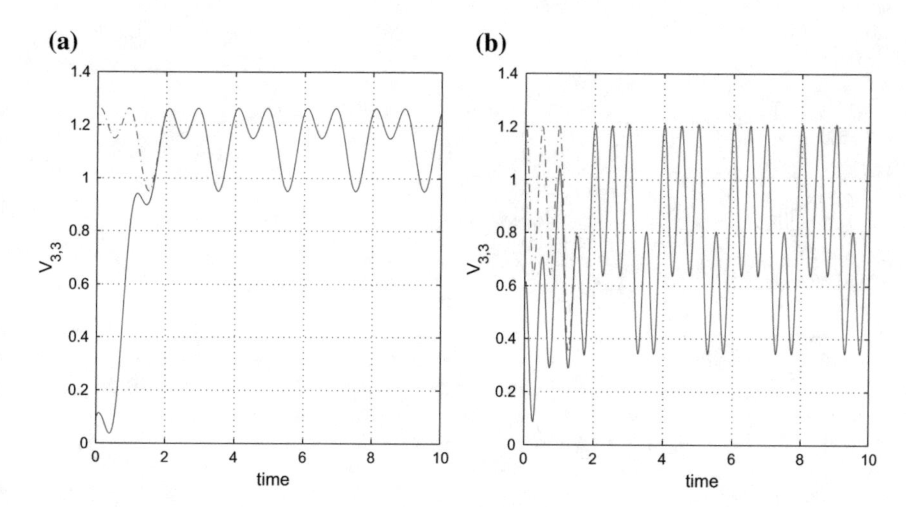

Fig. 14.2 2-factors commodities price PDE: tracking of reference setpoint (*dashed red line*) by the value of the PDE system (*blue line*) at the final grid point V_{NN} (**a**) Setpoint 3 (**b**) Setpoint 4

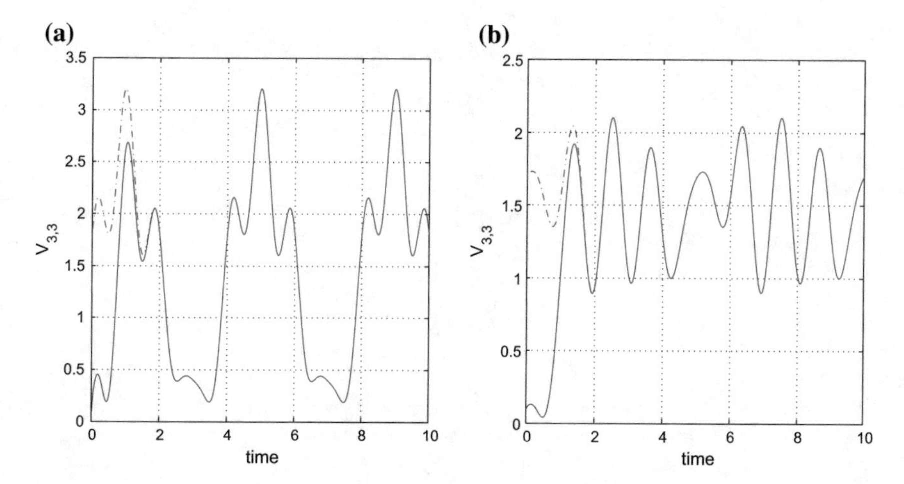

Fig. 14.3 2-factors commodities price PDE: tracking of reference setpoint (*dashed red line*) by the value of the PDE system (*blue line*) at the final grid point V_{NN} (**a**) Setpoint 5 (**b**) Setpoint 6

and has a moderate range of variation. The accuracy of tracking of the reference setpoints was quite satisfactory. The proposed method shows that stabilization of financial systems dynamics is possible through feedback control.

Chapter 15
Stabilization of Mortgage Price Dynamics Using Differential Flatness Theory

15.1 Outline

The previous results on the control of the multi-asset Black–Scholes PDE and on control of the commodities price PDE can be generalized towards stabilization of the mortgage price PDE. The latter is shown again to be a multi-asset PDE. One can demonstrate the differential flatness properties of this PDE and can apply to it flatness-based control implemented in multiple loops.

The value of a mortgage depends not only on the value of the residence (property) but is also determined by the varying interest rates provided by lenders and banks. Thus, in the recent years elaborated methods for the pricing of mortgages have been developed [91, 182, 183, 251]. A part of the associated results are based on models of the mortgage price variation in the form of stochastic differential equations, while another part uses equivalent models in the form of partial differential equations [49, 53, 119, 242, 252]. The chapter considers a model of variation of mortgages price defined by a partial differential equation. This PDE model is shown to be equivalent to the two asset Black–Scholes PDE, where the first asset is the residence's (or property's) value, and the second asset is the interest rate [80, 81, 143, 224]. For this PDE model it is shown that it is possible to apply boundary feedback control that is capable of making the mortgage's price converge to desirable setpoints. As already analyzed, results on stabilization of financial systems dynamics can be found in [157, 165, 177, 275]. Moreover, results on feedback control of PDE dynamics are given in [10, 20, 26, 82, 249]. The proposed control scheme is based on proving and exploiting the differential flatness properties of the mortgage price PDE.

Following the procedure applied in previous chapters, and after implementing semi-discretization and the finite differences' method, a states-space description of the PDE pricing model is obtained [179, 184, 189]. It is proven that this PDE model is a differentially flat one, which means that all its state variables and its control

G.G. Rigatos, *State-Space Approaches for Modelling and Control in Financial Engineering*, Intelligent Systems Reference Library 125,
DOI 10.1007/978-3-319-52866-3_15

inputs can be expressed as differential functions of specific algebraic variables which constitute's the model's flat output [133, 191, 225, 231]. Actually, it is shown that the PDE model is written in the triangular (backstepping integral) form for which differential flatness properties have been demonstrated and confirmed. Differential flatness is known to indicate that the system's nonlinear state-space description can be transformed through diffeomorphisms to an equivalent linear canonical form [78, 170, 222, 277]. Moreover, in this chapter a novel proof of the differential flatness properties of the mortgage price PDE and of the associated state-space model is given. By examining independently each row of the state-space model it is shown that this stands again for a differentially flat system, for which a virtual control input can be computed as in the case of flatness-based control for the trivial system. The virtual control input is chosen such that the ODE subsystem dynamics is linearized and the tracking error is eliminated. The boundary condition that appears in the nonlinear ODE subsystem that comprises the last row of the state-space description, stands for the aggregate control input.

The computation of the boundary control input for the mortgage price PDE uses recursively all virtual control inputs mentioned above, moving from the last row of the state-space model to the first one [222, 224]. This stands for implementation of flatness-based control in successive (cascading) loops. Thus, by tracing the rows of the state-space model backwards, at each iteration of the control algorithm, one can finally obtain the control input that should be applied to the multi-factor mortgage PDE system so as to assure that all its state vector elements will converge to the desirable setpoints. The stability of the control loop is proven in two manners. First, convergence to zero is proven for the tracking error of all subsystems into which the PDE's state-space model is decomposed. Next, with the use of Lyapunov analysis it is reconfirmed that this control scheme is asymptotically stable.

15.2 Options Theory-Based PDE Model of Mortgage Valuation

A mortgage is a financial contract in which a property (real estate asset) is reserved as a collateral for the repayment of a loan. This pledge is cancelled when the loan has been paid off. On the one side of the contract there is the lender (mortgager) and on the other side there is the borrower (mortgagee). The loan is paid off in consecutive installments, which are carried out at regular intervals, until a fixed date in the future which is called maturity. The borrower can take over the real estate asset either after the maturity date (when the last installment has been paid off) or before that date if the borrower is in position to pay the entire amount due. If the borrower cannot pay and cannot comply with the terms of the contract then a default of the borrower takes place.

The option theory-based PDE model of mortgage valuation is given by

$$\begin{aligned}\frac{\partial V}{\partial t} = -\frac{1}{2}h^2\sigma_h^2\frac{\partial^2 V}{\partial h^2} - \rho h\sqrt{\tau}\sigma_h\sigma_r\frac{\partial^2 V}{\partial h\partial r} - \frac{1}{2}r\sigma_v^2\frac{\partial^2 V}{\partial r^2} - \\ -\kappa(\theta - r)\frac{\partial V}{\partial r} - (r-\delta)h\frac{\partial V}{\partial h} + rV\end{aligned} \tag{15.1}$$

In the previous equation, the mortgage value is denoted by V and is a function of time t, of the property (e.g. residence) price h and of the interest rate r. This valuation of the PDE is defined in the time interval $T_{i-1} < t \leq T_i$ where T_i is the time instant at which the i-th installment is paid. For the property's value it holds $0 < h < \infty$ and for the interest rate one has $0 < r < \infty$.

15.3 Computation of the Mortgage Price PDE

Using Taylor series expansion the variation in the mortgage's price is computed as follows:

$$dV = \frac{\partial V}{\partial t}dt + \frac{\partial V}{\partial h}dh + \frac{1}{2}\left(\frac{\partial^2 V}{\partial h^2}dh^2 + 2\frac{\partial^2 V}{\partial h\partial r}dhdr + \frac{\partial^2 V}{\partial r^2}dr^2\right) + \cdots \tag{15.2}$$

Using that the value of the property h and the interest rate r vary in time according to stochastic processes one has

$$\begin{aligned}dh^2 \rightarrow \sigma_h^2 h^2 dX_h^2 \rightarrow \sigma_h^2 h^2 dt \\ dr^2 \rightarrow \sigma_r^2 r dX_r^2 \\ dhdr \rightarrow \sigma_h\sigma_r h\sqrt{r}dX_h dX_r = \rho\sigma_h\sigma_r h\sqrt{r}dt\end{aligned} \tag{15.3}$$

Moreover, the Wiener process X_h and X_r satisfies the following relation

$$dX_h dX_r = \rho dt \tag{15.4}$$

Using Eqs. (15.3) and (15.4) the previous Taylor series expansion is written as

$$dV = \frac{\partial V}{\partial t}dt + \frac{\partial V}{\partial h}dh + \frac{1}{2}\left(\sigma_h^2 h^2\frac{\partial^2 V}{\partial h^2} + 2\rho\sigma_h\sigma_r h\sqrt{r}\frac{\partial^2 V}{\partial h\partial r}dhdr + \sigma_r^2 r\frac{\partial^2 V}{\partial r^2}\right)dt + \cdots \tag{15.5}$$

Next, a portfolio Π is constructed, consisting of the mortgage $V_1(h, r, t)$ that has maturity T_1, $-\Delta_1$ units of another mortgage $V_2(h, r, t)$ that has maturity T_2, and $-\Delta_2$ units of a residence with price h. Thus, the portfolio is given by

$$\Pi = V_1 - \Delta_1 V_2 - \Delta_2 h \tag{15.6}$$

By considering that the infinitesimal variation for V_2 is ΔV_2 and that the infinitesimal variation for h is $dh + \delta h dt$, the differential variation of the portfolio in time is given by

$$d\Pi = dV_1 - \Delta_1 dV_2 - \Delta_2(dh + \delta h dt) \quad (15.7)$$

Moreover, by choosing the mortgage possession units to be

$$\Delta_1 = \frac{\frac{\partial V_1}{\partial r}}{\frac{\partial V_2}{\partial r}} \\ \Delta_2 = \frac{\partial V_1}{\partial h} - \Delta_1 \frac{\partial V_2}{\partial h} \quad (15.8)$$

the infinitesimal variation of the portfolio becomes

$$d\Pi = \frac{\partial V_1}{\partial t}dt + \frac{1}{2}(\sigma_h^2 h^2 \frac{\partial^2 V_1}{\partial h^2} + 2\rho\sigma_h\sigma_r h\sqrt{r}\frac{\partial^2 V_1}{\partial h \partial r} + \sigma_r^2 r \frac{\partial^2 V_1}{\partial r^2})dt - \delta h \frac{\partial V_1}{\partial h}dt \\ -\frac{\frac{\partial V_1}{\partial r}}{\frac{\partial V_2}{\partial r}}[\frac{\partial V_2}{\partial t}dt + \frac{1}{2}(\sigma_h^2 h^2 \frac{\partial^2 V_2}{\partial h^2} + 2\rho\sigma_h\sigma_r h\sqrt{r}\frac{\partial^2 V_2}{\partial h \partial r} + \sigma_r^2 r \frac{\partial^2 V_2}{\partial r^2})dt - \delta h \frac{\partial V_2}{\partial h}dt] \quad (15.9)$$

Next, it is assumed that

$$\delta\Pi = r\Pi dt \quad (15.10)$$

Moreover, using Eq. (15.8) in Eq. (15.6) one obtains

$$\Pi = V_1 - \frac{\frac{\partial V_1}{\partial r}}{\frac{\partial V_2}{\partial r}}V_2 - \frac{\partial V_1}{\partial h}h + \frac{\frac{\partial V_1}{\partial r}}{\frac{\partial V_2}{\partial r}}\frac{\partial V_2}{\partial h}h \quad (15.11)$$

and after substituting this expression in Eq. (15.9) one gets

$$\frac{1}{\frac{\partial V_1}{\partial r}}[\frac{\partial V_1}{\partial t} + \frac{1}{2}\sigma_h^2 h^2 \frac{\partial^2 V_1}{\partial h^2} + \rho\sigma_h\sigma_r h\sqrt{r}\frac{\partial^2 V_1}{\partial h \partial r} + \frac{1}{2}\sigma_r^2 r \frac{\partial^2 V_1}{\partial r^2} + (r-\delta)h\frac{\partial V_1}{\partial h} - rV_1] = \\ \frac{1}{\frac{\partial V_2}{\partial r}}[\frac{\partial V_2}{\partial t} + \frac{1}{2}\sigma_h^2 h^2 \frac{\partial^2 V_2}{\partial h^2} + \rho\sigma_h\sigma_r h\sqrt{r}\frac{\partial^2 V_2}{\partial h \partial r} + \frac{1}{2}\sigma_r^2 r \frac{\partial^2 V_2}{\partial r^2} + (r-\delta)h\frac{\partial V_2}{\partial h} - rV_2] \quad (15.12)$$

By denoting the right part of Eq. (15.12) as $a(h, r, t)$ and by omitting the subscript from V_1 in the left part of the above equation, one arrives at

$$-\frac{1}{\frac{\partial V}{\partial r}}[\frac{\partial V}{\partial t} + \frac{1}{2}\sigma_h^2 h^2 \frac{\partial^2 V}{\partial h^2} + \rho\sigma_h\sigma_r h\sqrt{r}\frac{\partial^2 V}{\partial h \partial r} + \frac{1}{2}\sigma_r^2 r \frac{\partial^2 V}{\partial r^2} + (r-\delta)h\frac{\partial V}{\partial h} - rV] = a(h, r, t) \quad (15.13)$$

where $a(h, r, t)$ indicates a market price of risk. If one finally uses the following equation to describe the market price of risk

$$a(h, r, t) = -\kappa(\theta - r) \quad (15.14)$$

then the PDE for the mortgage price dynamics is given by Eq. (15.1), that is

$$\frac{\partial V}{\partial t} = -\frac{1}{2}h^2\sigma_h^2\frac{\partial^2 V}{\partial h^2} - \rho h\sqrt{r}\sigma_h\sigma_r\frac{\partial^2 V}{\partial h \partial r} - \frac{1}{2}r\sigma_r^2\frac{\partial^2 V}{\partial r^2} - \kappa(\theta - r)\frac{\partial V}{\partial r} - (r-\delta)\frac{\partial V}{\partial h} + rV \quad (15.15)$$

15.4 Boundary Control of the Multi-factor Mortgage Price PDE

Next, the options pricing model based on the multi-asset Black–Scholes PDE is considered [157]:

$$\frac{\partial V}{\partial t}=\sum_{i=1}^{N}\sum_{j=1}^{N}\rho\sigma_i\sigma_j S_i S_j\frac{\partial^2 V}{\partial S_i\partial S_j}+\sum_{i=1}^{N}rS_i\frac{\partial V}{\partial S_i}-rV \tag{15.16}$$

Without loss of generality the two-asset Black–Scholes PDE is considered [224]

$$\frac{\partial V}{\partial t}=\frac{1}{2}\sigma_1^2S_1^2\frac{\partial V}{\partial^2 S_1^2}+\frac{1}{2}\sigma_2^2S_2^2\frac{\partial V}{\partial^2 S_2^2}+\rho\sigma_1\sigma_2S_1S_2\frac{\partial^2 V}{\partial S_1\partial S_2}+rS_1\frac{\partial V}{\partial S_1}+rS_2\frac{\partial V}{\partial S_2}-rV \tag{15.17}$$

The above 2-asset Black–Scholes PDE is shown to be equivalent to the 2-factor mortgage price PDE that was described in Eq. (15.15). This is demonstrated through the change of variable $S_1 = h$ that is S_1 is equal to the residence's (or property's) price, $S_2 = \sqrt{r}$ that is S_2 is dependent on the square root of the interest rate. After the coefficients of the three last partial derivative terms appearing in the right of Eq. (15.17) are suitably modified one arrives at the form:

$$\begin{aligned}\frac{\partial V}{\partial t}=-[&\tfrac{1}{2}\sigma_1^2S_1^2\tfrac{\partial V}{\partial^2 S_1^2}+\tfrac{1}{2}\sigma_2^2S_2^2\tfrac{\partial V}{\partial^2 S_2^2}+\rho\sigma_1\sigma_2S_1S_2\tfrac{\partial^2 V}{\partial S_1\partial S_2}+\\ &+(r-\delta)\tfrac{\partial V}{\partial S_1}+(\kappa(\theta-r)\tfrac{\partial V}{\partial S_2}-rV]\end{aligned} \tag{15.18}$$

Semi-discretization and the finite differences method is applied to the PDE model of Eq. (15.15). To this end the partial derivatives appearing in Eq. (15.17) are computed as follows:

$$\frac{\partial V}{\partial S_1}=\frac{V(S_{1,i+1,S_{2,j}})-V(S_{1,i,S_{2,j}})}{\Delta S_1} \tag{15.19}$$

$$\frac{\partial^2 V}{\partial^2 S_1}=\frac{V(S_{1,i+1},S_{2,j})-2V(S_{1,i},S_{2,j})+V(S_{1,i-1},S_{2,j})}{\Delta S_1{}^2} \tag{15.20}$$

$$\frac{\partial V}{\partial S_2}=\frac{V(S_{1,i,S_{2,j+1}})-V(S_{1,i,S_{2,j}})}{\Delta S_2} \tag{15.21}$$

$$\frac{\partial^2 V}{\partial^2 S_2}=\frac{V(S_{1,i},S_{2,j+1})-2V(S_{1,i},S_{2,j})+V(S_{1,i},S_{2,j-1})}{\Delta S_1{}^2} \tag{15.22}$$

$$\frac{\partial^2 V}{\partial S_1\partial S_2}=\frac{V(S_{1,i+i},S_{2,j+i})-V(S_{1,i+1},S_{2,j})-V(S_{1,i},S_{2,j+1})+V(S_{1,i},S_{2,j})}{\Delta S_1\Delta S_2} \tag{15.23}$$

Using the previous semi-discretization, for grid point (i, j) it holds

$$\frac{\partial V(S_{1,i},S_{2,j})}{\partial t} = -[\tfrac{1}{2}\sigma_1^2 S_{1,i}^2[\frac{V(S_{1,i+1},S_{2,j})-2V(S_{1,i},S_{2,j})+V(S_{1,i-1},S_{2,j})}{\Delta S_1^2}]+$$
$$\tfrac{1}{2}\sigma_2^2 S_{2,j}^2[\frac{V(S_{1,i},S_{2,j+1})-2V(S_{1,i},S_{2,j})+V(S_{1,i},S_{2,j-1})}{\Delta S_1^2}]+$$
$$\rho\sigma_1\sigma_2 S_{1,i}S_{2,j}[\frac{V(S_{1,i+1},S_{2,j+1})-V(S_{1,i+1},S_{2,j})-V(S_{1,i},S_{2,j+1})+V(S_{1,i},S_{2,j})}{\Delta S_1 \Delta S_2}]+$$
$$(r-\delta)[\frac{V(S_{1,i+1},S_{2,j})-V(S_{1,i},S_{2,j})}{\Delta S_1}]+(\kappa(\theta-S_{2,j}^2))[\frac{V(S_{1,i},S_{2,j+1})-V(S_{1,i},S_{2,j})}{\Delta S_2}]-rV(S_{1,i},S_{2,j})] \tag{15.24}$$

The boundary conditions of the PDE are taken to be $V_{i,0}{\neq}0$ only if $i = 1$, $V_{0,j}{\neq}0$ only if $j = 1$ and $V(i, j) = ct$ (constant) if $i > N$ or $j > N$.
Considering that $i = 1, 2, \cdots, N$ and $j = 1, 2, \cdots, N$ the option's values at the grid points (i, j) are denoted as $V_{i,j}$. Using this notation, the semi-discretized model of the PDE takes the following form:
At grid point $i = 1$ and $j = 1$

$$\frac{\partial V_{1,1}}{\partial t} = -[\tfrac{1}{2}\sigma_1^2 S_{1,1}^2[\frac{V_{2,1}-2V_{1,1}+V_{0,1}}{\Delta S_1^2}]+$$
$$\tfrac{1}{2}\sigma_2^2 S_{2,1}^2[\frac{V_{1,2}-2V_{1,1}+V_{1,0}}{\Delta S_2^2}]+\rho\sigma_1\sigma_2 S_{1,1}S_{2,1}[\frac{V_{2,2}-V_{2,1}-V_{1,2}+V+1,1}{\Delta S_1 \Delta S_2}]+ \tag{15.25}$$
$$(r-\delta)[\frac{V_{2,1}-V_{1,1}}{\Delta S_1}]+(\kappa(\theta-S_{2,1}^2))[\frac{V_{1,2}-V_{1,1}}{\Delta S_2}]-rV(S_{1,i},S_{2,j})]$$

At grid point $i > 1$ and $j > 1$ it holds

$$\frac{\partial V_{i,j}}{\partial t} = -[\tfrac{1}{2}\sigma_1^2 S_{1,i}^2[\frac{V_{i+1,j}-2V_{i,j}+V_{i-1,j}}{\Delta S_1^2}]+$$
$$\tfrac{1}{2}\sigma_2^2 S_{2,j}^2[\frac{V_{i,j+1}-2V_{i,j}+V_{i,j-1}}{\Delta S_2^2}]+\rho\sigma_1\sigma_2 S_{1,i}S_{2,j}[\frac{V_{i+1,j+1}-V_{i+1,j+1}-V_{i,j+1}+V+i,i}{\Delta S_1 \Delta S_2}]+ \tag{15.26}$$
$$(r-\delta)[\frac{V_{i+1,j}-V_{i,j}}{\Delta S_1}]+(\kappa(\theta-S_{2,j}^2))[\frac{V_{i,j+1}-V_{i,j}}{\Delta S_2}]-rV(S_{1,i},S_{2,j})]$$

Next, the following state vector variables are defined $x_{(i-1)N+j} = V_{i,j}$, $i = 1, 2, \cdots, N$ and $j = 1, 2, \cdots, N$. Using this notation of state variables Eq. (15.25) becomes

$$\dot{x}_1 = -[\tfrac{1}{2}\sigma_1^2 S_{1,1}^2[\tfrac{x_{N+1}-2x_1}{\Delta S_1^2}]+\tfrac{1}{2}\sigma_2^2 S_{2,1}^2[\tfrac{x_2-2x_1}{\Delta S_1^2}]+$$
$$\rho\sigma_1\sigma_2 S_{1,1}S_{1,2}[\tfrac{x_{N+2}-x_2-x_{N+1}+x_1}{\Delta S_1 \Delta S_2}]+(r-\delta)[\tfrac{x_{N+1}-x_1}{\Delta S_1}]+(\kappa(\theta-S_{2,1}^2))[\tfrac{x_2-x_1}{\Delta S_2}]-rx_1+$$
$$[\tfrac{1}{2}\sigma_1^2 S_{1,1}^2 \tfrac{V_{0,1}}{\Delta S_1^2}+\tfrac{1}{2}\sigma_2^2 S_{2,1}^2 \tfrac{V_{1,0}}{\Delta S_1^2}]] \tag{15.27}$$

Thus, by defining the control input associated with the boundary conditions as $u = [V_{0.1}, V_{1,0}]^T$ and $c_1 u = [\tfrac{1}{2}\sigma_1^2 S_{1,1}^2 \tfrac{V_{0,1}}{\Delta S_1^2} + \tfrac{1}{2}\sigma_2^2 S_{2,1}^2 \tfrac{V_{1,0}}{\Delta S_1^2}]$ one obtains a description for Eq. (15.27) in the form

$$\dot{x}_1 = f_1(x) + c_1 u \tag{15.28}$$

Equivalently for Eq. (15.26) one obtains

$$\begin{aligned}
\dot{x}_{(i-1)N+j} = -[\tfrac{1}{2}\sigma_1^2 S_{1,i}^2 [\tfrac{x_{iN+j}-2x_{(i-1)N+j}+x_{(i-2)N+j}}{\Delta S_1^2}]+ \\
\tfrac{1}{2}\sigma_2^2 S_{2,j}^2 [\tfrac{x_{(i-1)N+(j+1)}-2x_{(i-1)N+j}+x_{(i-1)N+(j-1)}}{\Delta S_2^2}]+ \\
\rho\sigma_1\sigma_2 S_{1,i} S_{2,j} [\tfrac{x_{(i)N+(j+1)}-x_{(i-1)N+(j+1)}-x_{iN+j}+x_{(i-1)N+j}}{\Delta S_1 \Delta S_2}] - r x_{(i-1)N+j}+ \\
(r-\delta)[\tfrac{x_{iN+j}-x_{(i-1)N+j}}{\Delta S_1}] + (\kappa(\theta - S_{2,j}^2) - \lambda)[\tfrac{x_{(i-1)N+(j+1)}-x_{(i-1)N+j}}{\Delta S_2}]]
\end{aligned} \tag{15.29}$$

Eq. (15.29) can be also written in the form

$$\begin{aligned}
\dot{x}_{(i-1)N+j} = \tfrac{1}{2}\sigma_1^2 S_{1,i}^2 [\tfrac{x_{iN+j}-2x_{(i-1)N+j}+x_{(i-2)N+j}}{\Delta S_1^2}]+ \\
\tfrac{1}{2}\sigma_2^2 S_{2,j}^2 [\tfrac{x_{(i-1)N+(j+1)}-2x_{(i-1)N+j}}{\Delta S_2^2}]+ \\
\rho\sigma_1\sigma_2 S_{1,i} S_{2,j} [\tfrac{x_{(i)N+(j+1)}-x_{(i-1)N+(j+1)}-x_{iN+j}+x_{(i-1)N+j}}{\Delta S_1 \Delta S_2}]+ \\
(r-\delta)[\tfrac{x_{iN+j}-x_{(i-1)N+j}}{\Delta S_1}] + (\kappa(\theta - S_{2,j^2}) - \lambda)[\tfrac{x_{(i-1)N+(j+1)}-x_{(i-1)N+j}}{\Delta S_2}] - r x_{(i-1)N+j}+ \\
[\tfrac{1}{2}\sigma_2^2 S_{2,j}^2 \tfrac{1}{\Delta S_2^2}] x_{(i-1)N+(j-1)}
\end{aligned} \tag{15.30}$$

Eq. (15.30) is finally written in the form

$$\dot{x}_{(i-1)N+j} = f_{(i-1)N+j}(x) + c_{(i-1)N+j} x_{(i-1)N+(j-1)} \tag{15.31}$$

where functions $f_{(i-1)N+j}(x)$ and $c_{(i-1)N+j}$ are given by

$$\begin{aligned}
f_{(i-1)N+j}(x) = \tfrac{1}{2}\sigma_1^2 S_{1,i}^2 [\tfrac{x_{iN+j}-2x_{(i-1)N+j}+x_{(i-2)N+j}}{\Delta S_1^2}]+ \\
\tfrac{1}{2}\sigma_2^2 S_{2,j}^2 [\tfrac{x_{(i-1)N+(j+1)}-2x_{(i-1)N+j}}{\Delta S_2^2}]+ \\
\rho\sigma_1\sigma_2 S_{1,i} S_{2,j} [\tfrac{x_{(i)N+(j+1)}-x_{(i-1)N+(j+1)}-x_{iN+j}+x_{(i-1)N+j}}{\Delta S_1 \Delta S_2}]+ \\
(r-\delta)[\tfrac{x_{iN+j}-x_{(i-1)N+j}}{\Delta S_1}] + (\kappa(\theta - S_{2,j^2})[\tfrac{x_{(i-1)N+(j+1)}-x_{(i-1)N+j}}{\Delta S_2}] - r x_{(i-1)N+j}
\end{aligned} \tag{15.32}$$

$$c_{(i-1)N+j} = [\frac{1}{2}\sigma_2^2 S_{2,j}^2 \frac{1}{\Delta S_2^2}] \tag{15.33}$$

Considering that $i = 1, 2, \cdots, N$ and $j = 1, 2, \cdots, N$ there are N^2 state-space equations. Thus, the dynamics of the PDE model is written as [224]

$$\begin{aligned}
\dot{x}_{N^2} &= f_{N^2}(x) + c_{N^2} x_{N^2-1} \\
\dot{x}_{N^2-1} &= f_{N^2-1}(x) + c_{N^2-1} x_{N^2-2} \\
&\cdots\cdots \\
\dot{x}_{(i-1)N+j} &= f_{(i-1)N+j} + c_{(i-1)N+j} x_{(i-1)N+(j-1)} \\
&\cdots\cdots \\
\dot{x}_2 &= f_2(x) + c_2 x_1 \\
\dot{x}_1 &= f_1(x) + c_1 u
\end{aligned} \tag{15.34}$$

15.5 Flatness-Based Control of the Multi-factor Mortgage Price PDE

First, it can be proven that the state-space description of the mortgage price PDE, given in Eq. (15.34), is a differentially flat one, with flat output $y = x_{N^2}$ [224]. Solving the i-th row of the state space model with respect to x_{i+1} one finds that state variables x_{i+1} is a differential function of the flat output y. Moreover, from the last row of Eq. (15.34) it holds that u is a function of the flat output and its derivatives. Next, the following virtual control inputs are defined

$$\begin{array}{ccc} \alpha_1 = x_{N^2-1}, & \alpha_2 = x_{N^2-2}, & \cdots \\ \alpha_{N^2-(i-1)N-(j-1)} = x_{(i-1)N+(j-1)}, & \cdots & \alpha_{N^2-1} = x_1 \end{array} \tag{15.35}$$

Using the virtual control inputs of Eq. (15.35) in the state-space model of Eq. (15.34) one gets

$$\begin{aligned} \dot{x}_{N^2} &= f_{N^2}(x) + c_{N^2}\alpha_1 \\ \dot{x}_{N^2-1} &= f_{N^2-1}(x) + c_{N^2-1}\alpha_2 \\ &\cdots\cdots \\ \dot{x}_{(i-1)N+j} &= f_{(i-1)N+j} + c_{(i-1)N+j}\alpha_{N^2-(i-1)N-(j-1)} \\ &\cdots\cdots \\ \dot{x}_2 &= f_2(x) + c_2\alpha_{N^2-1} \\ \dot{x}_1 &= f_1(x) + c_1 u \end{aligned} \tag{15.36}$$

By examining independently each nonlinear ODE of the previous state-space description of Eq. (15.36) and by defining as local flat output for the i-th ODE the state variable x_i it can been shown that the i-th row of the state-space description stands again for a differentially flat system. Actually, one has now N^2 subsystems, each one of them related to a row of the state-space model and the local flat outputs for these subsystems are

$$Y = [x_1, x_2, \cdots, x_{(i-1)N+j}, \cdots, x_{N^2-1}, x_{N^2}] \tag{15.37}$$

From the i-th row of the state-space model it can be seen that the virtual control input α_i is a differential function of the local flat output x_i, which shows again that the i-th subsystem, if independently examined, is also differentially flat.

The virtual control input for the i-th row of the state space model is chosen such that the ODE subsystem dynamics is linearized and the tracking error is eliminated. The boundary condition that appears in the nonlinear ODE subsystem that comprises the last rows of the state-space description, was used as the aggregate control input [224].

One can find the values α_1^* that the virtual control inputs should have, so as to eliminate the tracking error for each one of the subsystems that are obtained from the per-row decomposition of Eq. (15.36)

$$\alpha_1^* = \frac{1}{c_{N^2}}[\dot{x}_N^* - f_{N^2}(x) - k_{p_1}(x_{N^2} - x_{N^2}^*)] \tag{15.38}$$

with $k_{p_1} > 0$, while it also holds that $\alpha_1^* = x_{N^2-1}^*$. Continuing with the rest of the rows of Eq. (15.36), the associated virtual control input can be obtained:

$$\alpha_2^* = \frac{1}{c_{N^2-1}}[\dot{x}_{N^2-1}^* - f_{N^2-1}(x) - k_{p_2}(x_{N^2-1} - x_{N^2-1}^*)] \tag{15.39}$$

with $k_{p_2} > 0$, while it also holds that $\alpha_2^* = x_{N^2-2}^*$. Continuing in a similar manner, for the i-th row of Eq. (15.36), the associated virtual control input is

$$\begin{aligned}\alpha_{N^2-(i-1)N-(j-1)} = & \frac{1}{c_{(i-1)N+j}}[\dot{x}_{(i-1)N+j}^* - f_{(i-1)N+j}(x) - \\ & -k_{pN^2-(i-1)N-(j-1)}(x_{(i-1)N+j} - x_{(i-1)N+j}^*)]\end{aligned} \tag{15.40}$$

with $k_{pN^2-(i-1)N-(j-1)} > 0$, where $\alpha_{N^2-(i-1)N-(j-1)}^*$. By applying the same procedure, the virtual control input for the $N^2 - 1$ row of Eq. (15.36) is found

$$\alpha_{N^2-1}^* = \frac{1}{c_2}[\dot{x}_2^* - f_2(x) - k_{pN^2-1}(x_2 - x_2^*)] \tag{15.41}$$

where $\alpha_{N^2-1}^* = x_1^*$. Finally, from the N^2-th row of Eq. (15.36) one computes the boundary control input that is really exerted on the system

$$u = \frac{1}{c_1}[\dot{x}_1^* - f_1(x) - k_{pN^2}(x_1 - x_1^*)] \tag{15.42}$$

where $\alpha_{N^2}^* = u$. Using the previous definitions, the virtual control inputs can be written as

$$\alpha_1^* = \frac{1}{c_{N^2}}[\dot{x}_N^* - f_{N^2}(x) - k_{p_1}(x_{N^2} - x_{N^2}^*)] \tag{15.43}$$

$$\begin{aligned}\alpha_2^* = & \frac{1}{c_{N^2-1}}[\dot{\alpha}_1^* - f_{N^2-1}(x) - k_{p_2}(x_{N^2-1} - \alpha_1^*)] \\ & \cdots\cdots\end{aligned} \tag{15.44}$$

$$\begin{aligned}\alpha_{N^2-(i-1)N-(j-1)} = & \frac{1}{c_{(i-1)N+j}}[\dot{\alpha}_{N^2-(i-1)N-(j-1)-1}^* - f_{(i-1)N+j}(x) - \\ & -k_{pN^2-(i-1)N-(j-1)}(x_{(i-1)N+j} - \alpha_{N^2-(i-1)N-(j-1)-1}^*)]\end{aligned} \tag{15.45}$$

$$\begin{aligned}\alpha_{N^2-1}^* = & \frac{1}{c_2}[\dot{\alpha}_{N^2-2}^* - f_2(x) - k_{pN^2-1}(x_2 - \alpha_{N^2-2}^*)] \\ & \cdots\cdots\end{aligned} \tag{15.46}$$

$$u = \frac{1}{c_1}[\dot{\alpha}_{N^2-1}^* - f_1(x) - k_{pN^2}(x_1 - \alpha_{N^2-1}^*)] \tag{15.47}$$

15.6 Stability Analysis of the Control Loop of the Multi-factor Mortgage Price PDE

The dynamics of the multi-factor mortgage PDE system has been shown to be

$$\dot{x}_{N^2} = f_{N^2}(x) + c_{N^2}\alpha_1 \quad (15.48)$$

$$\dot{x}_{N^2-1} = f_{N^2-1}(x) + c_{N^2-1}\alpha_2 \quad (15.49)$$
$$\cdots\cdots$$

$$\dot{x}_{(i-1)N+j} = f_{(i-1)N+j} + c_{(i-1)N+j}\alpha_{N^2-(i-1)N-(j-1)} \quad (15.50)$$
$$\cdots\cdots$$

$$\dot{x}_2 = f_2(x) + c_2\alpha_{N^2-1} \quad (15.51)$$

$$\dot{x}_1 = f_1(x) + c_1 u \quad (15.52)$$

From Eqs. (15.47) and (15.52) one gets

$$\begin{array}{c}\dot{x}_1 = f_1(x) + c_1\frac{1}{c_1}[\dot{\alpha}^*_{N^2-1} - f_1(x) - k_{p_{N^2}}(x_1 - \alpha^*_{N^2-1})] \Rightarrow \\ (\dot{x}_1 - \dot{\alpha}^*_{N^2-1}) + k_{p_{N^2}}(x_1 - \alpha^*_{N^2-1})] = 0\end{array} \quad (15.53)$$

By defining $z_1 = x_1 - \alpha^*_{N^2-1}$ and taking $k_{p_{N^2}} > 0$ one obtains

$$\begin{array}{c}\dot{z}_1 + k_{p_{N^2}}z_1 = 0 \Rightarrow lim_{t\to\infty}z_1 = 0 \\ \Rightarrow lim_{t\to\infty}x_1 = \alpha^*_{N^2-1} \Rightarrow lim_{t\to\infty}x_1 = x_1^*\end{array} \quad (15.54)$$

From Eqs. (15.46) and (15.51), and using that $\alpha_{N^2-1} \to \alpha^*_{N^2-1}$ one gets

$$\begin{array}{c}\dot{x}_2 = f_2(x) + c_2\frac{1}{c_2}[\dot{\alpha}^*_{N^2-2} - f_2(x) - k_{p_{N^2-1}}(x_2 - \alpha^*_{N^2-2})] \Rightarrow \\ (\dot{x}_2 - \dot{\alpha}^*_{N^2-2}) + k_{p_{N^2-1}}(x_2 - \alpha^*_{N^2-2}) = 0\end{array} \quad (15.55)$$

By defining $z_2 = x_2 - \alpha^*_{N^2-1}$ and taking $k_{p_{N^2}} > 0$ one obtains

$$\begin{array}{c}\dot{z}_2 + k_{p_{N^2}}z_2 = 0 \Rightarrow lim_{t\to\infty}z_2 = 0 \\ \Rightarrow lim_{t\to\infty}x_2 = \alpha^*_{N^2-2} \Rightarrow lim_{t\to\infty}x_2 = x_2^*\end{array} \quad (15.56)$$

This procedure is also applied to the rest of the rows of the PDE's state-space description. From Eqs. (15.45) and (15.50) one gets

$$\begin{aligned}
\dot{x}_{(i-1)N+j} = & f_{(i-1)N+j}(x) + c_{(i-1)N+j}\frac{1}{c_{(i-1)N+j}}[\dot{\alpha}^*_{N^2-(i-1)N-(j-1)-1} - \\
& -f_{(i-1)N+j}(x) - k_{p_{N^2-(i-1)N-(j-1)}}(x_{(i-1)N+j} - \alpha^*_{N^2-(i-1)N-(j-1)-1})] \Rightarrow \\
& (\dot{x}_{(i-1)N+j} - \dot{\alpha}^*_{N^2-(i-1)N-(j-1)-1}) + k_{p_{N^2-(i-1)N-(j-1)}}(x_{(i-1)N+j} - \\
& -\alpha^*_{N^2-(i-1)N-(j-1)-1}) = 0
\end{aligned} \tag{15.57}$$

By defining $z_{N^2-(i-1)N-(j-1)} = x_{(i-1)N+j} - \alpha^*_{N^2-(i-1)N-(j-1)-1} - 1$ and taking the control gain $k_{p_{N^2-(i-1)N-(j-1)}} > 0$ one obtains

$$\begin{aligned}
& \dot{z}_{N^2-(i-1)N-(j-1)} + k_{p_{N^2-(i-1)N-(j-1)}} z_{N^2-(i-1)N-(j-1)} = 0 \Rightarrow \\
& lim_{t\to\infty} z_{N^2-(i-1)N-(j-1)} = 0 \Rightarrow \\
& lim_{t\to\infty} x_{(i-1)N+j} = \alpha^*_{N^2-(i-1)N-(j-1)-1} \Rightarrow \\
& lim_{t\to\infty} x_{(i-1)N+j} = x^*_{(i-1)N+j}
\end{aligned} \tag{15.58}$$

One processes in a similar manner the rest of the rows of the state-space description of the PDE given in Eq. (15.36). From Eqs. (15.44) and (15.49), and using that $\alpha_2 \to \alpha_2^*$ one gets

$$\begin{aligned}
\dot{x}_{N^2-1} = & f_{N^2-1}(x) + c_{N^2-1}\frac{1}{c_{N^2-1}}[\dot{\alpha}_1^* - f_{N^2-1}(x) - k_{p_2}(x_{N^2-1} - \alpha_1^*)] \Rightarrow \\
& (\dot{x}_{N^2-1} - \dot{\alpha}_1^*) + k_{p_2}(x_{N^2-1} - \dot{\alpha}_1^*) = 0
\end{aligned} \tag{15.59}$$

By defining $z_{N^2-1} = x_{N^2-1} - \alpha_1$ and taking $k_{p_2} > 0$ one obtains

$$\begin{aligned}
& \dot{z}_{N^2-1} + k_{p_2} z_{N^2-1} = 0 \Rightarrow lim_{t\to\infty} z_{N^2-1} = 0 \Rightarrow \\
& lim_{t\to\infty} x_{N^2-1} = \alpha_1^* \Rightarrow lim_{t\to\infty} x_{N^2-1} = x^*_{N^2-1}
\end{aligned} \tag{15.60}$$

Finally, from Eqs. (15.43) and (15.48), and using that $\alpha_1 \to \alpha_1^*$ one gets

$$\begin{aligned}
\dot{x}_{N^2} = & f_{N^2}(x) + c_{N^2}\frac{1}{c_{N^2}}[\dot{x}_N^* - f_{N^2}(x) - k_{p_1}(x_{N^2} - x^*_{N^2})] \Rightarrow \\
& (\dot{x}_{N^2} - \dot{x}^*_{N^2}) + k_{p_1}(x_{N^2} - \dot{x}^*_{N^2}) = 0
\end{aligned} \tag{15.61}$$

By defining $z_{N^2} = x_{N^2} - x^*_{N^2}$ and taking $k_{p_1} > 0$ one obtains

$$\begin{aligned}
& \dot{z}_{N^2} + k_{p_1} z_{N^2} = 0 \Rightarrow lim_{t\to\infty} z_{N^2} = 0 \Rightarrow \\
& lim_{t\to\infty} x_{N^2} = x^*_{N^2}
\end{aligned} \tag{15.62}$$

Through this procedure, it is proven that the tracking error for the individual control loops into which the PDE model is decomposed converges asymptotically to 0. From the previous analysis one can also demonstrate the stability of the control loop by applying the Lyapunov method. It holds that

$$\begin{aligned} &\dot{z}_{N^2} + k_{p_1} z_{N^2} = 0 \\ &\dot{z}_{N^2-1} + k_{p_2} z_{N^2-1} = 0 \\ &\cdots\cdots \\ &\dot{z}_{N^2-(i-1)N-(j-1)} + k_{p_{N^2-(i-1)N-(j-1)}} z_{N^2-(i-1)N-(j-1)} = 0 \\ &\cdots\cdots \\ &\dot{z}_2 + k_{p_{N^2-1}} z_2 = 0 \\ &\dot{z}_1 + k_{p_{N^2}} z_1 = 0 \end{aligned} \tag{15.63}$$

The following Lyapunov function is defined

$$V = \sum_{i=1}^{N^2} \frac{1}{2} z_i^2 \tag{15.64}$$

Setting $k_{p_i} > 0$ for $i = 1, 2, \cdots, N^2$, the first derivative of this Lyapunov function is

$$\begin{aligned} &\dot{V} = \sum_{i=1}^{N^2} \tfrac{1}{2} 2 z_i \dot{z}_i \Rightarrow \dot{V} = \sum_{i=1}^{N^2} z_i(-k_{p_i} z_i) \Rightarrow \\ &\dot{V} = -\sum_{i=1}^{N^2} k_{p_i} z_i^2 \Rightarrow \dot{V} < 0 \end{aligned} \tag{15.65}$$

The above result confirms the asymptotic stability of the multi-factor mortgage price PDE control loop, that has been based on differential flatness theory.

15.7 Simulation Tests

Simulation examples about the proposed control method for distributed parameter systems are provided for the case of the multi-factor mortgage price PDE that was given in Eq. (15.17). The boundary condition served as the control input and was computed as each iteration of the control algorithm according to the procedure described in Sect. 15.5. The obtained results are depicted in Figs. 15.1, 15.2, 15.3 and 15.4. The spatial discretization of the PDE model consisted of $N^2 = 16$ points. The boundary condition $V_{0,1}$ served as the control input, while the boundary condition $V_{1,0}$ was set equal to zero.

The numerical simulation experiments have confirmed the theoretical findings of this chapter. It has been shown that by applying the proposed control method, the multi-factor mortgage PDE dynamics can be modified so as to converge to the desirable reference profile. The control input that succeeds this changes smoothly and has a moderate range of variation. The accuracy of tracking of the reference setpoints was quite satisfactory. The proposed method shows that stabilization of financial systems dynamics is possible through feedback control.

The proposed control method confirms that by varying the residences' objective price (as defined by state services and authorized assessors) as well as the loan's

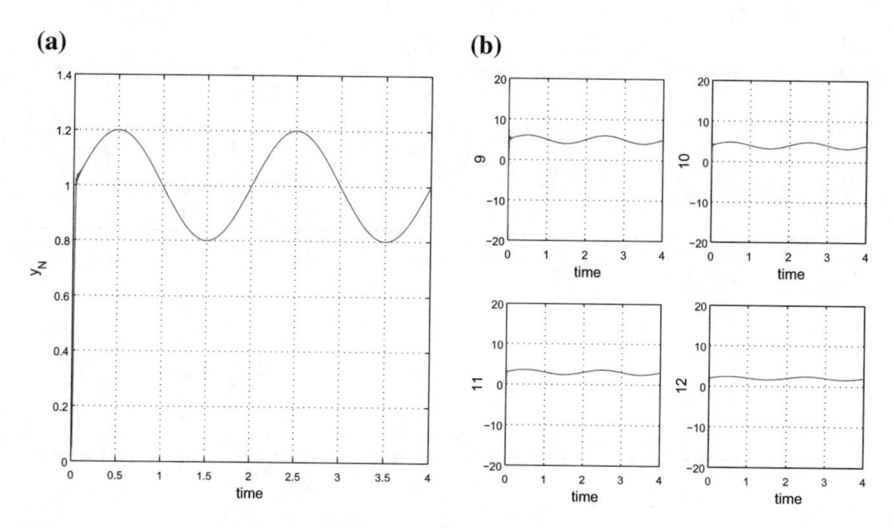

Fig. 15.1 Tracking of setpoint 1 by the 2-factors mortgage price PDE: **a** Convergence of the final grid point value V_{NN} of the PDE system (*blue line*) to the associated reference value (*dashed red line*) **b** Setpoint tracking at grid points p_9 to p_{12}

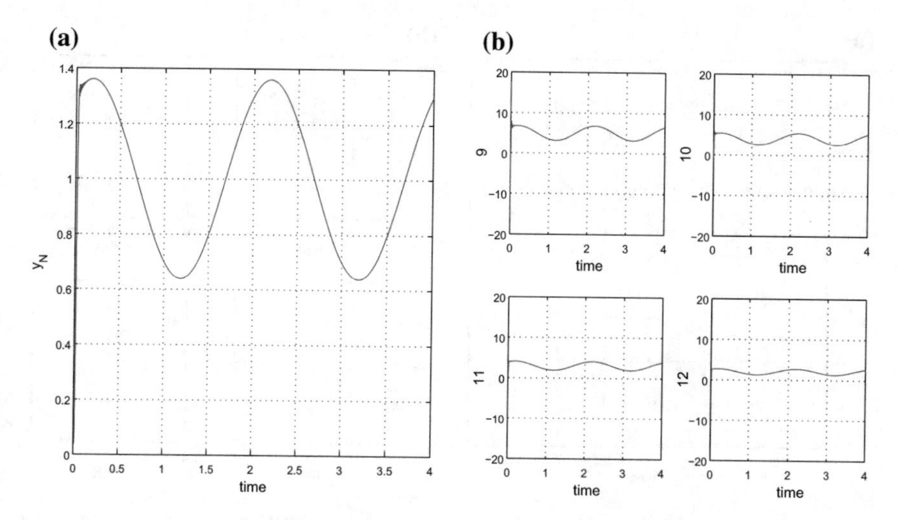

Fig. 15.2 Tracking of setpoint 2 by the 2-factors mortgage price PDE: **a** Convergence of the final grid point value V_{NN} of the PDE system (*blue line*) to the associated reference value (*dashed red line*) **b** Setpoint tracking at grid points p_9 to p_{12}

interest rates (as defined by banks and lenders) one can make the mortgage's value track and converge to specific reference values. This also means that mortgage-based markets can be controlled and stabilized through regulatory policies.

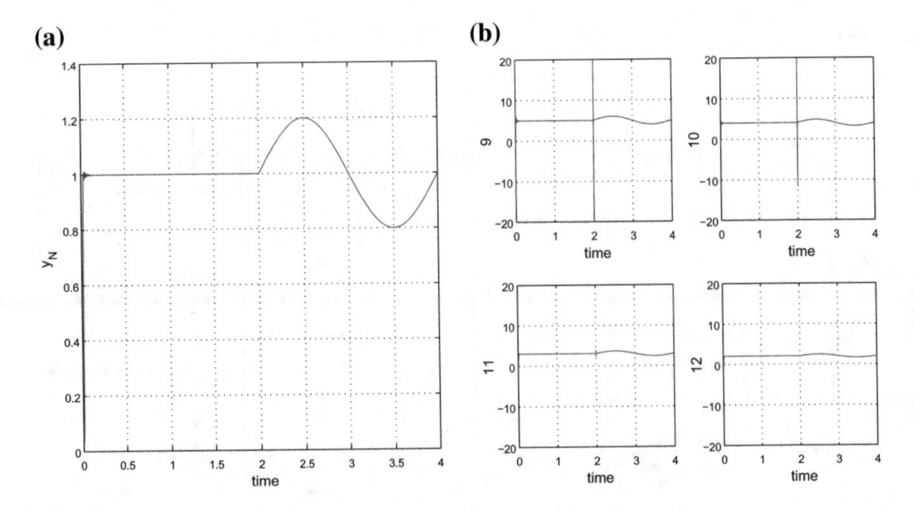

Fig. 15.3 Tracking of setpoint 3 by the 2-factors mortgage price PDE: **a** Convergence of the final grid point value V_{NN} of the PDE system (*blue line*) to the associated reference value (*dashed red line*) **b** Setpoint tracking at grid points p_9 to p_{12}

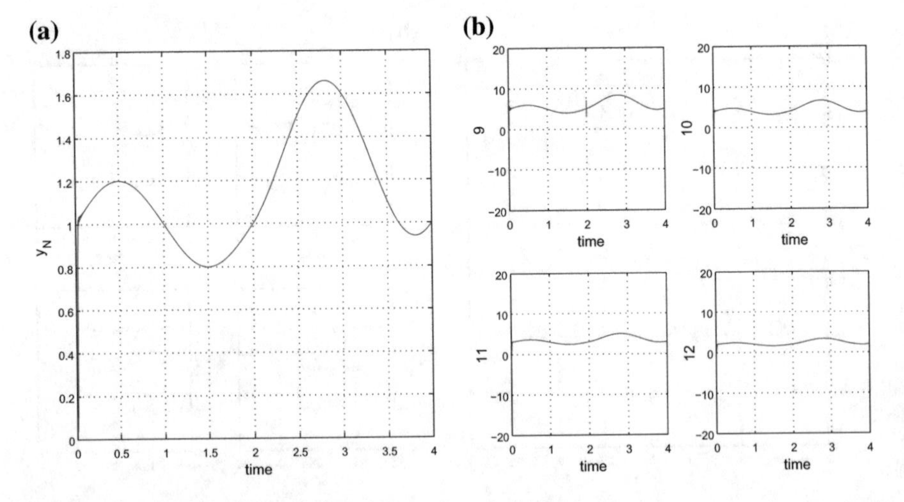

Fig. 15.4 Tracking of setpoint 4 by the 2-factors mortgage price PDE: **a** Convergence of the final grid point value V_{NN} of the PDE system (*blue line*) to the associated reference value (*dashed red line*) **b** Setpoint tracking at grid points p_9 to p_{12}

Through the previous theoretical analysis and test results the significance of state-space models for control, stabilization and estimation in financial engineering has been confirmed. The presented methods are generic can be extended to more financial engineering problems and applications.

References

1. Aid, R.: A review of optimal investment rules in electricity generation. Quantitative Energy Finance. Springer, Berlin (2014)
2. Amari, S., Murata, N., Müller, K.-R., Finke, M., Yang, H.H.: Asymptotic statistical theory of overtraining and cross-validation. IEEE Trans. Neural Netw. **8**(5), 985–996 (1997)
3. Amilon, H.: A neural network vs Black Scholes: a comparison of pricing and hedging performances. J. Forecast. J. Wiley **22**, 317–335 (2003)
4. Andrievskii, B., Fradkov, A.: Control of Chaos: Methods and Applications. II. Applications, Automation and Remote Control, vol. 65, 4th edn., pp. 505–533. Springer, Berlin (2004)
5. Arulampalam, S., Maskell, S.R., Gordon, N.J., Clapp, T.: A tutorial on particle filters for on-line nonlinear/non-Gaussian Bayesian tracking. IEEE Trans. Signal Process. **50**, 174–188 (2002)
6. Atiya, A.F.: Bankruptcy prediction for credit risk using neural networks: a survey and new results. IEEE Trans. Neural Netw. **12**(4), 929–935 (2001)
7. Balogh, A., Kristic, M.: Infinite dimensional backstepping style feedback transformations for a heat equation with an arbitrary level of instability. Eur. J. Control **8**, 165–175 (2002)
8. Barmish, B.R.: On performance limits of feedback control-based stock-trading strategies. In: Proceedings of the American Control Conference, 2011 American Control Conference, ACC 2011, San Francisco, California, USA
9. Bar-Shalom, Y., Rong Li, X., Kirubarajan, T.: Estimation with Applications to Tracking and Navigation. Wiley, Ney York (2001)
10. Basseville, M., Nikiforov, I.: Detection of Abrupt Changes: Theory and Applications. Prentice-Hall, Englewood Cliffs (1993)
11. Basseville, M.: On-board component fault detection and isolation using the statistical local approach. Autom. Elsevier **34**(11), 1391–1415 (1998)
12. Basseville, M., Benveniste, A., Zhang, Q.: Surveilliance d' installations industrielles : démarche générale et conception de l' algorithmique. IRISA Publication Interne No 1010 (1996)
13. Basseville, M., Abdelghani, M., Benveniste, A.: Sub-space-based fault detection and isolation methods - application to vibration monitoring. Publication Interne IRISA No 1143 (1997)
14. Basseville, M., Benveniste, A., Mathis, G., Zhang, Q.: Monitoring the combustion set of a gas turbine. In: Proceedings of the Safeprocess '94, Helsinki, Finland
15. Basseville, M., Benveniste, A., Goursat, M., Mevel, L.: In-flight vibration monitoring of aeronautical structures. IEEE Control Syst. Mag. **27**(5), 27–42 (2007)

G.G. Rigatos, *State-Space Approaches for Modelling and Control in Financial Engineering*, Intelligent Systems Reference Library 125,
DOI 10.1007/978-3-319-52866-3

16. Basseville, M., Benveniste, A., Zhang, Q.: Towards the handling of uncertainties in statistical FDI. In: 18th IFAC World Congress Milano, Italy (2011)
17. Basseville, M.: Information criteria for residual generation and fault detection and isolation. Autom. Elsevier **33**(5), 783–803 (1997)
18. Basseville, M.: On fault detectability and isolability. Eur. J. Control **7**(6), 625–637 (2001)
19. Basseville, M., Benveniste, A., Gach-Devauchelle, B., Goursat, M., Bonnecase, D., Dorey, P., Prevosto, M., Olagnon, M.: Damage monitoring in vibration mechanics: issues in diagnostics and predictive maintenance. Mech. Syst. Signal Process. **7**(5), 401–423 (1993)
20. Bensoussan, A., Prato, G.D., Delfour, M.C., Mitter, S.K.: Representation and Control of Infinite Dimensional Systems. Birkahaüser, Boston (2006)
21. Benveniste, A., Basseville, M., Moustakides, G.: The asymptotic local approach to change detection and model validation. IEEE Trans. Autom. Control **32**(7), 583–592 (1987)
22. Berg, D.: Bankruptcy prediction by generalized additive models. J. Appl. Stoch. Models Bus. Ind. **28**(2), 129–143 (2007)
23. Berkowitz, J., Christofersen, P., Pelletier, D.: Evaluating value-at-risk models with desk-level data. Manag. Sci. **57**, 2213–2227 (2011)
24. Berkowitz, J.: OBrien, J.: How accurate are value-at-risk models at commercial banks? J. Financ. **57**, 1093–1111 (2002)
25. Bernhart, M., Pham, H., Tankov, P., Warin, X.: Swing options valuation: a BSDE with constrained jumps approach. In: Carmona, R., Del Moral, P., Hu, P., Oudjane, N. (eds.) Numerical Methods in Finance, Springer Proceedings in Mathematics, vol. 12, pp. 379–400. Springer, Berlin (2012)
26. Bernhard, P.: The robust control approach to option pricing and interval models: an overview. In: Breton, N., Ben-Ameur, H. (eds.) Numerical Methods in Finance, pp. 91–108. Springer, Berlin (2005)
27. Bertsekas, D.P.: Incremental least squares methods and the extended Kalman filter. SIAM J. Optim. **6**(3), 807–822 (1998)
28. Bharath, S.T., Shumway, T.: Forecasting default with the KMV-Merton model. J. Int. Financ. Mark. Inst. Money (2014)
29. Black, F., Scholes, M.: The pricing of options and corporate liabilities. J. Polit. Econ. **81**, 637–654 (1973)
30. Bodea, A.: Valuation of swing options in electricity commodity markets. Ph.D. thesis, Ruprecht-Kurls-Universitat, Heidelberg (2012)
31. Bolić, M., Djurić, P.M., Hong, H.: Resampling algorithms and architectures for distributed particle filters. IEEE Trans. Signal Process. **53**, 2242–2450 (2005)
32. Borak, S., Detlefsen, K., Hardle, W.: FFT-based option pricing. In: Cizek, R., Hardle, w., Weron, R. (eds.), Statistical Tools for Finance and Insurance, pp. 183–200. Springer, Berlin (2005)
33. Boskovic, D.M., Krstic, M., Liu, W.J.: Boundary control of an unstable heat equation via measurement of domain averaged temperature. IEEE Trans. Autom. Control **46**, 2022–2028 (2002)
34. Bououden, S., Boutat, D., Zheng, G., Barbot, J.P., Kratz, F.: A triangular canonical form for a class of 0-flat nonlinear systems. Int. J. Control Taylor Francis **84**(2), 261–269 (2011)
35. Boussaada, I., Cela, A., Mounier, H., Niculescu, S.I.: Control of drilling vibrations: a time-delay system-based approach. In: 11th Workshop on Time Delay Systems, France (2013)
36. Brémaud, P.: Markov-Chains, Gibbs Fields. Monte-Carlo Simulation and Queues. Texts in Applied Mathematics. Springer, Berlin (1999)
37. Brianis, M.: Numerical methods for option pricing in jump-diffusion markets. Ph.D. thesis, University of Rome, La Sapienza (2003)
38. Brumback, B.D., Srinath, M.D.: A chi-square test for fault-detection in Kalman filters. IEEE Trans. Autom. Control **32**(6), 552–554 (1987)
39. Bunnin, F.O., Guo, Y., Ren, Y.: Option pricing under model and parameter uncertainty using predictive densities. Stat. Comput. Springer **12**(1), 37–44 (2002)

40. Cai, G., Yao, L., Hu, P., Fang, X.: Adaptive full state hybrid function projective synchronization of financial hyperchaotic systems with uncertainty parameters. Discret. Contin. Dyn. Syst. Ser. B **18**(8), 2019–2028 (2013)
41. Cai, G., Hu, P., Li, Y.: Modified function lag projective synchronization of a financial hyperchaotic system. Nonlinear Dyn. **69**, 1457–1464 (2012)
42. Campillo, F., Cérou, F., LeGland, F., Rakotozafy, R.: Particle and cell approximations for nonlinear filtering, Rapport de Recherche INRIA. No. 2567 (1995)
43. Campillo, F.: Particulaire & Modèles de Markov Cachés, Master Course Notes "Filtrage et traitement des doneées". Université de Sud-Toulon Var, France (2006)
44. Carmona, R., Del Moral, P., Hu, P., Oudjane, N.: In: Carmona, R., Del Moral, P., Hu, P., Oudjane, N. (eds.), An Introduction to Particle Methods with Financial Applications. Springer Proceedings in Mathematics, vol. 12, pp. 3–50. Springer, Berlin (2012)
45. Chak, C.K., Feng, G., Mia, J.: An adaptive fuzzy neural network for MIMO system model approximation in high dimensional spaces, IEEE Trans. on Syst. Man and Cybern. - Part B. Cybernetics **28**(3), 436–446 (1998)
46. Chao, L.U.: Empirical Research on Credit Risk Evaluation of Chinese Listed Corporation Based on Modified KMV Model. Advances in information Sciences and Service Sciences(AISS), vols. 5, 10 (2013)
47. Chauvin, J.: Observer design for a class of wave equations driven by an unknown periodic input. In: 18th World Congress. Milano, Italy (2011)
48. Chen, G., Moiola, J.L., Wang, H.O.: Bifurcation control: theories. Methods Appl. Int. J. Bifurc. Chaos **10**(3), 511–548 (2000)
49. Chen, J.: Simulation-based pricing of mortgage-backed securities. In: IEEE 2004 Simulation Conference, vol. 2, pp. 1589–1595 (2004)
50. Chen, M.Y., Linkens, D.A.: A systematic neuro-fuzzy modeling framework with application to material property prediction. IEEE Trans. Syst. Man Cybern. - Part B: Cybern. **31**(5), 781–790 (2001)
51. Chen, W.C.: Dynamics and control of a financial system with time-delayed feedbacks. Chaos Solitons Fractals Elsevier **37**, 1198–1207 (2008)
52. Chen, X., Liu, H., Xu, C.: The new result on delayed finance system. Nonlinear Dyn. Elsevier **78**, 1989–1998 (2014)
53. Chen, Y., Connoly, M., Tang, W., Su, T.: The value of mortgage prepayment and default options. J. Futur. Mark. J. Wiley **28**(9), 840–861 (2009)
54. Chian, A.C.: Nonlinear dynamics and chaos in macroeconomics. Int. J. Theor. Appl. Financ. **3**, 601–613 (2000)
55. Chian, A., Rempel, E., Rogers, C.: Complex economic dynamics: chaotic saddle, crisis and intermittency. Chaos Solitons Fractals Elsevier **29**, 1194–1218 (2006)
56. Cipolla, A., Missaglia, G.: Forecasting industry sector default rates through dynamic factor models. J. Risk Model Valid. **2**(3), 59–79 (2008)
57. Colletaz, G., Hurlin, C., Perignon, C.: The Risk Map: a new tool for validating risk models. J. Bank. Financ. Elsevier **37**(10), 38433854 (2013)
58. Cont, R.: Model uncertainty and its impact on the pricing of derivative instruments. Math. Financ. J. Wiley **16**(3), 519–547 (2006)
59. Cortazar, G., Swartz, E.: Implementing a stochastic model for oil futures prices. Energy Econ. Elsevier **25**(3), 215–238 (2003)
60. Crassidis, J.L., Junkins, J.L.: Optimal Estimation of Dynamic Systems, 2nd edn. CRC Press, Boca Raton (2012)
61. Danca, M., Garrappa, R., Tang, W., Chen, G.: Sustaining stable dynamics of a fractional-order chaotic financial system by parameter switching. Comput. Math. Appl. Elsevier **66**, 702–716 (2013)
62. Dannehl, J., Fuchs, F.W.: Flatness-based control of an induction machine fed via voltage source inverter - concept, control design and performance analysis, IECON 2006. In: IEEE Conference on Industrial Electronics, pp. 5125–5130 (2006)

63. Demetriou, M.A.: Design of consensus and adaptive consensus filters for distributed parameter systems. Autom. Elsevier **46**, 300–311 (2010)
64. Derbali, A., Hallara, S.: Haw the default probability is defined by the credit portfolio models: a comparative theoretical analysis between the structural models? Global J. Manag. Bus. Res. vol. 13, 1 Version 1.0 (2013)
65. Doucet, A., De Freitas, N., Gordon, N. (eds.): Sequential Monte Carlo Methods in Practice. Springer, Berlin (2001)
66. Doyle, J.C., Glover, K., Khargonekar, P.P., Francis, B.A.: State-space solutions to standard H_2 and H_∞ control problems. IEEE Trans. Autom. Control **34**, 831–847 (1989)
67. Dragonescu, A., Soane, A.M.: Multigrid solution of a distibuted optimal control problem constrained by the Stokes equations. Appl. Math. Comput. Elsevier **219**, 5622–5634 (2013)
68. Dubois, D., Prade, H., Ughetto, L.: Checking the coherence and redundancy of fuzzy knowledge bases. IEEE Trans. Fuzzy Syst. **5**(3), 398–417 (1997)
69. Dubois, D., Foulloy, L., Mauris, G., Prade, H.: Probability-possibility transformations. Triangular fuzzy sets and probabilistic inequalities. Reliab. Comput. **10**, 273–297 (2004)
70. Efimov, D.V., Fradkov, A.L.: Adaptive tuning to bifurcation for time-varying nonlinear systems. Autom. Elsevier **42**, 417–425 (2006)
71. Fanti, L., Manfredi, P.: Chaotic business cylces and fiscal policy: an IS-LM model with distributed tax collection lags. Chaos Solitons Fractals Elsevier **32**, 736–744 (2007)
72. Farinwata, S., Filev, D., Langari, R.: Fuzzy Control: Synthesis and Analysis. Wiley, New York (2000)
73. Fatone, L., Mariani, F., Recchioni, M.C., Zirilli, F.: The use of statistical tests to calibrate the Black-Scholes asset dynamics model applied to pricing options with uncertain volatility. J. Probab. Stat. Hindawi Publ. (2012). doi:10.1155/2012/93609
74. Fatone, L., Mariani, F., Recchioni, M.C., Zirilli, F.: Calibration of a multiscale stochastic volatility model using European option prices. Math. Methods Econ. Financ. Springer **3**(1), 49–61 (2008)
75. Fedorov, S., Mikhailov, S.: Option Pricing for Incomplete Markers Via Stochastic Optimization: Transaction Costs. Adaptive Control and Forecast, International Journal of Theoretical and Applied Finance, vol. 1, 1st edn., pp. 179–195. World Scientific, Singapore (2001)
76. Fischer, T., Riedler, J.: Prices, debt and market structure in an agent-based model of the financial market. J. Econ. Dyn. Control Elsevier **48**, 95120 (2014)
77. Fliess, M., Mounier, H.: Tracking control and π-freeness of infinite dimensional linear systems. In: Picci, G., Gilliam, D.S. (eds.), Dynamical Systems, Control, Coding and Computer Vision, vol. 258, pp. 41–68. Birkhaüser (1999)
78. Fliess, M., Mounier, H.: An algebraic framework for infinite-dimensional linear systems. In: Proceedings of International School on Automatic Control of Lille, "Control of Distributed Parameter Systems: Theory and Applications", Grenoble, France, 2002
79. Fliess, M., Joins, C.: A mathematical proof of the existence of trends in financial time series. In: International Conference on Systems Theory, Modelling, Analysis and Control, Fes Morocco, 2009
80. Fliess, M., Joins, C.: Delta hedging in financial engineering: towards a model-free approach. In: 18th Mediterranean Conference on Control and Automation, MED '10, Marrakech, Morocco, 2010
81. Fliess, M., Joins, C.: Preliminary remarks on option pricing and dynamic hedging. In: 1st IEEE International Conference on Systems and Computer Science, Lille, Villeneuve d' Ascq, France, June 2012
82. Forsyth nad, P.A., Labahn, G.: Numerical methods for controlled Hamilton-Jacobi-Bellman PDE in finance. J. Comput. Financ. **11**(2), 1 (2007)
83. Fliess, M., Lévine, J., Martin, P., Rouchon, P.: A Lie-Backlünd approach to equivalence and flatness of nonlinear systems. IEEE Trans. Autom. Control **44**(5), 922–937 (1999)
84. Fukumizu, K.: A regularity condition of the information matrix of a multilayer perceptron network. Neural Netw. **9**(5), 871–879 (1996)

85. Fukumizu, K.: Statistical analysis of unidentifiable models and its application to multilayer neural networks. In: Post-Conference of Bernoulli-Riken BSI 2000 on Neural Networks and Learning (2000)
86. Fukumizu, K.: Statistical active learning in multilayer perceptrons. IEEE Trans. Neural Netw. **11**(1), 17–26 (2000)
87. Gerdts, M., Greif, G., Pesch, H.J.: Numerical optimal control of the wave equation: optimal boundary control of a string to rest in finite time. Math. Comput. Simul. Elsevier **79**(4), 1020–1032 (2008)
88. Gibbs, B.P.: Advanced Kalman Filtering. Least Squares and Modelling. A Practical Handbook. Wiley, New York (2011)
89. Gibbs, R.G.: New Kalman filter and smoother consistency tests. Autom. Elsevier **49**(10), 3141–3144 (2013)
90. Glorennec, P.Y.: Algorithmes d' apprentissage pour systemes d' inference floue. Hermes Science Publications, Paris (1999)
91. Goncharev, Y.: An intensity-based approach to the valuation of mortgage contracts and computation of the endogenous mortgage rate. Int. J. Theor. Appl. Financ. World Sci. **9**(6), 889–914 (2006)
92. Gordon, N.J., Salmond, D.J., Smith, A.F.M.: A novel approach to nonlinear/non-Gaussian Bayesian state estimation. IEE Proc. Radar Signal Process. **140**, 107–113 (1993)
93. Guegan, D.: Chaos in economics and finance. Annu. Rev. Control Elsevier **33**(1), 89–93 (2009)
94. Guo, B.Z., Xu, C.Z., Hammouri, H.: Output feedback stabilization of a one-dimensional wave equation with an arbitrary time-delay in boundary observation, ESAIM: Control. Optim. Calc. Var. **18**, 22–25 (2012)
95. Guo, L., Billings, S.A.: State-space reconstruction and spatio-temporal prediction of lattice dynamical systems. IEEE Trans. Autom. Control **52**(4), 622–632 (2007)
96. Gustafsson, F., Gunnarsson, F., Bergman, N., Forssel, U.: Particle filters for positioning. navigation and tracking. IEEE Trans. Signal Process. - Special Issue Monte Carlo Methods Stat. Signal Process. **50**, 425–437 (2002)
97. Haas, L.: Stabilizing chaos in a dynamic macroeconomic model. J. Econ. Behavior Organ. **33**(3–4), 313–332 (1998)
98. Haine, G.: Observateurs en dimension infinie. Application à l étude de quelques problèmes inverses, Thèse de doctorat, Institut Elie Cartan Nancy (2012)
99. Hambly, B., Howison, S., Kluge, T.: Modeling spikes and pricing swing options in electricity markets, pp. 937–949. Quantitative Finance, Taylor and Francis (2009)
100. Hamid, S.A., Habib, A.: Can neural networks learn the Black-Scholes model? A Simplified Approach, Working Paper, Southern University of New Hampshire, 2005
101. Harris, C.J., Gan, Q.: State estimation and multi-sensor data fusion using data-based neuro-fuzzy local linearization process models. Inf. Fus. Elsevier **2**, 17–29 (2001)
102. Harvey, A., Jan, S.: Koopman. Unobserved components models in economics and finance: the role of Kalman Filter in time series econometrics. IEEE Control Syst. Mag. **29**(6), 71–81 (2009)
103. Haschke, R., Steil, J.J.: Input-space bifurcation manifolds of recurrent neural networks. Neurocomput. Elsevier **64**, 25–38 (2005)
104. Haykin, S.: Adaptive Filter Theory. Prentice-Hall, Englewood Cliff (1991)
105. Haykin, S.: Neural Networks: A Comprehensive Foundation. McMillan, New York (1994)
106. Hidayat, Z., Babuska, R., de Schutter, B., Nunez, A.: Decentralized Kalman Filter comparison for distributed parameter systems: a case study for a 1D heat conduction process. In: Proceedings of the 16th IEEE International Conference on Emerging Technologies and Factory Automation, ETFA 2011, Toulouse, France (2011)
107. Holyst, J., Hagel, T., Haag, G., Weidlich, W.: How to control a chaotic economy? J. Evolut. Econ. Springer **6**, 31–42 (1996)
108. Holyst, J., Zebrowska, M., Urbanowicz, K.: Observations of deterministic chaos in nancial time series by recurrence plots, can one control chaotic economy? Eur. Phys. J. B **20**, 531–535 (2001)

109. Hristache, M., Juditsky, A., Polzelh, J., Spokoiny, V.: Structure adaptive approach for dimension reduction. Ann. Stat. **29**(6), 1537–1566 (2001)
110. Huang, F., Sheng, Y.: Evaluation of default risk based on KMV model for ICBC (Commercial Bank of China), CCB (China Construction Bank) and BOC (Bank of Chine). Int. J. Econ. Financ. **2**(1) (2010)
111. Hull, J.: Options. Futures and Other Derivatives. Prentice Hall, Boston (2012)
112. Iazzolino, G., Fortino, A.: Credit risk analysis and the KMV Black and Scholes model: a proposal of correction and an empirical analysis. Invest. Manag. Financ. Innov. **9**(2), 167–181 (2012)
113. Isermann, R., Schreiber, A.: Identification methods for experimental modelling of nonlinear combustion processes. In: Collection of Technical Papers - 4th International Energy Conversion Engineering Conference, vol. 2, pp. 1468–1485 (2006)
114. Isermann, R.: Local basis function networks for identification of a turbocharger. Proc. IEE Conf. **427**(1), 7–12 (1996)
115. Jang, J.S.R., Sun, C.-T., Mizutani, E.: Neurofuzzy and Soft-Computing. A Computational Approach to Learning and Machine Intelligence. Prentice-Hall, Englewood Cliffs (1997)
116. Julier, S., Uhlmann, J., Durrant-Whyte, H.F.: A new method for the nonlinear transformations of means and covariances in filters and estimators. IEEE Trans. Autom. Control **45**(3), 477–482 (2000)
117. Julier, S.J., Uhlmann, J.K.: Unscented filtering and nonlinear estimation. Proc. IEEE **92**, 401–422 (2004)
118. Jwo, D.J., Cho, T.S.: A practical note on evaluating Kalman filter performance optimality and degradation. Appl. Math. Comput. Elsevier **45**(3), 193–482 (2007)
119. Kalotay, A., Yang, D., Fabozzi, F.: An option-theoretic prepayment model for mortgages and mortgage-backed securities. Int. J. Theor. Appl. Financ. World Scientific **7**(8), 949–978 (2004)
120. Kamen, E.W., Su, J.K.: Introduction to Optimal Estimation. Springer, Berlin (1999)
121. Khalil, H.K.: Nonlinear Systems, 2nd edn. Prentice Hall, Englewood Cliffs (1996)
122. Kitagawa, G.: Monte-Carlo filter and smoother for non-Gaussian non-linear state-space models. J. Comput. Gr. Stat. **5**(1), 1–25 (1996)
123. Kjaer, M.: Pricing of swings options in a mean reverting model with jumps. Appl. Math. Financ. Taylor Francis **15**(5–6), 471–502 (2008)
124. Kröner, A.: Adaptive finite element methods for optimal control second order hyberbolic equations. Comput. Methods Appl. Math. **11**(2), 214–240 (2011)
125. Kumar, P.R., Ravi, V.: Bankruptcy prediction in banks and firms via statistical and intelligent techniques: a review. Eur. J. Op. Res. Elsevier **180**(1), 1–28 (2007)
126. Kurikawa, T., Kaneko, K.: Learning to memorize input-output mapping as bifurcation in neural dynamics: relevance of multiple time-scales for synaptic changes. Neural Comput. Appl. Springer **31**, 725–734 (2012)
127. Laroche, B., Martin, P., Rouchon, P.: Motion planning of the heat equation. Int. J. Robust Nonlinear Control J. Wiley **40**(8), 629–643 (2000)
128. Laroche, B., Martin, Ph, Petit, N.: Commande par platitude: Équations différentielles ordinaires et aux derivées partielles. École Nationale Supérieure des Techniques Avancées, Paris (2007)
129. Larsson, E., Ahlander, K., Hall, A.: Multi-dimensional option pricing using radial basis functions and the generalized Fourier transform. J. Comput. Appl. Math. **222**, 175–192 (2008)
130. Lee, D.J.: Unscented information filtering for distributed estimation and multiple sensor fusion. AIAA Guidance, Navigation and Control Conference and Exhibit. Hawai, USA (2008)
131. Lee, D.J.: Nonlinear estimation and multiple sensor fusion using unscented information filtering. IEEE Signal Process. Lett. **15**, 861–864 (2008)
132. Leisch, F., Jain, L.C., Hornik, K.: Cross-validation with active pattern selection for neural-network classifiers. IEEE Trans. Neural Netw. **9**(1), 35–41 (1998)
133. Lévine, J.: Analysis and Control of Nonlinear Systems: A Flatness-Based Approach. Springer, Berlin (2009)

134. Lévine, J.: On necessary and sufficient conditions for differential flatness. Appl. Algebra Eng. Commun. Comput. Springer **22**(1), 47–90 (2011)
135. Liao, X., Wong, K.W., Wu, Z.: Bifurcation analysis of a two-neuron system with distributed delays. Phys. D Elsevier **149**, 123–141 (2001)
136. Lin, Z., Qiao, H., Dong, X.: Electric power enterprise financial risk evaluation based on rough set and BP networks. In: IEEE SOLI 2008. IEEE International Conference on Service Operation and Logistics, Beijing, China (2008)
137. Lindstrom, E., Ströjby, J., Brodén, M.: Sequential calibration of options. Comput. Stat. Data Anal. Elsevier **52**(6), 2877–2891 (2008)
138. Liu, J., Li, D., Xiong, Z.: Research on an improved residual Chi-square fault detection method for federated unscented Kalman filter. Chin. J. Sci. Instrum. **30**(12), 2568–2573 (2009)
139. Liu, W.J.: Boundary stabilization of an unstable heat equation. SIAM J. Control Optim. **42**, 1033–1043 (2003)
140. Liu, M., Zhang, S.: An LMI approach to design H_∞ controllers for discrete-time nonlinear systems based on unified models. Int. J. Neural Syst. World Scientific **18**(5), 443–452 (2008)
141. Lopez, J.A., Saidenberg, M.R.: Evaluating credit risk models. J. Bank. Financ. Elsevier **24**(1–2), 151–165 (2000)
142. Lorenz, H.W.: Nonlinear Dynamical Economics and Chaotic Motion. Spring, Berlin (1993)
143. Lotstedt, P., Persson, J., von Sydov, L., Tysk, J.: Space-time adaptive finite difference method for European multi-asset options. Int. J. Comput. Math. Appl. **33**, 1159–1180 (2007)
144. Lu, C., Li, X., Pan, H.B.: Empirical research on credit condition of Chinese listed corporations with KMV default models. In: 2006 IEEE International Conference on Machine Learning and Cybernetics, Dalian, China (2006)
145. Lublin, L., Athans, M.: An experimental comparison of and designs for interferometer testbed. In: Francis, B., Tannenbaum, A. (eds.) Lectures Notes in Control and Information Sciences: Feedback Control, Nonlinear Systems and Complexity, pp. 150–172. Springer, Berlin (1995)
146. Analysis of commodity prices with the particle filter: Lucene Aiube, F.A., Nanda Baidya, T.K., Haesaya Tito, E.A. Energy Econ. Elsevier **70**, 577–605 (2008)
147. Lucena Aiube, F.A.: On the comparison of Schwartz and Smith's two and three-factors models on commodity prices. Appl. Econ. Taylor Francis **40**(30), 3736–3748 (2014)
148. Ma, J.H., Chen, Y.S.: Study of the bifurcation topological structure and the global complicated character of a kind of nonlinear finance system. Appl. Math. Mech. **22**(11), 1240–1251 (2001)
149. Mathis, G.: Outils de detection des ruptures et de Diagnostic - Application a la surveillance de Turbines a Gaz. Universite de Rennes I, Thèse de Doctorat (1994)
150. Mackey, M.C., Glass, L.: Oscillation and chaos in physiological control systems. Sci. New Ser. **197**(4300), 287–289 (1997)
151. Maidi, A., Corriou, J.P.: Distributed control of nonlinear diffusion systems by input-output linearization. Int. J. Robust Nonlinear Control **26**, 389–405 (2014)
152. Makarenko, A., Durrany-Whyte, H.: Decentralized Bayesian algorithms for active sensor networks. In. Fus. Elsevier **7**, 418–433 (2006)
153. Manyika, J., Durrant-Whyte, H.: Data Fusion and Sensor Management: A Decentralized Information Theoretic Approach. Prentice Hall, Englewood Cliffs (1994)
154. Marino, R.: Adaptive observers for single output nonlinear systems. IEEE Trans. Autom. Control **35**(9), 1054–1058 (1990)
155. Marino, R., Tomei, P.: Global asymptotic observers for nonlinear systems via filtered transformations. IEEE Trans. Autom. Control **37**(8), 1239–1245 (1992)
156. Martin, P., Rouchon, P.: Two remarks on induction motors. In: CESA 96 IMACS Multiconference Lille, France vol. 1, pp. 76–79 (1996)
157. Martins-Vaquero, J., Khaliq, A.Q.M., Kleefeld, B.: Stabilized explicit Range-Kutta methods for multi-asset American options. Comput. Math. Appl. **67**, 1293–1308 (2014)
158. Martin, Ph, Rouchon, P.: Systèmes plats: planification et suivi des trajectoires. Journées X-UPS, École des Mines de Paris, Centre Automatique et Systèmes, Mai (1999)
159. Mauris, G., Berrah, L., Foulloy, L., Haurat, A.: Fuzzy handling of measurement errors in instrumentation. IEEE Trans. Meas. Instrum. **49**, 89–93 (2000)

160. Merton, R.: On the pricing of corporate debt: the risk structure of interest rates. J. Financ. **28**, 449–470 (1974)
161. van der Merwe, R. , Wan, E.A., Julier, S.I.: Sigma-Point Kalman filters for nonlinear estimation and sensor-fusion applications to intergrated navigation. In: Proceedings of the AIAA Guidance, Navigation and Control Conference, Providence, RI, USA, August, 2004
162. McGinnity, S., Irwin, G.: Nonlinear Kalman filtering using fuzzy local linear models. In: Proceedings of the American Control Conference, pp. 3299–3300. Albuquerque, New Mexico (1997)
163. Míguez, J.: Analysis of parallelizable resampling algorithms for particle filtering. Signal Process. Elsevier **87**, 3155–3174 (2007)
164. Miller, N., Weller, P.: Stochastic saddlepoint systems stabilization policy and the stock-market. J. Econ. Dyn. Control Elsevier **19**(1–2), 279–302 (1995)
165. Moon, K.S., Kim, H.: A multi-dimensional loop average lattice method for multi-asset models. Quant. Financ. (Spec. Issue Option Pricing Hedging) **13**(6), 873–884 (2013)
166. Moon, K.S., Kim, W.J., Kim, H.: Adaptive lattice methods for multi-asset methods. J. Comput. Math. Appl. **56**, 352–366 (2008)
167. Mounier, H.: Document de synthèse pour l' habilitation à diriger des recherches. Université de Paris XI, Juillet (2005)
168. Miller, N., Weller, F.: Stochastic saddlepoint systems stabilization policy and the stock-market. J. Econ. Dyn. Control Elsevier **19**(1–2), 279–302 (1998)
169. Mounier, H., Rudolph, J.: Trajectory tracking for π-flat nonlinear delay systems with a motor example. In: Isidori, A., Lamnabhi-Lagarrigue, F., Respondek, W. (eds.), Nonlinear Control in the Year 2000, vol. 1. Lecture Notes in Control and Information Science, vol. 258, pp. 339–352. Springer, Berlin (2001)
170. Mounier, H., Rudolph, J., Woittenneck, F.: Boundary value problems and convolutional systems over rings of ultradistributions. Advances in the theory of Control, Signal and Systems with Physical Modelling. Lecture Notes in Control an Information Sciences, pp. 179–188. Springer, Berlin (2010)
171. Musso, C., Oudjane, N., Le Gland, F.: Imrpoving regularized particle filters. In: Doucet, A., de Freitas, N., Gordon, N. (eds.) Sequential Monte Carlo Methods in Practice, pp. 247–272. Springer, Berlin (2001)
172. Nelles, O., Fischer, M.: Local linear model trees (LOLIMOT) for linear system identification of a cooling blast. In: Proceedings of EUFIT 96 Conference, Aachen, Germany, 1996
173. Nettleton, E., Durrant-Whyte, H., Sukkarieh, S.: A robust architecture for decentralized data fusion, ICAR03. In: Proceedings of 11th International Conference on Advanced Robotics, Coimbra, Portugal, 2003
174. Nguyen, M.H., Ehrhardt, M.: Modelling and numerical valuation of power derivatives in energy markets. Adv. Appl. Math. Mech. **4**(3), 259–293 (2012)
175. Nguyen, L.H., Hong, K.S.: Hopf bifurcation control via a dynamic state-feedback control. Phys. Lett. A **376**, 442–446 (2012)
176. Nguyen Huu, A.: Valorisation financière sur les marchés d' électricité, Thèse de Doctorat, Université Paris IX Dauphine, 2012
177. Oksendal, B., Sulem, A.: Applied Stochastic Control of Jump Diffusions. Springer, Berlin (2006)
178. Oliveira, J.V.: Towards neuro-linguistic modelling: constraints for optimization of membership functions. Fuzzy Sets Syst. Elsevier **106**, 357–380 (1999)
179. Olivier, F., Sedoglavic, A.: A generalization of flatness to nonlinear systems of partial differential equations: application to the control of a flexible rod. In: Proceedings of the 5th IFAC Symposium on Nonlinear Control Systems, Saint-Petersbourg, 2001
180. Ong, L.L., Bailey, T., Durrant-Whyte, H., Upcroft, B.: Decentralized particle filtering for multiple target tracking in wireless sensor networks. In: Fusion 2008, the 11th International Conference on Information Fusion, Cologne, Germany, 2008
181. Ong, L.L., Upcroft, B., Bailey, T., Ridley, M., Sukkarieh, S., Durrant-Whyte, H.: A decentralized particle filtering algorithm for multi-target tracking across multiple flight vehicles. In:

2006 IEEE/RSJ International Conference on Intelligent Robots and Systems, Beijing, China, 2006
182. Papi, M., Briani, M.: A PDE-based approach for pricing mortgage-backed securities. In: di Nunno, G., Oksendal, B. (eds.) Advanced Mathematical Models for Finance. Springer, Berlin (2011)
183. Parshad, R., Bayazit, D., Barlow, N., Prasad, V.: On the strong solution of a class of partial differential equations that arise in the pricing of mortgage backed securities. Commun. Math. Sci. Int. Press **9**(4), 1033–1050 (2011)
184. Pascucci, A.: PDE and Martingale Methods in Option Pricing. Springer, Berlin (2011)
185. Pedrycz, W.: Why triangular membership functions? Fuzzy Sets Syst. **64**, 21–30 (1994)
186. Pilloy, E., O' Hara, J.G.: FFT based option pricing under a mean reverting process with stochastic volatility and jumps. J. Comput. Appl. Math. Elsevier **235**, 3378–3384 (2011)
187. Pinsky, M.: Partial Differential Equations and Boundary Value Problems. Prentice-Hall, Englewood Cliffs (1991)
188. Pirrong, G.: Commodity Price Dynamics: A Structural Approach. Cambridge University Press, Cambridge (2014)
189. Platen, E., Heath, D.: A Benchmark Approach to Quantitative Finance. Springer, Berlin (2006)
190. Psichogios, D.C., Ungar, L.H.: SVD-NET: an algorithm that automatically selects network structure. IEEE Trans. Neural Netw. **5**(3), 513–515 (1994)
191. Sira-Ramirez, H., Agrawal, S.: Differentially Flat Systems. Marcel Dekker, New York (2004)
192. Rao, B.S., Durrant-Whyte, H.F.: Fully decentralized algorithm for multisensor Kalman filtering. IET Proc: Control Theory Appl. **138**(5), 413–451 (1991)
193. Refregier, A.: Shapelets - I. A method for image analysis. Mon. Not. R. Astron. Soc. **338**, 35–47 (2003)
194. Reisz, A., Perlich, C.: A market-based framework for bankruptcy prediction. J. Financ. Stab. Elsevier **3**(2), 85–131 (2007)
195. Rigatos, G., Zhang, Q.: Fuzzy Model Validation Using the Local Statistical Approach, Publication Interne IRISA No 1417. Rennes, France (2001)
196. Rigatos, G.G., Tzafestas, S.G.: Adaptive fuzzy control for the ship steering problem. J. Mech. Elsevier **16**(6), 479–489 (2006)
197. Rigatos, G.G., Tzafestas, S.G.: Neural Structures Using the Eigenstates of the Quantum Harmonic Oscillator. Open Systems and Information Dynamics, vol. 13, pp. 27–41. Springer, Berlin (2006)
198. Rigatos, G.G., Tzafestas, S.G.: Extended Kalman Filtering for Fuzzy Modelling and Multi-Sensor Fusion. Mathematical and Computer Modelling of Dynamical Systems. Taylor & Francis vol. 13, pp. 251–266 (2007)
199. Rigatos, G.G.: Adaptive fuzzy control with output feedback for H_∞ tracking of SISO nonlinear systems. Int. J. Neural Syst. World Scientific **18**(4), 1–16 (2008)
200. Rigatos, G.G.: Particle filtering for state estimation in industrial robotic systems. IMeche J. Syst. Control Eng. **222**, 437–455 (2008)
201. Rigatos, G., Zhang, Q.: Fuzzy Model Validation Using the Local Statistical Approach, IEEE SMC '02 Conference. Hammamet, Tunisia (2002)
202. Rigatos, G.G., Siano, P., Piccolo, A.: A neural network-based approach for early detection of cascading events in electric power systems. IET J. Gener. Transm. Distrib. **3**(7), 650–665 (2009)
203. Rigatos, G.G.: Model-based and model-free control of flexible-link robots: a comparison between representative methods. Appl. Math. Model. Elsevier **33**, 3906–3925 (2009)
204. Rigatos, G.G.: Particle filtering for state estimation in nonlinear industrial systems. IEEE Trans. Instrum. Meas. **58**, 3885–3900 (2009)
205. Rigatos, G., Zhang, Q.: Fuzzy model validation using the local statistical approach. Fuzzy Sets Syst. Elsevier **60**(7), 882–904 (2009)
206. Rigatos, G.G.: Adaptive fuzzy control of DC motors using state and output feedback. Electr. Power Syst. Res. Elsevier **79**(11), 1579–1592 (2009)

207. Rigatos, G.G.: Sigma-point Kalman Filters and Particle Filters for integrated navigation of unmanned aerial vehicles. In: International Workshop on Robotics for Risky Interventions and Environmental Surveillance, RISE 2009, Brussels, Belgium, 2009
208. Rigatos, G.G.: Extended Kalman and Particle Filtering for sensor fusion in motion control of mobile robots. Math. Comput. Simul. Elsevier **81**(3), 590607 (2010)
209. Rigatos, G.G.: A derivative-free Kalman Filtering approach for sensorless control of nonlinear systems. In: IEEE ISIE 2010. IEEE International Symposium on Industrial Electronics, Bari, Italy (2010)
210. Rigatos, G.G.: Flatness-based adaptive fuzzy control for nonlinear dynamical systems. In: AIM 2011, IEEE/ASME, International Conference on Advanced Intelligent Mechatronics, Budapest, Hungary, 2011
211. Rigatos, G.G.: Distributed nonlinear filtering under packet drops and variable delays for robotic visual servoing. In: Robot Arms, In-Tech Publications, Vienna, Austria, 2011
212. Rigatos, G., Siano, P.: Validation of fuzzy Kalman Filters using the local statistical approach to fault diagnosis. In: IMACS Mascot, : Annual Conference of the Italian Institute for Calculus Applications. Gran Canaria, Spain (2012). 2012
213. Rigatos, G., Siano, P.: Sensorless nonlinear control of induction motors using unscented Kalman filtering. In: IEEE IECON, : 38th International Conference of the Industrial Electronics Society. Montreal, Canada (2012). 2012
214. Rigatos, G., Siano, P.: DFIG control using Differential flatness theory and extended Kalman filtering. In: IFAC INCOM 2012, 14th IFAC International Conference on Information Control Problems in Manufacturing, Bucharest, Romania, 2012
215. Rigatos, G.G.: A derivative-free Kalman Filtering approach to state estimation-based control of nonlinear dynamical systems. IEEE Trans. Ind. Electron. **59**(10), 3987–3997 (2012)
216. Rigatos, G.: A Kalman Filtering approach of improved precision for fault diagnosis in distributed parameter systems. Internal Report, Industrial Systems Institute (2013). arXiv.org/cs/arXiv:1310.3358
217. Rigatos, G.: A Kalman Filtering approach for detection of option mispricing in the Black-Scholes PDE model. In: IEEE Computational Intelligence for Financial Engineering and Economics, London, UK (2014)
218. Rigatos, G.: A Kalman Filtering approach to the detection of option mispricing in electric power markets. In: IEEE SCCI, Orlando, Florida, USA (2014)
219. Rigatos, G., Siano, P., Zervos, N.: An approach to fault diagnosis of nonlinear systems using neural networks with invariance to Fourier transform. J. Ambient Intell. Humaniz. Comput. Springer **4**, 621–639 (2013)
220. Rigatos, G., Siano, P., Zervos, N.: Sensorless control of distributed power generators with the derivative-free nonlinear Kalman Filter. IEEE Trans. Ind. Electron. **61**(11), 6369–6382 (2014)
221. Rigatos, G., Siano, P.: A new nonlinear H-infinity feedback control approach to the problem of autonomous robot navigation. J. Intell. Ind. Syst. Springer **1**(3), 179–186 (2015)
222. Rigatos, G.: Parameter change detection in financial options models using neural networks with invariance to Fourier transform. Industrial Systems Institute, Internal Report (2015)
223. Rigatos, G.: Stabilization of option price dynamics through feedback control of the Black-Scholes PDE, IFAC MICNON 2015. In: 1st IFAC Conference on Modelling, Identification and Control of Nonlinear Systems, St. Petersburg, Russia, 2015
224. Rigatos, G., Siano, P.: Feedback control of the multi-asset Black-Scholes PDE using differential flatness theory. World Scientific, J. Financ. Eng (2015)
225. Rigatos, G.: Modelling and Control for Intelligent Industrial Systems: Adaptive Algorithms in Robotics and Industrial Engineering. Springer, Berlin (2011)
226. Rigatos, G.: Advanced Models of Neural Networks: Nonlinear Dynamics and Stochasticity in Biological Neurons. Springer, Berlin (2013)
227. Rigatos, G.: Nonlinear Control and Filtering Using Differential Flatness Approaches: Applications to Electromechanical Systems. Springer, Berlin (2015)

228. Rigatos, G.: Intelligent Renewable Energy Systems: Modelling and Control. Springer, Berlin (2017)
229. Rikkens, F., Thibeault, A.E.: A structural firm default prediction model for SMEs: evidence from the Dutch market. Multinatl. Financ. J. **13**(3–4), 229–264 (2009)
230. Ristic, B., Arulampalam, M.S., Gordon, N.J.: Beyond the Kalman Filter: Particle Filters for Tracking Applications. Artech House (2004)
231. Rudolph, J.: Flatness Based Control of Distributed Parameter Systems. Examples and Computer Exercises from Various Technological Domains. Shaker Verlag, Aachen (2003)
232. Salberg, S.A., Maybeck, P.S., Oxley, M.E.: Infinite-dimensional sampled-data Kalman Filtering and stochastic heat equation. In: 49th IEEE Conference on Decision and Control, Atlanta, Georgia, USA, 2010
233. Saporta, G.: Probabilités, analyse des données et statistique, Editions Technip, 1990
234. Särrkä, S.: On unscented Kalman Filtering for state estimation of continuous-time nonlinear systems. IEEE Trans. Autom. Control **52**(9), 1631–1641 (2007)
235. Schmutz, A., Gnansounou, E., Sarlas, G.: Economic performance of contracts in electricity markets: a fuzzy and multiple criteria approach. IEEE Trans. Power Syst. **17**(4), 966–973 (2002)
236. Schön, T.B., Törnqvist, D., Gustafsson, F.: Fast particle fitlers for multi-rate sensors. In: EUSIPCO 2007 Conference, Proceedings of the 15th European Signal Processing Conference, Poznan, Poland, 2007
237. Semmler, W., Bernard, L., Robert, M.: Credit risk, credit derivatives and firm value based models. Invest. Manag. Financ. Innov. **5**(4) (2008)
238. Serrano, S., Barrio, R., Dena, A., Rodrguez, M.: Crisis curves in nonlinear business cycles. Commun. Nonlinear Sci. Numer. Simul. Elsevier **17**, 788–794 (2012)
239. Setnes, M., Babuska, R.: Fuzzy modeling for predictive control. In: Farinwata, Filev, Langari (eds.), Fuzzy Control - Synthesis and Analysis, pp. 23–46. Wiley, New York (2000)
240. Setnes, M., Kaymak, U.: Extended fuzzy c-means with volume prototypes and cluster merging. In: Proceedings of the EUFIT'98, Aachen, Germany, 1998
241. Setnes, M., Babuska, R., Kaymak, U., van Nauta Lemke, H.R.: Similarity measures in fuzzy rule base simplification. IEEE Trans. Syst. Man Cybern. - Part B. Cybern. **28**(3), 376–386 (1998)
242. Sharp, N., Newton, D., Duck, P.: An improved fixed-rate mortgage valuation methodology with prepayment and default options. J. Real Estate Financ. Econ. **36**, 307–342 (2008)
243. Simandl, M., Straka, O.: Sampling densities of particle: a survey and comparison. In: Proceedings of the 2007 American Control Conference, pp. 4437–4442. New York, USA, 2007
244. Simon, D.: A game theory approach to constrained minimax state estimation. IEEE Trans. Signal Process. **54**(2), 405–412 (2006)
245. Simon, D.: Training fuzzy systems with the Extended Kalman Filter. Fuzzy Sets Syst. Elsevier **132**, 189–199 (2002)
246. Simon, D.: Sum normal optimization of fuzzy membership functions. Int. J. Uncertain. Fuzziness. Knowl.-Based Syst. **10**, 363–384 (2002)
247. Sira-Ramirez, H., Fliess, M.: On the output feedback control of a synchronous generator. In: 43rd IEEE Conference on Decision and Control, Bahamas, 2004
248. Slotine, J.J.: Applied Nonlinear Control. Prentice-Hall, Englewood Cliffs (1991)
249. Smyshlyaev, A., Krstic, M.: Adaptive Control of Parabolic PDEs. Princeton University Press, Princeton (2010)
250. Sorrentino, G.: Option pricing in a path integral framework. Ph.D. thesis, Victoria University, Melbourne, Australia, 2009
251. Stanton, R.: Rational prepayment and the valuation of mortgage-backed securities. Rev. Financ. Stud. **8**, 672–708 (1995)
252. Stokes, J.R., Britch, B.M.: Valuing agricultural mortgage-based securities. J. Agric. Appl. Econ. **33**(3), 493–511 (2001)
253. Sundureshar, S.: A review of Merton's model of the the firm's capital structure with wide applications. Annu. Rev. Financ. Econ. **5**(5), 1–5 (2013)

254. Schwartz, E.: The stochastic behavior of commodity prices: implications for volatility snd hedging. J. Financ. J. Wiley **53**(3), 927–973 (1997)
255. Schwartz, E.: Valuing long-term commodity assets. Financ. Manag. J. Wiley **27**(1), 57–66 (1998)
256. Takagi, T., Sugeno, M.: Fuzzy identification of systems and its applications to modeling and control. IEEE Trans. Syst. Man Cybern. **15**, 116–132 (1985)
257. Thrun, S., Burgard, W., Fox, D.: Probabilistic Robotics. MIT Press, Cambridge (2005)
258. Tong, S., Li, H.-X., Chen, G.: Adaptive fuzzy decentralized control for a class of large-scale nonlinear systems. IEEE Trans. Syst. Man Cybern. - Part B: Cybern. **34**(1), 770–775 (2004)
259. Toussaint, G.J., Basar, T., Bullo, F.: H_∞ optimal tracking control techniques for nonlinear underactuated systems. In: Proceedings of the IEEE CDC 2000, 39th IEEE Conference on Decision and Control, Sydney Australia, 2000
260. Tudela, M., Young, G.: A Merton-model approach to assessing the default risk of UK public companies. Bank Engl. Q. Bull. **43**(3), 328 (2003)
261. Utz, T., Meurer, T., Kugi, A.: Trajectory planning for two-dimensional quasi-linear parabolic PDE based on finite difference semi-discretization. In: 18th IFAC World Congress. Milano, Italy (2011)
262. Valestrand, R., Naevdal, G., Stordal, A.S.: Evaluation of EnKF and variants on the PUNQ-S3 case, Oil & gas science and technology rev. IFP Energies Nouv. **67**(5), 841–855 (2012)
263. Vargas, J., Grzeidak, E., Hemerly, E.: Robust adaptive synchronization of a hyperchaotic finance system. Nonlinear Dyn. Springer **80**, 239–248 (2015)
264. Vercauteren, T., Wang, X.: Decentralized sigma-point information filters for target tracking in collaborative sensor networks. IEEE Trans. Signal Process. **53**(8), 2997–3009 (2005)
265. Villagra, J., d'Andrea-Novel, B., Mounier, H., Pengov, M.: Flatness-based vehicle steering control strategy with SDRE feedback gains tuned via a sensitivity approach. IEEE Trans. Control Syst. Technol. **15**, 554–565 (2007)
266. Wade, B.A., Khaliq, A.Q.M., Yousuf, M., Vigo-Aguiar, J., Deininger, R.: On smoothing of the Crank-Nicolson scheme and higher-order schemes for pricing barrier options. J. Comput. Appl. Math. Elsevier **204**, 144–158 (2007)
267. Wang, L.X.: Adaptive Fuzzy Systems and Control: Design and Stability Analysis. Prentice Hall, Englewood Cliffs (1994)
268. Wang, L.X.: A Course in Fuzzy Systems and Control. Prentice-Hall, Upper Saddle River (1998)
269. Wang, Z., Huang, X., Shi, G.: Analysis of nonlinear dynamics and chaos in a fractional order financial system with time delay. Comput. Math. Appl. Elsevier **62**, 1531–1539 (2011)
270. Wang, Z., Huang, X., Shen, H.: Control of an uncertain fractional order economic system via adaptive sliding mode. Neurocomputing Elsevier **83**, 8388 (2012)
271. Wang, G., Li, C., Wen, F., Xu, N.: Risk assessment and control for distribution companies in electricity market environment. In: IEEE DRPT 2008, 3rd IEEE International Conference on Electric Utility Deregulation and Restructuring and Power Technologies, Nanjuing, China, 2008
272. Watanabe, K., Pathiranage, C.D., Izumi, K.: A fuzzy Kalman filter approach to the SLAM problem of nonholonomic mobile robots. In: Proceedings of the 17th IFAC World Congress, pp. 4600–4605. Seoul, Korea, 2008
273. Wei, D.Q., Luo, X.S.: Bo Zhang, and Ying Hua Qin, Controlling chaos in spaceclamped FitzHugh-Nagumo neuron by adaptive passive method. Nonlinear Anal.: Real World Appl. Elsevier **11**(3), 1752–1759 (2010)
274. Wiggins, S.: Introduction to Applied Nonlinear Dynamical Systems and Chaos. Series: Texts in Applied Mathematics, vol. 2. Springer, Berlin (2003)
275. Windcliff, H., Forsyth, P.A., Vetzal, K.R.: Analysis of the stability of the linear boundary condition for the Black-Scholes equation. J. Comput. Financ. **8**, 65–92 (2004)
276. Winkler, F.J., Lohmann, B.: Design of a decoupling controller structure for first order hyperbolic PDEs with distributed control action. In: 2010 American Control Conference. Baltimore, MD USA (2010)

277. Woitteneck, F., Mounier, H.: Controllability of networks of statially one-dimensional second order PDEs - an algebraic approach. SIAM J. Control Optim. **48**(6), 3882–3902 (2010)
278. Woittennek, F., Rudolph, J.: Controller canonical forms and flatness-based state feedback for 1D hyperbolic systems. In: 7th Vienna International Conference on Mathematical Modelling, MATHMOD, 2012
279. Wong, T.C., Hu, C.H., Lo, C.F.: Discriminatory power and predictions of default of structural credit risk models. J. Risk Model Valid. **3**(4), 39–60 (2010)
280. Worrel, C.A., Brady, S.M., Bala, J.W.: Comparison of data classification methods for predictive ranking of banks exposed to risk of failure. In: IEEE CIFEr 2012, IEEE International Conference on Computational Intelligence for Financial Engineering & Economics, 2012
281. Wu, H.N., Wang, J.W., Li, H.K.: Design of distributed H_∞ fuzzy controllers with constraint for nonlinear hyperbolic PDE systems. Autom. Elsevier **48**, 2535–2543 (2012)
282. Wu, Z.G., Shi, P., Su, H., Chu, J.: Sampled-data fuzzy control of chaotic systems based on T-S Fuzzy model. IEEE Trans. Fuzzy Syst. **22**(1), 153–163 (2014)
283. Xin, B., Zhang, J.: Finite-time stabilizing a fractional-order chaotic financial system with market confidence. Nonlinear Dyn. Springer **79**, 1399–1409 (2015)
284. Xu, X., Hu, H.Y., Wang, H.L.: Stability switches. Hopf bifurcations and chaos of a neuron model with delay dependent parameters. Phys. Lett. A **354**, 126–136 (2006)
285. Yang, X.S., Huang, Y.: Complex dynamics in simple Hopfield networks. AIP Chaos **16**, 033114-1–033114-7 (2006)
286. Yen, J., Wang, L.: Simplifying fuzzy rule-based models using orthogonal transformation methods. IEEE Trans. Syst. Man Cybern. - Part B **29**, 13–24 (1999)
287. Yousef, N.M., Jaffer, M.: The analysis of KMV-Merton model in forecasting default probability. In: 2012 IEEE Symposium on Humanities Science and Engineering. Kuala Lumpur, Malaysia (2012)
288. Yu, H., Cai, G., Li, Y.: Dynamic analysis and control of a new hyperchaotic finance system. Nonlinear Dyn. Elsevier **67**, 2171–2182 (2012)
289. Yu, D., Chakravotry, S.: A randomly perturbed iterative proper orthogonal decomposition technique for filtering distributed parameter systems. In: American Control Conference, Montreal, Canada, 2012
290. Zhang, Q., Basseville, M., Benveniste, A.: Fault detection and isolation in nonlinear dynamic systems : a combined input-output and local approach. Autom. Elsevier **34**(11), 1359–1373 (1998)
291. Zhang, Q.: Fault detection and isolation with nonlinear black-box models. In: Proceedings of the SYSID '97, Kitakyushu, Japan
292. Zhang, Q.: Nonlinear system identification with an integrable continuous time nonlinear ARX model. Journal Européen des Systèmes Automatisés **46**(6–7), 691–710 (2012)
293. Zhang, Q., Basseville, M., Benveniste, A.: Early warning of slight changes in systems. Autom. Spec. Issue Stat. Methods Signal Process. Control **30**(1), 95–113 (1994)
294. Zhang, Q., Basseville, M.: Statistical fault detection and isolation for linear time-varying systems. In: Proceedings of the 16th IFAC Conference on Systems Identification, Brussels, Belgium, 2012
295. Zhang, Q., Basseville, M.: Advanced numerical computation of chi2-tests for fault detection and isolation. Safeprocess '03 - 5th IFAC/IMACS Symposium on Fault Detection. Supervision and Safety of Technical Processes, Washington DC (2003)
296. Zhang, Q., Campillo, F., Cérou, F., Legland, F.: Nonlinear fault detection and isolation based on bootstrap particle filters. In: Proceedings of the 44th IEEE Conference on Decision and Control, and European Control Conference, Seville Spain, 2005
297. Zhang, S., Wang, L.: Fast Fourier transform option pricing with stochastic interest rate, stochastic volatility and double jumps. Appl. Math. Comput. Elsevier **219**, 10928–10933 (2013)
298. Zhang, B., Oosterlee, C.W.: Pricing of early-exercise Asian options under Lévy processes based on Fourier transform. Appl. Numer. Math. **78**, 14–30 (2014)
299. Zhang, B., Oosterlee, C.W.: An efficient pricing algorithm for swing options based on Fourier-cosine expansions. J. Comput. Financ. **16**(4), 1–31 (2013)

300. Zhang, B., Oosterlee, C.W.: Efficient pricing of European-style Asian options under exponential Lévy Processes based on Fourier-cosine expansions. SIAM J. Financ. Math. **4**, 399–426 (2013)
301. Zhang, P., Zhou, H.: The credit risk measurement of chinas listed companies based on the KMV model. In: Wu, D.D. (ed.) Quantitative Financial Risk Management. Springer, Berlin (2011)
302. Zheng, P., Tang, W., Zhang, J.: Some novel double-scroll chaotic attractors in Hopfield networks. Neurocomputing Elsevier **73**(10–12), 2280–2285 (2010)
303. Zhao, M., Wang, J.: H-infinity control of a chaotic finance system in the presence of external disturbance and input time-delay. Appl. Math. Comput. Elsevier **233**, 320–327 (2014)
304. Zhao, X., Li, Z., Li, S.: Synchronization of a chaotic finance system. Appl. Math. Comput. Elsevier **217**, 6031–6039 (2011)
305. Zhou, W., Yang, M., Han, L.: Black-Scholes versus artificial neural networks in pricing call warrants: the case of China market. In: Third International Conference on Natural Computation, ICNC 2007, Haikou, China, 2007
306. Zhuang, C., Fu, L., Fan, Y.: Multiple fading Kalman filter based on hypothesis testing. J. Beijing Univ. Aeronaut. Astronaut. **30**(1), 18–22 (2004)
307. Zielinski, T.: Merton's and KMV models in credit risk management. Econ. Stud. **127**, 123–135 (2013)
308. Zimmermann, H.G., Neuneier, R., Grothmann, R.: Multiagent modeling of multiple FX-markets by neural networks. IEEE Trans. Neural Netw. **12**(4), 735–743 (2001)
309. Zimmermann, H.G., Neuneier, R., Grothmann, R.: Multi-agent FX-Market Modeling by Neural Networks, Operations Research Proceedings 2001, Volume 2001 of the Series Operations Research Proceedings 2001, pp 413–420. Springer, Berlin (2001)
310. Zolghadri, A.: An algorithm for real-time failure detection in Kalman filters. IEEE Trans. Autom. Control **41**(10), 1537–1539 (1996)
311. Zwart, H., Le Gorrec, Y., Maschke, B.: Linking hyperbolic and parabolic PDEs. In: 2011 50th IEEE Conference on Decision and Control and European Control Conference, CDC-ECC, Orlando, Florida, USA, 2011

Index

G.G. Rigatos, *State-Space Approaches for Modelling and Control in Financial Engineering*, Intelligent Systems Reference Library 125,
DOI 10.1007/978-3-319-52866-3

Printed in the United States
By Bookmasters